Introduction to
Finite Element Analysis Using MATLAB® and Abaqus

Introduction to
Finite Element Analysis Using MATLAB® and Abaqus

Amar Khennane

CRC Press
Taylor & Francis Group
Boca Raton London New York

CRC Press is an imprint of the
Taylor & Francis Group, an **informa** business

CRC Press
Taylor & Francis Group
6000 Broken Sound Parkway NW, Suite 300
Boca Raton, FL 33487-2742

© 2013 by Taylor & Francis Group, LLC
CRC Press is an imprint of Taylor & Francis Group, an Informa business

No claim to original U.S. Government works

Version Date: 20130220

International Standard Book Number-13: 978-1-4665-8020-6 (Hardback)

Library of Congress Cataloging-in-Publication Data

Khennane, Amar.
    Introduction to finite element analysis using MATLAB and Abaqus / author, Amar Khennane.
    pages cm
    Includes bibliographical references and index.
    ISBN 978-1-4665-8020-6 (hardback)
    1. Finite element method--Data processing. 2. Engineering mathematics. 3. MATLAB. 4. Abaqus (Electronic resource) I. Title.

TA347.F5K54 2013
518'.25028553--dc23
                                                                                2013003260

Visit the Taylor & Francis Web site at
http://www.taylorandfrancis.com

and the CRC Press Web site at
http://www.crcpress.com

Contents

List of Figures xiii

List of Tables xxv

Preface xxvii

Author... xxix

Chapter 1 Introduction 1

    1.1   Prologue 1
    1.2   Finite Element Analysis and the User 1
    1.3   Aim of the Book... 2
    1.4   Book Organization 2

Chapter 2 Bar Element 5

    2.1   Introduction.. 5
    2.2   One-Dimensional Truss Element 5
            2.2.1   Formulation of the Stiffness Matrix: The Direct Approach 5
            2.2.2   Two-Dimensional Truss Element.................................... 7
    2.3   Global Stiffness Matrix Assembly... 9
            2.3.1   Discretization 9
            2.3.2   Elements' Stiffness Matrices in Local Coordinates..................... 9
            2.3.3   Elements' Stiffness Matrices in Global Coordinates.................. 10
                    2.3.3.1   Element 1 11
                    2.3.3.2   Element 2 11
                    2.3.3.3   Element 3 12
             2.3.4   Global Matrix Assembly 12
                    2.3.4.1   Only Element 1 Is Present................................... 13
                    2.3.4.2   Only Element 2 Is Present................................... 13
                    2.3.4.3   Only Element 3 Is Present................................... 13
             2.3.5   Global Force Vector Assembly 14
    2.4   Boundary Conditions.. 15
            2.4.1   General Case... 15
    2.5   Solution of the System of Equations 16
    2.6   Support Reactions 17
    2.7   Members' Forces 18
    2.8   Computer Code: truss.m 19
            2.8.1   Data Preparation 20
                    2.8.1.1   Nodes Coordinates... 20
                    2.8.1.2   Element Connectivity.. 20
                    2.8.1.3   Material and Geometrical Properties 20
                    2.8.1.4   Boundary Conditions 20
                    2.8.1.5   Loading 21
             2.8.2   Element Matrices 21
                    2.8.2.1   Stiffness Matrix in Local Coordinates 21
                    2.8.2.2   Transformation Matrix 22

                    2.8.2.3   Stiffness Matrix in Global Coordinates 22
                    2.8.2.4   "Steering" Vector 22
          2.8.3   Assembly of the Global Stiffness Matrix 23
          2.8.4   Assembly of the Global Force Vector 23
          2.8.5   Solution of the Global System of Equations 23
          2.8.6   Nodal Displacements 23
          2.8.7   Element Forces 23
          2.8.8   Program Scripts 24
    2.9   Problems 27
          2.9.1   Problem 2.1 27
          2.9.2   Problem 2.2 32
    2.10 Analysis of a Simple Truss with Abaqus 35
          2.10.1 Overview of Abaqus 35
          2.10.2 Analysis of a Truss with Abaqus Interactive Edition 36
                    2.10.2.1 Modeling 36
                    2.10.2.2 Analysis 51
          2.10.3 Analysis of a Truss with Abaqus Keyword Edition 57

Chapter 3   Beam Element 63

    3.1   Introduction 63
    3.2   Stiffness Matrix 63
    3.3   Uniformly Distributed Loading 67
    3.4   Internal Hinge 71
    3.5   Computer Code: beam.m 73
          3.5.1   Data Preparation 73
                    3.5.1.1   Nodes Coordinates 73
                    3.5.1.2   Element Connectivity 74
                    3.5.1.3   Material and Geometrical Properties 74
                    3.5.1.4   Boundary Conditions 74
                    3.5.1.5   Internal Hinges 74
                    3.5.1.6   Loading 75
                    3.5.1.7   Stiffness Matrix 76
          3.5.2   Assembly and Solution of the Global System
                  of Equations 76
          3.5.3   Nodal Displacements 76
          3.5.4   Element Forces 77
    3.6   Problems 81
          3.6.1   Problem 3.1 81
          3.6.2   Problem 3.2 84
          3.6.3   Problem 3.3 87
    3.7   Analysis of a Simple Beam with Abaqus 90
          3.7.1   Interactive Edition 90
          3.7.2   Analysis of a Beam with Abaqus Keyword Edition 103

Chapter 4   Rigid Jointed Frames 107

    4.1   Introduction 107
    4.2   Stiffness Matrix of a Beam–Column Element 107
    4.3   Stiffness Matrix of a Beam–Column Element in the Presence
          of Hinged End 107

4.4 Global and Local Coordinate Systems 108
4.5 Global Stiffness Matrix Assembly and Solution for Unknown Displacements 109
4.6 Computer Code: frame.m 109
    4.6.1 Data Preparation 109
           4.6.1.1 Nodes Coordinates.. 110
           4.6.1.2 Element Connectivity..................................... 110
           4.6.1.3 Material and Geometrical Properties 110
           4.6.1.4 Boundary Conditions 110
           4.6.1.5 Internal Hinges.. 111
           4.6.1.6 Loading 111
    4.6.2 Element Matrices 112
           4.6.2.1 Stiffness Matrix in Local Coordinates 112
           4.6.2.2 Transformation Matrix 113
           4.6.2.3 Stiffness Matrix in Global Coordinates 113
           4.6.2.4 "Steering" Vector 113
           4.6.2.5 Element Loads 113
    4.6.3 Assembly of the Global Stiffness Matrix........................... 113
    4.6.4 Solution of the Global System of Equations 114
    4.6.5 Nodal Displacements 114
    4.6.6 Element Forces 114
4.7 Analysis of a Simple Frame with Abaqus.................................. 124
    4.7.1 Interactive Edition 124
    4.7.2 Keyword Edition.. 132

Chapter 5 Stress and Strain Analysis....................................... 135

5.1 Introduction... 135
5.2 Stress Tensor.. 135
    5.2.1 Definition 135
    5.2.2 Stress Tensor–Stress Vector Relationships......................... 137
    5.2.3 Transformation of the Stress Tensor.............................. 139
    5.2.4 Equilibrium Equations 139
    5.2.5 Principal Stresses 140
    5.2.6 von Mises Stress 141
    5.2.7 Normal and Tangential Components of the Stress Vector 141
    5.2.8 Mohr's Circles for Stress... 143
    5.2.9 Engineering Representation of Stress 144
5.3 Deformation and Strain 144
    5.3.1 Definition 144
    5.3.2 Lagrangian and Eulerian Descriptions............................. 145
    5.3.3 Displacement 146
    5.3.4 Displacement and Deformation Gradients 147
    5.3.5 Green Lagrange Strain Matrix 148
    5.3.6 Small Deformation Theory... 149
           5.3.6.1 Infinitesimal Strain 149
           5.3.6.2 Geometrical Interpretation of the Terms of the Strain Tensor 150
           5.3.6.3 Compatibility Conditions 152
    5.3.7 Principal Strains 152

		5.3.8	Transformation of the Strain Tensor	153
		5.3.9	Engineering Representation of Strain	153
	5.4	Stress–Strain Constitutive Relations		154
		5.4.1	Generalized Hooke's Law	154
		5.4.2	Material Symmetries	155
			5.4.2.1 Symmetry with respect to a Plane	155
			5.4.2.2 Symmetry with respect to Three Orthogonal Planes	157
			5.4.2.3 Symmetry of Rotation with respect to One Axis	157
		5.4.3	Isotropic Material	158
			5.4.3.1 Modulus of Elasticity	160
			5.4.3.2 Poisson's Ratio	160
			5.4.3.3 Shear Modulus	160
			5.4.3.4 Bulk Modulus	160
		5.4.4	Plane Stress and Plane Strain	162
	5.5	Solved Problems		163
		5.5.1	Problem 5.1	163
		5.5.2	Problem 5.2	164
		5.5.3	Problem 5.3	167
		5.5.4	Problem 5.4	168
		5.5.5	Problem 5.5	170
		5.5.6	Problem 5.6	171
		5.5.7	Problem 5.7	172
		5.5.8	Problem 5.8	174

Chapter 6   Weighted Residual Methods 175

	6.1	Introduction	175
	6.2	General Formulation	175
	6.3	Galerkin Method	176
	6.4	Weak Form	178
	6.5	Integrating by Part over Two and Three Dimensions (Green Theorem)	179
	6.6	Rayleigh Ritz Method	183
		6.6.1 Definition	183
		6.6.2 Functional Associated with an Integral Form	183
		6.6.3 Rayleigh Ritz Method	183
		6.6.4 Example of a Natural Functional	185

Chapter 7   Finite Element Approximation 191

	7.1	Introduction	191
	7.2	General and Nodal Approximations	191
	7.3	Finite Element Approximation	193
	7.4	Basic Principles for the Construction of Trial Functions	195
		7.4.1 Compatibility Principle	195
		7.4.2 Completeness Principle	196
	7.5	Two-Dimensional Finite Element Approximation	197
		7.5.1 Plane Linear Triangular Element for C^0 Problems	197
			7.5.1.1 Shape Functions	197
			7.5.1.2 Reference Element	199
			7.5.1.3 Area Coordinates	202
		7.5.2 Linear Quadrilateral Element for C^0 Problems	203

		7.5.2.1	Geometrical Transformation	203
		7.5.2.2	Construction of a Trial Function over a Linear Quadrilateral Element	206
7.6		Shape Functions of Some Classical Elements for C^0 Problems		207
	7.6.1	One-Dimensional Elements		207
		7.6.1.1	Two-Nodded Linear Element	207
		7.6.1.2	Three-Nodded Quadratic Element	207
	7.6.2	Two-Dimensional Elements		207
		7.6.2.1	Four-Nodded Bilinear Quadrilateral	207
		7.6.2.2	Eight-Nodded Quadratic Quadrilateral	208
		7.6.2.3	Three-Nodded Linear Triangle	208
		7.6.2.4	Six-Nodded Quadratic Triangle	208
	7.6.3	Three-Dimensional Elements		208
		7.6.3.1	Four-Nodded Linear Tetrahedra	208
		7.6.3.2	Ten-Nodded Quadratic Tetrahedra	209
		7.6.3.3	Eight-Nodded Linear Brick Element	209
		7.6.3.4	Twenty-Nodded Quadratic Brick Element	210

Chapter 8 Numerical Integration ... 211

8.1		Introduction		211
8.2		Gauss Quadrature		211
	8.2.1	Integration over an Arbitrary Interval $[a, b]$		214
	8.2.2	Integration in Two and Three Dimensions		215
8.3		Integration over a Reference Element		216
8.4		Integration over a Triangular Element		217
	8.4.1	Simple Formulas		217
	8.4.2	Numerical Integration over a Triangular Element		218
8.5		Solved Problems		219
	8.5.1	Problem 8.1		219
	8.5.2	Problem 8.2		221
	8.5.3	Problem 8.3		226

Chapter 9 Plane Problems ... 231

9.1		Introduction		231
9.2		Finite Element Formulation for Plane Problems		231
9.3		Spatial Discretization		234
9.4		Constant Strain Triangle		235
	9.4.1	Displacement Field		236
	9.4.2	Strain Matrix		237
	9.4.3	Stiffness Matrix		237
	9.4.4	Element Force Vector		237
		9.4.4.1	Body Forces	238
		9.4.4.2	Traction Forces	238
		9.4.4.3	Concentrated Forces	239
	9.4.5	Computer Codes Using the Constant Strain Triangle		240
		9.4.5.1	Data Preparation	241
		9.4.5.2	Nodes Coordinates	243
		9.4.5.3	Element Connectivity	243
		9.4.5.4	Material Properties	243

          9.4.5.5   Boundary Conditions 243
          9.4.5.6   Loading 243
          9.4.5.7   Main Program 243
          9.4.5.8   Element Stiffness Matrix 245
          9.4.5.9   Assembly of the Global Stiffness Matrix 246
          9.4.5.10 Solution of the Global System of Equations 246
          9.4.5.11 Nodal Displacements 246
          9.4.5.12 Element Stresses and Strains 246
          9.4.5.13 Results and Discussion 247
          9.4.5.14 Program with Automatic Mesh Generation 249
     9.4.6   Analysis with Abaqus Using the CST 253
          9.4.6.1   Interactive Edition 253
          9.4.6.2   Keyword Edition 260
9.5   Linear Strain Triangle 263
     9.5.1   Displacement Field 264
     9.5.2   Strain Matrix 265
     9.5.3   Stiffness Matrix 266
     9.5.4   Computer Code: LST_PLANE_STRESS_MESH.m 266
          9.5.4.1   Numerical Integration of the Stiffness Matrix 270
          9.5.4.2   Computation of the Stresses and Strains 271
     9.5.5   Analysis with Abaqus Using the LST 272
          9.5.5.1   Interactive Edition 272
          9.5.5.2   Keyword Edition 278
9.6   The Bilinear Quadrilateral 279
     9.6.1   Displacement Field 280
     9.6.2   Strain Matrix 281
     9.6.3   Stiffness Matrix 282
     9.6.4   Element Force Vector 282
     9.6.5   Computer Code: Q4_PLANE_STRESS.m 284
          9.6.5.1   Data Preparation 284
          9.6.5.2   Main Program 287
          9.6.5.3   Integration of the Stiffness Matrix 289
          9.6.5.4   Computation of the Stresses and Strains 290
          9.6.5.5   Program with Automatic Mesh Generation 291
     9.6.6   Analysis with Abaqus Using the Q4 Quadrilateral 295
          9.6.6.1   Interactive Edition 295
          9.6.6.2   Keyword Edition 302
9.7   The 8-Node Quadrilateral 304
     9.7.1   Formulation 304
     9.7.2   Equivalent Nodal Forces 307
     9.7.3   Program Q8_PLANE_STRESS.m 307
          9.7.3.1   Data Preparation 307
          9.7.3.2   Main Program 311
          9.7.3.3   Integration of the Stiffness Matrix 314
          9.7.3.4   Results with the Coarse Mesh 314
          9.7.3.5   Program with Automatic Mesh Generation 315
     9.7.4   Analysis with Abaqus Using the Q8 Quadrilateral 321
9.8   Solved Problem with MATLAB® 326

9.8.1   Strip Footing with the CST Element 326

9.8.2   Strip Footing with the LST Element 331

9.8.3   Bridge Pier with the Q8 Element................................... 336

Chapter 10   Axisymmetric Problems 353

10.1 Definition 353

10.2 Strain–Displacement Relationship 353

10.3 Stress–Strain Relations 354

10.4 Finite Element Formulation 355

    10.4.1 Displacement Field 355

    10.4.2 Strain Matrix 355

    10.4.3 Stiffness Matrix 356

    10.4.4 Nodal Force Vectors 356

        10.4.4.1 Body Forces 356

        10.4.4.2 Surface Forces Vector...................... 356

        10.4.4.3 Concentrated Loads 357

        10.4.4.4 Example 357

10.5 Programming... 358

    10.5.1 Computer Code: AXI_SYM_T6.m 359

        10.5.1.1 Numerical Integration of the Stiffness

                 Matrix... 362

        10.5.1.2 Results 363

    10.5.2 Computer Code: AXI_SYM_Q8.m 365

        10.5.2.1 Numerical Integration of the Stiffness

                 Matrix... 368

        10.5.2.2 Results 370

10.6 Analysis with Abaqus Using the 8-Node Quadrilateral..................... 372

Chapter 11   Thin and Thick Plates 379

11.1 Introduction 379

11.2 Thin Plates 379

    11.2.1 Differential Equation of Plates Loaded

           in Bending.. 379

    11.2.2 Governing Equation in terms of Displacement

           Variables 382

11.3 Thick Plate Theory or Mindlin Plate Theory 383

    11.3.1 Stress–Strain Relationship................................ 384

11.4 Linear Elastic Finite Element Analysis of Plates........................... 385

    11.4.1 Finite Element Formulation for Thin Plates 385

        11.4.1.1 Triangular Element........................... 385

        11.4.1.2 Rectangular Element......................... 387

    11.4.2 Finite Element Formulation for Thick Plates......................... 388

11.5 Boundary Conditions 389

    11.5.1 Simply Supported Edge 389

    11.5.2 Built-in or Clamped Edge 390

    11.5.3 Free Edge... 390

11.6 Computer Program for Thick Plates Using the 8-Node

       Quadrilateral 390

11.6.1 Main Program: Thick_plate_Q8.m................................. 390
11.6.2 Data Preparation 395
11.6.2.1 Stiffness Matrices 395
11.6.2.2 Boundary Conditions 395
11.6.2.3 Loading.. 396
11.6.2.4 Numerical Integration of the Stiffness
Matrix.. 397
11.6.3 Results... 398
11.6.3.1 Determination of the Resulting Moments
and Shear Forces 398
11.6.3.2 Contour Plots.. 399
11.7 Analysis with Abaqus 400
11.7.1 Preliminary 400
11.7.1.1 Three-Dimensional Shell Elements 401
11.7.1.2 Axisymmetric Shell Elements........................... 401
11.7.1.3 Thick versus Thin Conventional Shell 401
11.7.2 Simply Supported Plate... 401
11.7.3 Three-Dimensional Shells 406

Appendix A: List of MATLAB® Modules and Functions 419

Appendix B: Statically Equivalent Nodal Forces 445

Appendix C: Index Notation and Transformation Laws for Tensors 447

References and Bibliography 453

Index.. 455

List of Figures

FIGURE 2.1   Truss structure..6

FIGURE 2.2   Bar element 6

FIGURE 2.3   Degrees of freedom of a rod element in a two-dimensional space. 7

FIGURE 2.4   Truss element oriented at an arbitrary angle θ 8

FIGURE 2.5   Model of a truss structure... 10

FIGURE 2.6   Free body diagram of the truss. 14

FIGURE 2.7   Free body diagram of element 3.. 18

FIGURE 2.8   Equilibrium of node 3. 19

FIGURE 2.9   Model of Problem 2.1.. 28

FIGURE 2.10   Model of Problem 2.2.. 32

FIGURE 2.11   Abaqus documentation. 36

FIGURE 2.12   Starting Abaqus. 36

FIGURE 2.13   Abaqus CAE main user interface. 37

FIGURE 2.14   Creating a part... 37

FIGURE 2.15   Choosing the geometry of the part. 37

FIGURE 2.16   Fitting the sketcher to the screen. 38

FIGURE 2.17   Drawing using the connected line button. 38

FIGURE 2.18   Drawing the truss geometry. 38

FIGURE 2.19   Finished part. 38

FIGURE 2.20   Material definition. 39

FIGURE 2.21   Material properties... 39

FIGURE 2.22   Create section window.. 40

FIGURE 2.23   Edit material window. 40

FIGURE 2.24   Section assignment... 40

FIGURE 2.25   Regions to be assigned a section. 41

FIGURE 2.26   Edit section assignment... 41

FIGURE 2.27   Loading the meshing menu.. 41

FIGURE 2.28   Selecting regions to be assigned element type............................. 42

FIGURE 2.29   Selecting element type. 42

FIGURE 2.30   Mesh. 43

FIGURE 2.31   Assembling the model. 43

FIGURE 2.32   Creating instances. 44

FIGURE 2.33   Numbering of the degrees of freedom. 44

FIGURE 2.34   Creating boundary conditions. 45

FIGURE 2.35   Type of boundary conditions. 45

FIGURE 2.36   Selecting a region to be assigned boundary conditions. 46

FIGURE 2.37   Edit boundary condition dialog box for pinned support. 46

FIGURE 2.38   Edit boundary condition dialog box for roller support. 47

FIGURE 2.39   Creating a step for load application. 47

FIGURE 2.40   Create step dialog box. 48

FIGURE 2.41   Edit step dialog box. 48

FIGURE 2.42   Creating a load. 49

FIGURE 2.43   Creating a concentrated load. 49

FIGURE 2.44   Selecting a joint for load application. 50

FIGURE 2.45   Entering the magnitude of a joint force. 50

FIGURE 2.46   Loaded truss. 50

FIGURE 2.47   Creating a job. 51

FIGURE 2.48   Naming a job. 51

FIGURE 2.49   Editing a job. 52

FIGURE 2.50   Submitting a job. 52

FIGURE 2.51   Monitoring of a job. 52

FIGURE 2.52   Opening the visualization module. 53

FIGURE 2.53   Common plot options. 53

FIGURE 2.54   Elements and nodes' numbering. 53

FIGURE 2.55   Deformed shape. 54

FIGURE 2.56   Field output dialog box. 54

FIGURE 2.57   Contour plot of the vertical displacement $U2$. 55

FIGURE 2.58   Viewport annotations options. 55

FIGURE 2.59   Normal stresses in the bars. 55

FIGURE 2.60   Selecting variables to print to a report. 56

FIGURE 2.61   Choosing a directory and the file name to which to write the report. 56

FIGURE 2.62   Running Abaqus from the command line. 61

FIGURE 3.1   Beam element 64

FIGURE 3.2   Differential element of a beam. 64

FIGURE 3.3   Nodal degrees of freedom 65

FIGURE 3.4   Statically equivalent nodal loads. 68

FIGURE 3.5   Loading, bending moment, and shear force diagrams. 68

FIGURE 3.6   Support reactions for individual members. 71

FIGURE 3.7   Beam with an internal hinge. 71

FIGURE 3.8   Beam elements with a hinge. 73

FIGURE 3.9   Example of a continuous beam. 73

FIGURE 3.10   Example 1: Continuous beam results. 81

FIGURE 3.11   Problem 3.1. 81

FIGURE 3.12   Problem 3.2 and equivalent nodal loads for elements 3 and 4. 84

FIGURE 3.13   Problem 3.3. 87

FIGURE 3.14   Continuous beam. 90

FIGURE 3.15   Beam cross section; dimensions are in mm. 90

FIGURE 3.16   Creating the Beam_Part. 91

FIGURE 3.17   Drawing using the connected line icon. 91

FIGURE 3.18   Material definition. 91

FIGURE 3.19   Creating a beam profile. 92

FIGURE 3.20   Entering the dimensions of a profile. 92

FIGURE 3.21   Creating a section. 93

FIGURE 3.22   Editing a beam section. 93

FIGURE 3.23   Editing section assignments. 94

FIGURE 3.24   Beam orientation. 94

FIGURE 3.25   Assigning beam orientation. 94

FIGURE 3.26   Rendering beam profile. 95

FIGURE 3.27   Rendered beam. 95

FIGURE 3.28   Selecting a beam element. 96

FIGURE 3.29   Seeding a mesh by size. 96

FIGURE 3.30   Node and element labels. 97

FIGURE 3.31   Creating a node set. 97

FIGURE 3.32   Selecting multiple nodes. 98

FIGURE 3.33   Creating element sets. 98

FIGURE 3.34   Imposing BC using created sets. 98

FIGURE 3.35   Selecting a node set for boundary conditions. 99

FIGURE 3.36   Editing boundary conditions. 99

FIGURE 3.37   Imposing BC using created sets. 100

FIGURE 3.38   Imposing a concentrated load using a node set. 100

FIGURE 3.39   Imposing a line load on an element set. 101

FIGURE 3.40   Field output. 101

FIGURE 3.41   Submitting a job in Abaqus CAE. 101

FIGURE 3.42   Plotting stresses in the bottom fiber. 102

FIGURE 4.1   Beam column element with six degrees of freedom. 108

FIGURE 4.2   Example 1: Portal frame. 110

FIGURE 4.3   Frame with an internal hinge. 119

FIGURE 4.4   Finite element discretization. 119

FIGURE 4.5   Statically equivalent nodal loads. 120

FIGURE 4.6   Portal frame. 124

FIGURE 4.7   Profiles' sections; dimensions are in mm. 125

FIGURE 4.8   Creating the Portal_frame part. 125

FIGURE 4.9   Material and profiles definitions. 126

FIGURE 4.10   Creating sections. 126

FIGURE 4.11   Editing section assignments. 127

FIGURE 4.12   Assigning beam orientation. 127

FIGURE 4.13   Rendering beam profile. 127

FIGURE 4.14   Seeding by number. 128

FIGURE 4.15   Mesh. 128

FIGURE 4.16   Creating the element set Rafters. 129

FIGURE 4.17   Imposing BC using created sets. 129

FIGURE 4.18   Imposing a line load in global coordinates. 130

FIGURE 4.19   Imposing a line load in local coordinates. 130

FIGURE 4.20   Analyzing a job in Abaqus CAE. 131

FIGURE 4.21   Plotting stresses in the bottom fiber (interactive edition). 131

FIGURE 4.22   Plotting stresses in the bottom fiber (keyword edition). 134

FIGURE 5.1   Internal force components. 136

FIGURE 5.2   Stress components at a point. 136

FIGURE 5.3   Stress components on a tetrahedron. 137

FIGURE 5.4   Equilibrium of an infinitesimal cube. 139

FIGURE 5.5   Principal directions of a stress tensor. 141

FIGURE 5.6   Tangential and normal components of the stress vector. 142

FIGURE 5.7   Mohr's circles. 143

FIGURE 5.8   Schematic representation of the deformation of a solid body. 145

FIGURE 5.9   Reference and current configurations. 146

FIGURE 5.10   Deformations of an infinitesimal element. 147

FIGURE 5.11   Geometrical representation of the components of strain at a point. 151

FIGURE 5.12   Monoclinic material. 155

FIGURE 5.13   Symmetry of rotation. 157

FIGURE 5.14   A state of plane stress. 162

FIGURE 5.15   State of plane strain. 163

FIGURE 5.16   Change of basis. 165

FIGURE 5.17   Displacement field (Problem 5.3). 167

FIGURE 5.18   Displacement field (Problem 5.5). 170

FIGURE 5.19   Strain rosette. 171

FIGURE 5.20   Problem 5.7. 172

FIGURE 5.21   Displacements without the rigid walls. 173

FIGURE 6.1   Graphical comparison of exact and approximate solution. 178

FIGURE 6.2   Integration by parts in two and three dimensions. 180

FIGURE 6.3   Infinitesimal element of the boundary. 180

FIGURE 6.4   Graphical comparison of the exact and approximate solutions. 186

FIGURE 7.1   Thick wall with embedded thermocouples. 192

FIGURE 7.2   Finite element discretization. 193

FIGURE 7.3   Finite element approximation. 195

FIGURE 7.4   Geometrical illustration of the compatibility principle 195

FIGURE 7.5   Linear triangle. 197

FIGURE 7.6   Geometrical transformation for a triangular element. 200

FIGURE 7.7   Three-node triangular element with an arbitrary point O. 202

FIGURE 7.8   Three-node triangular reference element. 204

FIGURE 7.9   Geometrical transformation. 204

FIGURE 7.10   One-dimensional elements. 207

FIGURE 7.11   Two-dimensional quadrilateral elements. 207

FIGURE 7.12 Two-dimensional triangular elements. 208

FIGURE 7.13 Three-dimensional tetrahedric elements. 209

FIGURE 7.14 Three-dimensional brick elements. 210

FIGURE 8.1 Positions of the sampling points for a triangle: Orders 1, 2, and 3. 219

FIGURE 8.2 Gauss quadrature over an arbitrary area. 219

FIGURE 8.3 Double change of variables. 220

FIGURE 8.4 Coarse mesh of two 8-nodded elements. 221

FIGURE 8.5 Eight elements finite element approximation with two 8-nodded elements. 222

FIGURE 8.6 Estimation of rainfall using finite element approximation. 226

FIGURE 9.1 Discretization error involving overlapping. 234

FIGURE 9.2 Discretization error involving holes between elements. 235

FIGURE 9.3 Plane elements with shape distortions. 235

FIGURE 9.4 Geometrical discretization error. 235

FIGURE 9.5 Linear triangular element. 236

FIGURE 9.6 Element nodal forces. 239

FIGURE 9.7 Analysis of a cantilever beam in plane stress. 240

FIGURE 9.8 Finite element discretization with linear triangular elements. 241

FIGURE 9.9 Deflection of the cantilever beam. 248

FIGURE 9.10 Stresses along the x-axis. 249

FIGURE 9.11 Automatic mesh generation with the CST element. 252

FIGURE 9.12 Deflection of the cantilever beam obtained with the fine mesh. 253

FIGURE 9.13 Stresses along the x-axis obtained with the fine mesh. 253

FIGURE 9.14 Creating the Beam_CST Part. 254

FIGURE 9.15 Drawing using the create-lines rectangle icon. 254

FIGURE 9.16 Creating a partition. 255

FIGURE 9.17 Creating a plane stress section. 255

FIGURE 9.18 Editing section assignments. 255

FIGURE 9.19 Mesh controls. 256

FIGURE 9.20 Selecting element type. 256

FIGURE 9.21 Seeding part by size. 256

FIGURE 9.22 Mesh. 257

FIGURE 9.23 Imposing BC using geometry. 257

FIGURE 9.24 Imposing a concentrated force using geometry. 257

FIGURE 9.25 Analyzing a job in Abaqus CAE. 258

FIGURE 9.26   Plotting displacements on deformed and undeformed shapes. 258

FIGURE 9.27   Generating a mesh manually in Abaqus. 261

FIGURE 9.28   Displacement contour. 263

FIGURE 9.29   Linear strain triangular element. 263

FIGURE 9.30   Automatic mesh generation with the LST element. 271

FIGURE 9.31   Deflection of the cantilever beam obtained with the LST element. 272

FIGURE 9.32   Stresses along the *x*-direction obtained with the LST element. 273

FIGURE 9.33   Aluminum plate with a hole. 273

FIGURE 9.34   Making use of symmetry. 273

FIGURE 9.35   Creating the Plate_LST Part. 274

FIGURE 9.36   Creating a plane stress section. 274

FIGURE 9.37   Editing section assignments. 275

FIGURE 9.38   Mesh controls. 275

FIGURE 9.39   Seeding edge by size and simple bias. 276

FIGURE 9.40   Creating a node set. 276

FIGURE 9.41   Creating a surface. 277

FIGURE 9.42   Imposing BC using node sets. 277

FIGURE 9.43   Imposing a pressure load on a surface. 278

FIGURE 9.44   Plotting the maximum in-plane principal stress (under tension). 279

FIGURE 9.45   Plotting the maximum in-plane principal stress (under compression). 279

FIGURE 9.46   Linear quadrilateral element. 280

FIGURE 9.47   Element loading. 283

FIGURE 9.48   Equivalent nodal loading. 284

FIGURE 9.49   Finite element discretization with 4-nodded quadrilateral elements. 285

FIGURE 9.50   Contour of the vertical displacement v_2. 290

FIGURE 9.51   Contour of the stress σ_{xx}. 291

FIGURE 9.52   Automatic mesh generation with the Q4 element. 295

FIGURE 9.53   Contour of the vertical displacement v_2. 295

FIGURE 9.54   Contour of the stresses along the *x*-axis σ_{xx}. 295

FIGURE 9.55   Creating the Beam_Q4 Part. 296

FIGURE 9.56   Creating a partition. 296

FIGURE 9.57   Creating a plane stress section. 297

FIGURE 9.58   Editing section assignments. 297

FIGURE 9.59   Mesh controls. 297

FIGURE 9.60   Selecting element type. 298

FIGURE 9.61   Seeding part by size. 298

FIGURE 9.62   Mesh. 298

FIGURE 9.63   Imposing BC using geometry. 299

FIGURE 9.64   Imposing a concentrated force using geometry. 299

FIGURE 9.65   Plotting displacements on deformed and undeformed shapes. 300

FIGURE 9.66   Generating a mesh manually in Abaqus. 302

FIGURE 9.67   Mesh generated with the keyword edition. 304

FIGURE 9.68   Displacement contour. 305

FIGURE 9.69   Eight-nodded isoparametric element. 305

FIGURE 9.70   Equivalent nodal loads. 307

FIGURE 9.71   Geometry and loading. 307

FIGURE 9.72   Coarse mesh. 308

FIGURE 9.73   Contour of the vertical displacement v_2. 314

FIGURE 9.74   Contour of the stress σ_{xx}. 314

FIGURE 9.75   Contour of the stress τ_{xy}. 315

FIGURE 9.76   Slender beam under 4-point bending. 315

FIGURE 9.77   Automatic mesh generation with the Q8 element. 319

FIGURE 9.78   Contour of the vertical displacement v_2. 320

FIGURE 9.79   Contour of the stress σ_{xx}. 320

FIGURE 9.80   Contour of the stress τ_{xy}. 320

FIGURE 9.81   Creating the Deep_Beam_Q8 Part. 321

FIGURE 9.82   Creating a plane stress section. 321

FIGURE 9.83   Editing section assignments. 322

FIGURE 9.84   Mesh controls and element type. 322

FIGURE 9.85   Mesh. 323

FIGURE 9.86   Creating the node set **Loaded_node**. 323

FIGURE 9.87   Creating the node set **Centerline**. 324

FIGURE 9.88   Creating the node set **Support**. 324

FIGURE 9.89   Imposing BC using a node set. 325

FIGURE 9.90   BC and loads. 325

FIGURE 9.91   Contour of the vertical displacement. 326

FIGURE 9.92   Contour of the horizontal stress σ_{xx}. 326

FIGURE 9.93   Strip footing. 327

FIGURE 9.94 Strip footing model. 328

FIGURE 9.95 Mesh with the CST element. 328

FIGURE 9.96 Computed result with the CST element. 332

FIGURE 9.97 Mesh with the LST element. 332

FIGURE 9.98 Statically equivalent loads for the LST element. 333

FIGURE 9.99 Computed result with the LST element. 336

FIGURE 9.100 Bridge pier. 337

FIGURE 9.101 Bridge pier model. 338

FIGURE 9.102 Element internal node numbering. 338

FIGURE 9.103 Finite element discretization of the pier model. 339

FIGURE 9.104 Contour of the vertical displacement. 350

FIGURE 9.105 Contour of the maximum principal stress σ_1. 350

FIGURE 9.106 Contour of the minimum principal stress σ_2. 351

FIGURE 10.1 Typical axisymmetric problem. 354

FIGURE 10.2 Strains and corresponding stresses in an axisymmetric solid. 354

FIGURE 10.3 Tangential strain. 354

FIGURE 10.4 Axisymmetric equivalent nodal loads. 356

FIGURE 10.5 Typical quadrilateral element on which axisymmetric loads are applied. 357

FIGURE 10.6 Circular footing on a sandy soil. 358

FIGURE 10.7 Geometrical model for the circular footing. 358

FIGURE 10.8 Finite element mesh using the 6-node triangle. 362

FIGURE 10.9 Contour plot of the vertical displacement. 363

FIGURE 10.10 Contour plot of the radial stress. 364

FIGURE 10.11 Contour plot of the vertical stress. 364

FIGURE 10.12 Contour plot of the shear stress. 365

FIGURE 10.13 Finite element mesh using the 8-node quadrilateral. 369

FIGURE 10.14 Contour plot of the vertical displacement. 370

FIGURE 10.15 Contour plot of the radial stress. 370

FIGURE 10.16 Contour plot of the vertical stress. 371

FIGURE 10.17 Contour plot of the shear stress. 371

FIGURE 10.18 Creating the FOOTING_Q8 Part. 372

FIGURE 10.19 Creating an axisymmetric section. 372

FIGURE 10.20 Editing section assignments. 373

FIGURE 10.21 Edge partition. 373

FIGURE 10.22   Mesh controls and element type. 374

FIGURE 10.23   Mesh.. 374

FIGURE 10.24   Imposing BC using geometry. 375

FIGURE 10.25   Imposing loads using geometry. 375

FIGURE 10.26   Contour of the vertical displacement. 376

FIGURE 10.27   Contour of the vertical stress σ_{yy}. 376

FIGURE 11.1   Deformed configuration of a thin plate in bending............................ 380

FIGURE 11.2   Internal stresses in a thin plate. Moments and shear forces due to internal stresses in a thin plate. 380

FIGURE 11.3   Moments and shear forces due to inernal stresses in a thin plate............. 380

FIGURE 11.4   Free body diagram of a plate element. 382

FIGURE 11.5   Deformed configuration of a thick plate in bending. 383

FIGURE 11.6   Three-node triangular plate bending element. 386

FIGURE 11.7   Four-node rectangular plate bending element. 387

FIGURE 11.8   Plate boundary conditions... 390

FIGURE 11.9   Simply supported plate on all edges. 391

FIGURE 11.10   Finite element mesh of one quadrant of the simply supported plate. 395

FIGURE 11.11   Contour plot of the vertical displacement. 399

FIGURE 11.12   Contour plot of the moment M_{xx}. 400

FIGURE 11.13   Contour plot of the moment M_{xy}. 400

FIGURE 11.14   Lifting of corners of a plate. 401

FIGURE 11.15   Creating the Slab_S4R Part. 402

FIGURE 11.16   Sketching the Slab_S4R Part.. 402

FIGURE 11.17   Creating a homogeneous shell section.. 402

FIGURE 11.18   Editing section assignments.. 403

FIGURE 11.19   Mesh controls and element type. 403

FIGURE 11.20   Mesh.. 404

FIGURE 11.21   Creating a node set. 404

FIGURE 11.22   Imposing BC **Edge_X0** using geometry....................................... 404

FIGURE 11.23   Imposing BC **Edge_Z18** using geometry...................................... 405

FIGURE 11.24   Imposing BC **Edge_Z0** using geometry. 405

FIGURE 11.25   Imposing BC **Edge_X9** using geometry....................................... 405

FIGURE 11.26   Imposing a concentrated force using a node set............................... 406

FIGURE 11.27   Plotting displacements on deformed shape..................................... 407

FIGURE 11.28   Castellated beam. 407

FIGURE 11.29   Base profile. 407

FIGURE 11.30   Castellated beam profile. 408

FIGURE 11.31   Geometrical details of the castellated beam. 408

FIGURE 11.32   Loading and boundary conditions. 408

FIGURE 11.33   Sketching the I profile. 409

FIGURE 11.34   Adding dimensions. 409

FIGURE 11.35   Finishing dimensioning the profile. 410

FIGURE 11.36   Editing shell extrusion. 410

FIGURE 11.37   Selecting a plane for an extruded cut. 410

FIGURE 11.38   Magnify View tool. 411

FIGURE 11.39   Sketching a hexagon. 411

FIGURE 11.40   Delete tool. 412

FIGURE 11.41   Dimension tool. 412

FIGURE 11.42   Linear pattern tool. 413

FIGURE 11.43   Editing a linear pattern. 413

FIGURE 11.44   Edit cut extrusion. 414

FIGURE 11.45   Creating a shell section. 414

FIGURE 11.46   Editing section assignments. 415

FIGURE 11.47   Mesh controls and element type. 415

FIGURE 11.48   Element type. 416

FIGURE 11.49   Mesh. 416

FIGURE 11.50   Imposing BC using geometry. 417

FIGURE 11.51   Applying a pressure load on a shell surface. 417

FIGURE 11.52   Contour of the vertical displacement. 418

FIGURE 11.53   Contour plot of the von Mises stress. 418

FIGURE B.1   Common beam loadings. 445

FIGURE C.1   Transformation of coordinates. 449

FIGURE C.2   Rotation around the third axis. 450

List of Tables

TABLE 5.1 Relationships between the Coefficients of Elasticity 161

TABLE 8.1 Abscissa and Weights for Gauss Quadrature 213

TABLE 8.2 Abscissae and Weights for a Triangle 218

TABLE 8.3 Coordinates of Rain Gages and Precipitations 227

Preface

The advent of the digital computer has revolutionized engineering curricula. In this day and age, the analysis of all but the simplest problem is carried out with the aid of a computer program that not only speeds up calculations but also allows the display of results in fancy graphics. For instance, when graduate engineers enter the design office, they encounter advanced commercial finite element software whose capabilities, and the theories behind their development, are far more superior to the training they have received during their university studies. These packages also come with a graphical user interface (GUI). Most of the time, this is the only component the user will interact with, and learning how to use the software is often a matter of trial and error assisted by the documentation that accompanies the software. However, proficiency in using the GUI is by no means related to the accuracy of the results. The latter depends very much on a deep understanding of the mathematics governing the theory. So, what is to be taught? This is the challenge facing experts and educators in engineering. Should only the theory be taught, with the practical aspects to be "picked up" later? Or, on the other hand, should the emphasis be on more "hands-on" applications using computer software at the expense of theory? The many textbooks that describe the theory of the finite element and/or its engineering applications fall into one of the following two categories: those that deal with the theory, assuming that the reader has access to some sort of software, and those that deal with the programming aspect, assuming that the reader has some theoretical knowledge of the method.

The theoretical approach is beneficial to students in the long term as it provides them with a deeper understanding of the mathematics behind the development of the finite element method. It also helps them prepare for postgraduate studies. However, it leaves very little time for practical applications, and as such it is not favored by employers, as they have to provide extra training for graduates in solving real-life problems. In addition, from my personal experience, it is often less attractive to students as it involves a lot of mathematics such as differential equations, matrix algebra, and advanced calculus. Indeed, finite element analysis subjects are usually taught in the two later years of the engineering syllabus, and at these later stages in their degree, most students expect that they have completed their studies in mathematics in the first two years. The "hands-on" approach, on the other hand, makes extensive use of the availability of computer facilities. Real-life problems are usually used as examples. It is very popular with students as it helps them solve problems quickly and efficiently with the results presented in attractive graphics. Students become experts at using the pre- and postprocessor abilities of the software and usually claim competency with a given computer package, which employers look well upon. However, this approach gives students a false sense of achievement. When faced with a novel problem, they usually do not know how to choose a suitable model and how to check the accuracy and the validity of the answers. In addition, modern packages have abilities beyond the student knowledge and experience. This is a serious cause for concern. In addition, given the many available computer software, it is also very unlikely that after graduating a student will use the same package on which he or she was trained.

The aim of this book, therefore, is to bridge this gap. It introduces the theory of the finite element method while keeping a balanced approach between its mathematical formulation, programming implementation and as its application using commercial software. The computer implementation is carried out using MATLAB®, while the practical applications are carried out in both MATLAB and Abaqus. MATLAB is a high-level language specially designed for dealing with matrices, making it particularly suited for programming the finite element method. In addition, it also allows the reader to focus on the finite element method by alleviating the programming burden. Experience has shown that books that include programming examples that can be implemented are of benefit to beginners. This book also includes detailed step-by-step procedures for solving problems with Abaqus interactive and keyword editions. Abaqus is one of the leading finite element packages and has much operational and verification experience to back it up, notwithstanding the quality of the pre- and postprocessing capabilities.

Finally, if you want to understand the introductory theory of the finite element method, to program it in MATLAB, and/or to get started with Abaqus, then this book is for you.

ABAQUS is a registered trade mark of Dassault Systèmes. For product information, please contact: Web: www.3ds.com

MATLAB® is a registered trademark of The MathWorks, Inc. For product information, please contact:

The MathWorks, Inc.
3 Apple Hill Drive
Natick, MA 01760-2098 USA
Tel: 508-647-7000
Fax: 508-647-7001
E-mail: info@mathworks.com
Web: www.mathworks.com

Author

Dr. Amar Khennane is a senior lecturer in the School of Engineering and Information Technology at the University of New South Wales, Canberra, Australian Capital Territory, Australia. He earned his PhD in civil engineering from the University of Queensland, Australia; a master of science in structural engineering from Heriot Watt University, United Kingdom; and a bachelor's degree in civil engineering from the University of Tizi-Ouzou, Algeria. His teaching experience spans 20 years and 2 continents. He has taught structural analysis, structural mechanics, and the finite element method at various universities.

1 Introduction

1.1 PROLOGUE

Undoubtedly, the finite element method represents one of the most significant achievements in the field of computational methods in the last century. Historically, it has its roots in the analysis of weight-critical framed aerospace structures. These framed structures were treated as an assemblage of one-dimensional members, for which the exact solutions to the differential equations for each member were well known. These solutions were cast in the form of a matrix relationship between the forces and displacements at the ends of the member. Hence, the method was initially termed matrix analysis of structures. Later, it was extended to include the analysis of continuum structures. Since continuum structures have complex geometries, they had to be subdivided into simple components or "elements" interconnected at nodes. It was at this stage in the development of the method that the term "finite element" appeared. However, unlike framed structures, closed form solutions to the differential equations governing the behavior of continuum elements were not available. Energy principles such as the theorem of virtual work or the principle of minimum potential energy, which were well known, combined with a piece-wise polynomial interpolation of the unknown displacement, were used to establish the matrix relationship between the forces and the interpolated displacements at the nodes numerically. In the late 1960s, when the method was recognized as being equivalent to a minimization process, it was reformulated in the form of weighted residuals and variational calculus, and expanded to the simulation of nonstructural problems in fluids, thermomechanics, and electromagnetics. More recently, the method is extended to cover multiphysics applications where, for example, it is possible to study the effects of temperature on electromagnetic properties that might affect the performance of electric motors.

1.2 FINITE ELEMENT ANALYSIS AND THE USER

Nowadays, in structural design, the analysis of all but simple structures is carried out using the finite element method. When graduate structural engineers enter the design office, they will encounter advanced commercial finite element software whose capabilities, and the theories behind its development, are far superior to the training they have received during their undergraduate studies. Indeed, current commercial finite element software is capable of simulating nonlinearity, whether material or geometrical, contact, structural interaction with fluids, metal forming, crash simulations, and so on. . . . Commercial software also come with advanced pre- and postprocessing abilities. Most of the time, these are the only components the user will interact with, and learning how to use them is often a matter of trial and error assisted by the documentation accompanying the software. However, proficiency in using the pre- and postprocessors is by no means related to the accuracy of the results. The preprocessor is just a means of facilitating the data input, since the finite element method requires a large amount of data, while the postprocessor is another means for presenting the results in the form of contour maps. The user must realize that the core of the analysis is what happens in between the two processes. To achieve proficiency in finite element analysis, the user must understand what happens in this essential part, often referred to as the "black box." This will only come after many years of high-level exposure to the fields that comprise FEA technology (differential equations, numerical analysis, and vector calculus). A formal training in numerical procedures and matrix algebra as applied in the finite element method would be helpful to the user, particularly if he/she is one of the many design engineers applying finite element techniques in their work without a prior training in numerical procedures.

1.3   AIM OF THE BOOK

The many textbooks that describe the theory of the finite element and/or its engineering applications can be split into two categories: those that deal with the theory, assuming that the reader has access to some sort of software, and those that deal with the programming aspect, assuming that the reader has some theoretical knowledge of the method. The aim of this book is to bridge this gap. It introduces the theory of the finite element method while keeping a balanced approach between its mathematical formulation, programming implementation, and its application using commercial software. The key steps are presented in sufficient details. The computer implementation is carried out using MATLAB®, while the practical applications are carried out in both MATLAB and Abaqus®.

MATLAB is a high-level language specially designed for dealing with matrices. This makes it particularly suited for programming the finite element method. In addition, MATLAB will allow the reader to focus on the finite element method by alleviating the programming burden. Experience has shown that books that include programming examples are of benefit to beginners. It should be pointed out, however, that this book is not about writing software to solve a particular problem. It is about teaching the first principles of the finite element method.

If the reader wishes to solve real-life problems, he/she will be better off using commercial software such as Abaqus rather than writing his/her own code. Home-written software may have serious bugs that can compromise the results of the analysis, while commercial software has much operational and verification experience to back it up, notwithstanding the quality of the pre- and postprocessing abilities. For this purpose, detailed step-by-step procedures for solving problems with Abaqus interactive and keyword editions are given in this book. Abaqus is a suite of commercial finite element codes. It consists of Abaqus Standard, which is a general purpose finite element software, and Abaqus Explicit for dynamic analysis. It is now owned by Dassault Systèms and is part of the SIMULIA range of products, http://www.simulia.com/products/unified_fea.html. Data input for a finite element analysis with Abaqus can be done either through Abaqus/CAE or CATIA, which are intuitive graphic user interfaces. They also allow monitoring and viewing of results. Data can be entered in or using an input file prepared with a text editor and executed through the command line, or using a script prepared with Python. Python is an object-oriented programming language and is included in Abaqus as Abaqus Python. The latter is an advanced option reserved for experienced users and will not be covered in this book.

1.4   BOOK ORGANIZATION

The organization of the book contents follows the historical development of the finite element method. After some introductory notes in Chapter 1, Chapters 2 through 4 introduce matrix structural analysis for trusses, beams, and frames. The matrix relationships between the forces and nodal displacements for each element type are derived using the direct approaches from structural mechanics. Using a truss as an example in Chapter 1, the different steps required in a finite element code; such as describing loads, supports, material, and mesh preparation, matrix manipulation, introduction of boundary condition, and equation solving are described succinctly. Indeed, a truss offers all the attributes necessary to illustrate the coding of a finite element code. Similar codes are developed for beams and rigid jointed frames in Chapters 3 and 4, respectively. The described procedures are implemented as MATLAB codes at the end of each chapter. In addition, detailed step-by-step procedures for solving similar problems with both the Abaqus interactive and keyword editions are provided at the end of each chapter.

Chapter 5 marks the change of philosophy between matrix structural analysis and finite element analysis of a continuum. In matrix analysis, there is only one dominant stress, which is the longitudinal stress. In a continuum, on the other hand, there are many stresses and strains at a point. Chapter 5 introduces the theories of stress and strain, and the relationships between them. It also includes many solved problems that would help the reader understand the developed theories.

Chapters 6 and 7 introduce, respectively, the weighted residual methods and finite element approximation, which include the various types of continuum elements and the different techniques used to construct the piece-wise polynomial interpolations of the unknown quantities. These methods are necessary to establish the matrix relationships between forces and nodal displacements for continuum elements of complicated geometry, and whose behavior is governed by differential equations for which closed form solutions cannot be easily established.

Chapter 8 is entirely devoted to numerical integration using the Gauss Legendre and Hammer formulae with many examples at the end of the chapter. Indeed, during the implementation of the finite element method, many integrals arise, as will be seen in Chapters 9 through 11. When the number of elements is large, and/or their geometrical shape is general, as is the case in most applications, the use of analytical integration is quite cumbersome and ill suited for computer coding.

In Chapter 9, the finite element formulation for plane stress/strain problems is presented. The stiffness matrices for the triangular and quadrilateral families of elements are developed in detail, enabling the reader to solve a wide variety of problems. The chapters also include a wide variety of solved problems with MATLAB and Abaqus.

Chapter 10 introduces axisymmetric problems while Chapter 11 is devoted to the theory of plates. The stiffness matrices for the most common elements are developed in detail, and numerous examples are solved at the end of each chapter using both MATLAB and Abaqus.

The appendices and http://www.crcpress.com/product/isbn9781466580206 contain all the MATLAB codes used in the examples.

2 Bar Element

2.1 INTRODUCTION

There is no better way of illustrating the steps involved in a finite element analysis than by analyzing a simple truss. Indeed, a truss is the first structural system introduced into the cursus of engineering studies. As early as the first year, the student becomes acquainted with a truss in engineering statics. A truss offers all the attributes needed to illustrate a finite analysis without the need to resort to advanced mathematical tools such as numerical integration and geometrical transformations that are required in the analysis of complicated structures.

A truss is a structure that consists of axial members connected by pin joints, as shown in Figure 2.1. The loads on a truss are assumed to be concentrated at the joints. The members of a truss support the external load through axial force as they do not undergo bending deformation. Therefore, no bending moments are present in truss members.

2.2 ONE-DIMENSIONAL TRUSS ELEMENT

2.2.1 FORMULATION OF THE STIFFNESS MATRIX: THE DIRECT APPROACH

A member of a truss is the simplest solid element, namely, an elastic rod with ends 1 and 2 referred to hereafter as nodes. Consider an element of length L, cross section A, and made of a linear elastic material having a Young's modulus E as represented in Figure 2.2a. If we apply a normal force N_1 at node 1, and at the same time maintaining node 2 fixed in space, the bar shortens by an amount u_1 as represented in Figure 2.2b.

The force N_1 is related to the displacement u_1 through the spring constant

$$N_1 = \frac{AE}{L} u_1 \tag{2.1}$$

In virtue of Newton's third law, there must be a reaction force R_2 at node 2 equal (in magnitude) and opposite (in direction) to the force N_1; that is,

$$R_2 = -\frac{AE}{L} u_1 \tag{2.2}$$

Similarly, if we apply a normal force N_2 at node 2, and at the same time maintaining node 1 fixed in space, the bar lengthens by an amount u_2 as represented in Figure 2.2c. In the same fashion, the force N_2 is related to the displacement u_2 through the spring constant

$$N_2 = \frac{AE}{L} u_2 \tag{2.3}$$

Again, in virtue of Newton's third law, there must be a reaction force R_1 at node 1 equal (in magnitude) and opposite (in direction) to the force N_2; that is,

$$R_1 = -\frac{AE}{L} u_2 \tag{2.4}$$

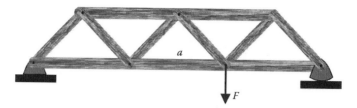

FIGURE 2.1 Truss structure.

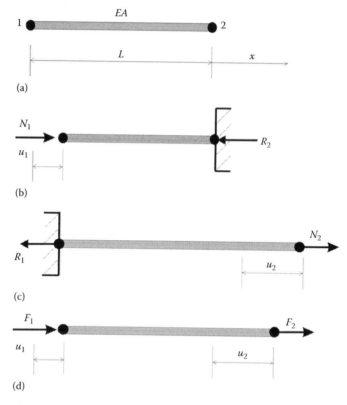

FIGURE 2.2 Bar element: (a) geometry, (b) nodal force applied at node 1, (c) nodal force applied at node 2, (d) nodal forces applied at both nodes.

When the bar is subjected to both forces N_1 and N_2 in virtue of the principle of superposition, the total forces F_1 and F_2 shown in Figure 2.2d will be

$$F_1 = N_1 - R_1 = \frac{AE}{L}u_1 - \frac{AE}{L}u_2$$
$$F_2 = N_2 - R_2 = -\frac{AE}{L}u_1 + \frac{AE}{L}u_2 \tag{2.5}$$

Rearranging Equations (2.5) in a matrix form yields

$$\begin{bmatrix} AE/L & -AE/L \\ -AE/L & AE/L \end{bmatrix} \begin{Bmatrix} u_1 \\ u_2 \end{Bmatrix} = \begin{Bmatrix} F_1 \\ F_2 \end{Bmatrix} \tag{2.6}$$

or simply as

$$[K_e]\{u_e\} = \{F_e\} \tag{2.7}$$

where
> the vector $\{u_e\}$ is the vector of nodal displacements
> the vector $\{F_e\}$ is the vector of nodal forces

The matrix $[K_e]$ is called the stiffness matrix; it relates the nodal displacements to the nodal forces.

Knowing the forces F_1 and F_2, one may be tempted to solve the system of Equation (2.6) to obtain the displacements u_1 and u_2. This is not possible, at least in a unique sense. Indeed, taking a closer look at the matrix $[K_e]$, it can be seen that its determinant is equal to zero; that is,

$$det([K_e]) = \left(\frac{AE}{L}\right)^2 - \left(\frac{AE}{L}\right)^2 = 0 \tag{2.8}$$

That is, any set of displacements u_1 and u_2 is a solution to the system. As odd as it may appear at this stage, this actually makes a lot of physical sense. In Figure 2.2d, the bar is subject to the forces F_1 and F_2. Under the action of these forces, the bar will experience a rigid body movement since it is not restrained in space. There will be many sets of displacements u_1 and u_2 that are solutions to the system (2.6). To obtain a unique solution, the bar must be restrained in space against rigid body movement. The state of restraints of the bar, or the structure in general, is introduced in the form of boundary conditions. This will be covered in detail in Section 2.4.

2.2.2 Two-Dimensional Truss Element

As shown in Figure 2.1, a plane truss structure consists of axial members with different orientations. A longitudinal force in one member may act at a right angle to another member. For example, the force F in Figure 2.1 acts at right angle to member **a**, and therefore causing it to displace in its transversal direction.

The nodal degrees of freedom (nodal displacements) of the rod element become four as represented in Figure 2.3, and they are given as

$$\{d_e\} = \{u_1, v_1, u_2, v_2\}^T \tag{2.9}$$

The corresponding stiffness matrix becomes

$$[K_e] = \begin{bmatrix} AE/L & 0 & -AE/L & 0 \\ 0 & 0 & 0 & 0 \\ -AE/L & 0 & AE/L & 0 \\ 0 & 0 & 0 & 0 \end{bmatrix} \tag{2.10}$$

Note that the second and fourth columns and rows associated with the transversal displacements are null since the truss member has axial deformation only.

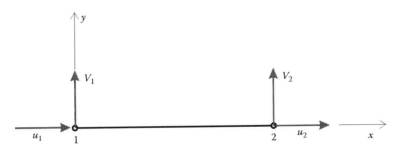

FIGURE 2.3 Degrees of freedom of a rod element in a two-dimensional space.

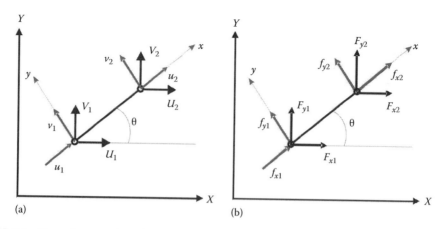

FIGURE 2.4   Truss element oriented at an arbitrary angle θ: (a) Nodal displacements, (b) Nodal forces.

Another problem that arises from the fact that all truss members do not have the same orientation is that when it comes to assemble the global stiffness, we need to have the element degrees of freedom (nodal displacements) given in terms of the common reference axes of the truss.

Figure 2.4 shows a truss element oriented at an arbitrary angle θ with respect to the horizontal axis (X, Y) of the structure. It also shows two sets of nodal displacements: The first set (u, v) is given in terms of the local set of axis (x, y) associated with the element, while the second set of displacements (U, V) is associated with the global set of axis (X, Y).

The element stiffness matrix is expressed in terms of the local displacements u and v. In order to be assembled with the stiffness matrices of the other elements to form the global stiffness matrix of the whole structure, it should be transformed such that it is expressed in terms of the global displacements U and V.

If we consider node 1, it can be seen that the displacements U_1 and V_1 can be written in terms of u_1 and v_1 as

$$U_1 = u_1 \cos \theta - v_1 \sin \theta$$
$$V_1 = u_1 \sin \theta + v_1 \cos \theta$$

(2.11)

In a similar fashion, U_2 and V_2 can be expressed in terms of u_2 and v_2 as

$$U_2 = u_2 \cos \theta - v_2 \sin \theta$$
$$V_2 = u_2 \sin \theta + v_2 \cos \theta$$

(2.12)

Grouping Equations (2.11) and (2.12) yields

$$\begin{Bmatrix} U_1 \\ V_1 \\ U_2 \\ V_2 \end{Bmatrix} = \begin{bmatrix} \cos \theta & -\sin \theta & 0 & 0 \\ \sin \theta & \cos \theta & 0 & 0 \\ 0 & 0 & \cos \theta & -\sin \theta \\ 0 & 0 & \sin \theta & \cos \theta \end{bmatrix} \begin{Bmatrix} u_1 \\ v_1 \\ u_2 \\ v_2 \end{Bmatrix}$$

(2.13)

or in a more compact form as

$$\{\overline{d_e}\} = [C]\{d_e\}$$

(2.14)

The matrix $[C]$ is called the transformation matrix. It is an orthonormal matrix with a determinant equal to one. Its inverse is simply equal to its transpose; that is,

$$[C]^{-1} = [C]^T \tag{2.15}$$

The vector of the global nodal forces $\{\overline{f_e}\} = \{F_{x1}, F_{y1}, F_{x2}, F_{y2}\}^T$ may be also obtained from the vector of local nodal forces $\{f_e\} = \{f_{x1}, f_{y1}, f_{x2}, f_{y2}\}^T$ as

$$\{\overline{f_e}\} = [C]\{f_e\} \tag{2.16}$$

In the local coordinate system, the force–displacement relation is given as

$$[K_e]\{d_e\} = \{f_e\} \tag{2.17}$$

Using $\{d_e\} = [C]^T\{\overline{d_e}\}$ and $\{f_e\} = [C]^T\{\overline{f_e}\}$, and substituting in (2.17) yields

$$[K_e][C]^T\{\overline{d_e}\} = [C]^T\{\overline{f_e}\} \tag{2.18}$$

Premultiplying both sides by $[C]$ yields

$$[C][K_e][C]^T\{\overline{d_e}\} = \{\overline{f_e}\} \tag{2.19}$$

which can be rewritten as

$$[\overline{K_e}]\{\overline{d_e}\} = \{\overline{f_e}\} \tag{2.20}$$

with

$$[\overline{K_e}] = [C][K_e][C]^T \tag{2.21}$$

The matrix $[\overline{K_e}]$ is called the element stiffness matrix in the global coordinate system; it relates the global nodal displacements to the global nodal forces.

2.3 GLOBAL STIFFNESS MATRIX ASSEMBLY

2.3.1 DISCRETIZATION

To illustrate how the elements' stiffness matrices are put together to form the global stiffness matrix, we proceed with a very simple example. Consider the truss represented in Figure 2.5.

First, we number all the elements and the nodes as well as identifying the nodal degrees of freedom (global displacement), as shown in Figure 2.5. In total, there are three nodes, three elements, and six degrees of freedom $[U_1, V_1, U_2, V_2, U_3, V_3]$.

2.3.2 ELEMENTS' STIFFNESS MATRICES IN LOCAL COORDINATES

Referring to Equation (2.10), it can be seen that the element stiffness matrix is a function of the material properties through the elastic modulus E, the cross-sectional area A of the element, and its length L. The elastic modulus refers to the material used to build the truss. If we assume that all the members of the truss are made of steel with an elastic modulus of 200000 MPa, and all the elements have the same cross-sectional area, say 2300 mm², then it is possible to evaluate each element stiffness matrix.

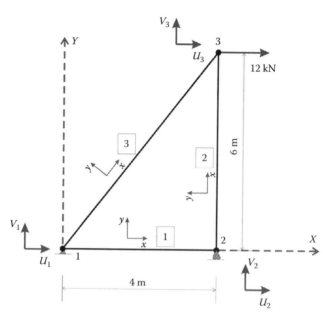

FIGURE 2.5   Model of a truss structure.

Element 1 has a length of 4000 mm. Substituting in Equation (2.10), its stiffness matrix in its local coordinates is obtained as

$$[K_1]_L = \begin{bmatrix} 115000 & 0 & -115000 & 0 \\ 0 & 0 & 0 & 0 \\ -115000 & 0 & 115000 & 0 \\ 0 & 0 & 0 & 0 \end{bmatrix} \tag{2.22}$$

Element 2 has a length of 6000 mm. Its stiffness matrix in its local coordinates is obtained as

$$[K_2]_L = \begin{bmatrix} 76666.67 & 0 & -76666.67 & 0 \\ 0 & 0 & 0 & 0 \\ -76666.67 & 0 & 76666.67 & 0 \\ 0 & 0 & 0 & 0 \end{bmatrix} \tag{2.23}$$

Element 3 has a length of 7211 mm, which can be calculated with the well-known Pythagoras formula. Its stiffness matrix in its local coordinates is obtained as

$$[K_3]_L = \begin{bmatrix} 63791.43 & 0 & -63791.43 & 0 \\ 0 & 0 & 0 & 0 \\ -63791.43 & 0 & 63791.43 & 0 \\ 0 & 0 & 0 & 0 \end{bmatrix} \tag{2.24}$$

2.3.3   ELEMENTS' STIFFNESS MATRICES IN GLOBAL COORDINATES

The elements' stiffness matrices, as respectively given by Equations (2.22) through (2.24), cannot be assembled into the global stiffness matrix of the truss because they are formulated in their respective local coordinate systems. In order to do so, they need to be transformed from their local coordinate systems (x, y) to the global coordinate system (X, Y).

2.3.3.1 Element 1

The local axis x of element 1 makes an angle of $0°$ with the global X axis of the structure. In virtue of Equation (2.13), its transformation matrix $[C]$ is given as

$$[C_1] = \begin{bmatrix} \cos(0) & -\sin(0) & 0 & 0 \\ \sin(0) & \cos(0) & 0 & 0 \\ 0 & 0 & \cos(0) & -\sin(0) \\ 0 & 0 & \sin(0) & \cos(0) \end{bmatrix} = \begin{bmatrix} 1 & 0 & 0 & 0 \\ 0 & 1 & 0 & 0 \\ 0 & 0 & 1 & 0 \\ 0 & 0 & 0 & 1 \end{bmatrix} \quad (2.25)$$

The transformation matrix $[C_1]$ of element 1 is an identity matrix. Therefore, as per Equation (2.19), premultiplying the matrix $[K_1]_L$ by $[C_1]$ and postmultiplying it by $[C_1]^T$, that is, $[C_1][K_1]_L[C_1]^T$, would not change anything, the reason being that the local axes of element 1 are co-linear with global axes (X, Y) of the structure. Therefore, the stiffness matrix of element 1 $[K_1]_G$ in the global coordinates system remains unchanged; that is,

$$[K_1]_G = \begin{array}{c} \\ U_1/u_1 \\ V_1/v_1 \\ U_2/u_2 \\ V_2/v_2 \end{array} \begin{array}{cccc} U_1/u_1 & V_1/v_1 & U_2/u_2 & V_2/v_2 \\ \begin{bmatrix} 115000 & 0 & -115000 & 0 \\ 0 & 0 & 0 & 0 \\ -115000 & 0 & 115000 & 0 \\ 0 & 0 & 0 & 0 \end{bmatrix} \end{array} \quad (2.26)$$

In its local coordinates (x, y), it has the degrees of freedom $\{u_1, v_1, u_2, v_2\}$, while in the global coordinates, as shown in Figure 2.5, it has the global degrees of freedom $\{U_1, V_1, U_2, V_2\}$. The top row and the left column outside the matrix show the correspondence between the local and the global degrees of freedom.

2.3.3.2 Element 2

The local axis x of element 2 makes an angle of $90°$ with the global X axis of the structure. In virtue of Equation (2.13), its transformation matrix $[C]$ is given as

$$[C_2] = \begin{bmatrix} 0 & -1 & 0 & 0 \\ 1 & 0 & 0 & 0 \\ 0 & 0 & 0 & -1 \\ 0 & 0 & 1 & 0 \end{bmatrix} \quad (2.27)$$

Premultiplying the matrix $[K_2]_L$ by $[C_2]$ and postmultiplying it by $[C_2]^T$ yields the stiffness matrix $[K_2]_G = [C_2][K_2]_L[C_2]^T$ of element 2 in the global system of axes:

$$[K_2]_G = \begin{array}{c} \\ U_2/u_1 \\ V_2/v_1 \\ U_3/u_2 \\ V_3/v_2 \end{array} \begin{array}{cccc} U_2/u_1 & V_2/v_1 & U_3/u_2 & V_3/v_2 \\ \begin{bmatrix} 0 & 0 & 0 & 0 \\ 76666.67 & 0 & -76666.67 & 0 \\ 0 & 0 & 0 & 0 \\ -76666.67 & 0 & 76666.67 & 0 \end{bmatrix} \end{array} \quad (2.28)$$

In its local coordinates (x, y), it has the degrees of freedom $\{u_1, v_1, u_2, v_2\}$, while in the global coordinates, as shown in Figure 2.5, it has the global degrees of freedom $\{U_2, V_2, U_3, V_3\}$.

2.3.3.3 Element 3

The local axis x of element 3 makes an angle of $\theta = \tan^{-1}(6/4) = 56.31°$ with the global X axis of the structure. Using Equation (2.13), its transformation matrix $[C_3]$ is given as

$$[C_3] = \begin{bmatrix} 0.554699 & -0.832051 & 0 & 0 \\ 0.832051 & 0.554699 & 0 & 0 \\ 0 & 0 & 0.554699 & -0.832051 \\ 0 & 0 & 0.832051 & 0.554699 \end{bmatrix} \quad (2.29)$$

Premultiplying the matrix $[K_3]_L$ by $[C_3]$ and postmultiplying it by $[C_3]^T$ yields the stiffness matrix $[K_3]_G = [C_3][K_3]_L[C_3]^T$ of element 3 in the global system of axes:

$$[K_3]_G = \begin{array}{c} \\ U_1/u_1 \\ V_1/v_1 \\ U_3/u_2 \\ V_3/v_2 \end{array} \begin{array}{c} \overset{\displaystyle U_1/u_1 \quad V_1/v_1 \quad U_3/u_2 \quad V_3/v_2}{} \\ \begin{bmatrix} 19628 & 29442 & -19628 & -29442 \\ 29442 & 44163 & -29442 & -44163 \\ -19628 & -29442 & 19628 & 29442 \\ -29442 & -44163 & 29442 & 44163 \end{bmatrix} \end{array} \quad (2.30)$$

In its local coordinates (x, y), it has the degrees of freedom $\{u_1, v_1, u_2, v_2\}$, while in the global coordinates, as shown in Figure 2.5, it has the global degrees of freedom $\{U_1, V_1, U_3, V_3\}$.

2.3.4 GLOBAL MATRIX ASSEMBLY

As shown in Figure 2.5, the truss has six degrees of freedom $\{U_1, V_1, U_2, V_2, U_3, V_3\}$; that is, two degrees of freedom per node. Its stiffness matrix must therefore have six lines and six columns each corresponding to a degree of freedom:

$$[\mathbf{K}] = \begin{array}{c} \\ U_1 \\ V_1 \\ U_2 \\ V_2 \\ U_3 \\ V_3 \end{array} \begin{array}{c} \overset{\displaystyle U_1 \quad V_1 \quad U_2 \quad V_2 \quad U_3 \quad V_3}{} \\ \begin{bmatrix} 0 & 0 & 0 & 0 & 0 & 0 \\ 0 & 0 & 0 & 0 & 0 & 0 \\ 0 & 0 & 0 & 0 & 0 & 0 \\ 0 & 0 & 0 & 0 & 0 & 0 \\ 0 & 0 & 0 & 0 & 0 & 0 \\ 0 & 0 & 0 & 0 & 0 & 0 \end{bmatrix} \end{array} \quad (2.31)$$

To populate the global stiffness matrix, imagine three hypothetical states:

- First, only element 1 is present
- Second, only element 2 is present
- Third, only element 3 is present

2.3.4.1 Only Element 1 Is Present

$$[\mathbf{K}] = \begin{array}{c} \\ U_1 \\ V_1 \\ U_2 \\ V_2 \\ U_3 \\ V_3 \end{array} \begin{array}{cccccc} U_1 & V_1 & U_2 & V_2 & U_3 & V_3 \\ \left[\begin{array}{cccccc} 115000 & 0 & -115000 & 0 & 0 & 0 \\ 0 & 0 & 0 & 0 & 0 & 0 \\ -115000 & 0 & 115000 & 0 & 0 & 0 \\ 0 & 0 & 0 & 0 & 0 & 0 \\ 0 & 0 & 0 & 0 & 0 & 0 \\ 0 & 0 & 0 & 0 & 0 & 0 \end{array}\right] \end{array} \quad (2.32)$$

Notice that only the cases corresponding to the global degrees of freedom of element 1 are populated.

2.3.4.2 Only Element 2 Is Present

$$[\mathbf{K}] = \begin{array}{c} \\ U_1 \\ V_1 \\ U_2 \\ V_2 \\ U_3 \\ V_3 \end{array} \begin{array}{cccccc} U_1 & V_1 & U_2 & V_2 & U_3 & V_3 \\ \left[\begin{array}{cccccc} 0 & 0 & 0 & 0 & 0 & 0 \\ 0 & 0 & 0 & 0 & 0 & 0 \\ 0 & 0 & 0 & 0 & 0 & 0 \\ 0 & 0 & 0 & 76666.67 & 0 & -76666.67 \\ 0 & 0 & 0 & 0 & 0 & 0 \\ 0 & 0 & 0 & -76666.67 & 0 & 76666.67 \end{array}\right] \end{array} \quad (2.33)$$

2.3.4.3 Only Element 3 Is Present

$$[\mathbf{K}] = \begin{array}{c} \\ U_1 \\ V_1 \\ U_2 \\ V_2 \\ U_3 \\ V_3 \end{array} \begin{array}{cccccc} U_1 & V_1 & U_2 & V_2 & U_3 & V_3 \\ \left[\begin{array}{cccccc} 19628 & 29442 & 0 & 0 & -19628 & -29442 \\ 29442 & 44163 & 0 & 0 & -29442 & -44163 \\ 0 & 0 & 0 & 0 & 0 & 0 \\ 0 & 0 & 0 & 0 & 0 & 0 \\ -19628 & -29442 & 0 & 0 & 19628 & 29442 \\ -29442 & -44163 & 0 & 0 & 29442 & 44163 \end{array}\right] \end{array} \quad (2.34)$$

By direct addition of the preceding matrices, the global structure stiffness matrix is obtained as

$$
[\mathbf{K}] = \begin{array}{c} \\ U_1 \\ V_1 \\ U_2 \\ V_2 \\ U_3 \\ V_3 \end{array}
\begin{array}{c} U_1 \qquad V_1 \qquad U_2 \qquad V_2 \qquad U_3 \qquad V_3 \end{array}
\begin{bmatrix}
115000+19628 & 29442 & -115000 & 0 & -19628 & -29442 \\
29442 & 44163 & 0 & 0 & -29442 & -44163 \\
-115000 & 0 & 115000 & 0 & 0 & 0 \\
0 & 0 & 0 & 76666.67 & 0 & -76666.67 \\
-19628 & -29442 & 0 & 0 & 19628 & 29442 \\
-29442 & -44163 & 0 & -76666.67 & 29442 & 44163+76666.67
\end{bmatrix} \qquad (2.35)
$$

2.3.5 GLOBAL FORCE VECTOR ASSEMBLY

Figure 2.6 shows a free body diagram where all the external forces acting on the truss are represented. At node 1, which is pinned, there are two reaction forces: R_{X1} and R_{Y1}. At node 2, which is a roller support, there is one reaction force R_{Y2}. Node 3 is free, but there is an external force of 12000 N acting in the positive x-direction.

The external forces can be grouped in the global force vector as

$$
\{F\} = \begin{Bmatrix} R_{X1} \\ R_{Y1} \\ 0 \\ R_{Y2} \\ 12000 \\ 0 \end{Bmatrix} \qquad (2.36)
$$

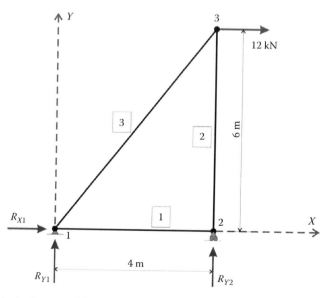

FIGURE 2.6 Free body diagram of the truss.

2.4   BOUNDARY CONDITIONS

2.4.1   General Case

Once the global stiffness matrix and the global force vector are assembled, the equilibrium equations of the truss are written as follows:

$$
\begin{bmatrix}
134628 & 29442 & -115000 & 0 & -19628 & -29442 \\
29442 & 44163 & 0 & 0 & -29442 & -44163 \\
-115000 & 0 & 115000 & 0 & 0 & 0 \\
0 & 0 & 0 & 76666.67 & 0 & -76666.67 \\
-19628 & -29442 & 0 & 0 & 19628 & 29442 \\
-29442 & -44163 & 0 & -76666.67 & 29442 & 120829.67
\end{bmatrix}
$$

$$
\times
\begin{Bmatrix}
U_1 \\ V_1 \\ U_2 \\ V_2 \\ U_3 \\ V_3
\end{Bmatrix}
=
\begin{Bmatrix}
R_{X1} \\ R_{Y1} \\ 0 \\ R_{Y2} \\ 12000 \\ 0
\end{Bmatrix}
$$

or in more compact form as

$$[K]\{\delta\} = \{F\} \tag{2.37}$$

As given by Equation (2.37), the system of equations cannot be solved in a unique fashion since the matrix $[K]$ is singular. Indeed, it is assembled from the elements' stiffness matrices, which are singular. In addition, the right-hand-side vector contains the unknown support reactions. To solve the system of equations, it is necessary to partition the matrix $[K]$ according to known and unknown quantities. The vector of displacements $\{\delta\}$ can be partitioned into known and unknown quantities. Node 1 is a pinned support; therefore, the displacements U_1 and V_1 are both equal to zero. Node 2 is a roller support; therefore, the displacement V_2 is also equal to zero. It follows therefore that the vector $\{\delta\}$ can be partitioned as follows:

$$
\{\delta\} =
\begin{Bmatrix}
U_1 = 0 \\
V_1 = 0 \\
V_2 = 0 \\
\cdots \\
U_2 \\
U_3 \\
V_3
\end{Bmatrix}
\tag{2.38}
$$

Similarly, the right-hand-side vector of global forces can be partitioned accordingly:

$$
\{F\} =
\begin{Bmatrix}
R_{X1} \\
R_{Y1} \\
R_{Y2} \\
\cdots \\
0 \\
12000 \\
0
\end{Bmatrix}
\tag{2.39}
$$

Note that unknown displacements correspond to known forces and known displacements correspond to unknown forces.

Finally, the matrix $[K]$ is partitioned as

$$
\begin{bmatrix}
134628 & 29442 & 0 & \vdots & -115000 & -19628 & -29442 \\
29442 & 44163 & 0 & \vdots & 0 & -29442 & -44163 \\
0 & 0 & 76666.67 & \vdots & 0 & 0 & -76666.67 \\
\cdots & \cdots & \cdots & \cdots & \cdots & \cdots & \cdots \\
-115000 & 0 & 0 & \vdots & 115000 & 0 & 0 \\
-19628 & -29442 & 0 & \vdots & 0 & 19628 & 29442 \\
-29442 & -44163 & -76666.67 & \vdots & 0 & 29442 & 120829.67
\end{bmatrix}
$$

$$
\times
\begin{Bmatrix}
U_1 = 0 \\
V_1 = 0 \\
V_2 = 0 \\
\cdots \\
U_2 \\
U_3 \\
V_3
\end{Bmatrix}
=
\begin{Bmatrix}
R_{X1} \\
R_{Y1} \\
R_{Y2} \\
\cdots \\
0 \\
12000 \\
0
\end{Bmatrix}
$$

As a result of the position of V_2 being interchanged with that of U_2 in the vector $\{\delta\}$, column 3 and line 3 have also been respectively interchanged with column 4 and line 4 in the matrix $[K]$. Finally, the partitioned system of equations can be rewritten in a compact form as

$$
\begin{bmatrix}
[K_{PP}] & \vdots & [K_{PF}] \\
\cdots & \cdots & \cdots \\
[K_{FP}] & \vdots & [K_{FF}]
\end{bmatrix}
\begin{Bmatrix}
\{\delta_P\} \\
\cdots \\
\{\delta_F\}
\end{Bmatrix}
=
\begin{Bmatrix}
\{F_P\} \\
\cdots \\
\{F_F\}
\end{Bmatrix}
\tag{2.40}
$$

where

The subscripts P and F refer respectively to the prescribed and free degrees of freedom
$\{\delta_P\}^T = \{0.\ 0.\ 0.\}$ the vector of the known prescribed displacements
$\{\delta_F\}^T = \{U_2\ U_3\ V_3\}$ the vector of the unknown free displacements
$\{F_P\}^T = \{R_{X1}\ R_{Y1}\ R_{Y2}\}$ the vector of the unknown reaction forces corresponding to the prescribed displacements
$\{F_F\}^T = \{0\ 12000\ 0\}$ the vector of the known applied external forces

2.5 SOLUTION OF THE SYSTEM OF EQUATIONS

Equation (2.40) can be expanded to yield

$$
[K_{PP}]\{\delta_P\} + [K_{PF}]\{\delta_F\} = \{F_P\}
\tag{2.41}
$$

$$
[K_{FP}]\{\delta_P\} + [K_{FF}]\{\delta_F\} = \{F_F\}
\tag{2.42}
$$

Since $\{\delta_P\}$ and $\{F_F\}$ are known quantities, it is then possible to obtain from Equation (2.42) the vector $\{\delta_P\}$ as

$$
\{\delta_F\} = [K_{FF}]^{-1}\{\{F_F\} - [K_{FP}]\{\delta_P\}\}
\tag{2.43}
$$

However, since $\{\delta_P\}^T = \{0.\ 0.\ 0.\}$, Equation (2.43) reduces to

$$\{\delta_F\} = [K_{FF}]^{-1} \{F_F\} \tag{2.44}$$

which is simply equivalent to eliminating the lines and the columns corresponding to the restrained degrees of freedom in the global matrix; that is,

$$\begin{Bmatrix} U_2 \\ U_3 \\ V_3 \end{Bmatrix} = \begin{bmatrix} 115000 & 0 & 0 \\ 0 & 19628 & 29442 \\ 0 & 29442 & 120829.67 \end{bmatrix}^{-1} \begin{Bmatrix} 0 \\ 12000 \\ 0 \end{Bmatrix}$$

Solving the system of equations yields

$$\{\delta_F\} = \begin{Bmatrix} U_2 \\ U_3 \\ V_3 \end{Bmatrix} = \begin{Bmatrix} 0 \\ 0.9635 \\ -0.2348 \end{Bmatrix} \text{ mm}$$

In summary, the vector of global displacements can be obtained as

$$\{\delta\} = \begin{Bmatrix} U_1 = 0. \\ V_1 = 0. \\ U_2 = 0. \\ V_2 = 0. \\ U_3 = 0.9635 \\ V_3 = -0.2348 \end{Bmatrix} \tag{2.45}$$

2.6 SUPPORT REACTIONS

Once $\{\delta_F\}$ is known, it is possible to obtain from Equation (2.41) the vector of the unknown reaction forces $\{F_P\}^T = \{R_{X1}\ R_{Y1}\ R_{Y2}\}$. Since $\{\delta_P\}^T = \{0.\ 0.\ 0.\}$, the vector $\{F_P\}$ is obtained as

$$\{F_P\} = [K_{PF}] \{\delta_F\}$$

That is,

$$\begin{Bmatrix} R_{X1} \\ R_{Y1} \\ R_{Y2} \end{Bmatrix} = \begin{bmatrix} -115000 & -19628 & -29442 \\ 0 & -29442 & -44163 \\ 0 & 0 & -76666.67 \end{bmatrix} \begin{Bmatrix} 0 \\ 0.9635 \\ -0.2348 \end{Bmatrix} = \begin{Bmatrix} -12 \\ -18 \\ 18 \end{Bmatrix} \text{ kN}$$

The obtained values for the support reactions can be easily checked using the equilibrium equations of a rigid body. Considering the free body diagram of the truss as shown in Figure 2.6, and taking moments with respect to node 1 yields

$$\Sigma_{/1} = R_{Y2} \times 4 - 12 \times 6 = 0 \Longrightarrow R_{Y2} = 18 \text{ kN}$$

Considering vertical equilibrium yields

$$\Sigma_Y = R_{Y2} + R_{Y1} = 0 \Longrightarrow R_{Y1} = -18 \text{ kN}$$

Considering horizontal equilibrium yields

$$\Sigma_X = 12 + R_{X1} = 0 \Longrightarrow R_{X1} = -12 \text{ kN}$$

2.7   MEMBERS' FORCES

Once all the displacements are known, the member forces can be easily obtained. For example, element 3 has the following vector of global displacements, $\{\overline{d_3}\}$, extracted from the global displacements vector $\{\delta\}$ Equation (2.45):

$$\{\overline{d_3}\} = \begin{Bmatrix} U_1 = 0 \\ V_1 = 0 \\ U_3 = 0.9635 \\ V_3 = -0.2348 \end{Bmatrix}$$

The vector of displacements in local coordinates $\{d_3\}$ is obtained using the inverse transformation $\{d_3\} = [C_3]^T\{\overline{d_3}\}$; that is,

$$\{d_3\} = \begin{bmatrix} 0.554699 & 0.832051 & 0 & 0 \\ -0.832051 & 0.554699 & 0 & 0 \\ 0 & 0 & 0.554699 & 0.832051 \\ 0 & 0 & -0.832051 & 0.554699 \end{bmatrix} \begin{Bmatrix} 0 \\ 0 \\ 0.9635 \\ -0.2348 \end{Bmatrix} = \begin{Bmatrix} 0 \\ 0 \\ 0.3391 \\ -0.9319 \end{Bmatrix}$$

Multiplying the local stiffness matrix of element 3, $[K_3]_L$, by the local displacement vector $\{d_3\}$ yields the local vector of forces $\{f_3\}$; that is,

$$\{f_3\} = \begin{bmatrix} 63791.43 & 0 & -63791.43 & 0 \\ 0 & 0 & 0 & 0 \\ -63791.43 & 0 & 63791.43 & 0 \\ 0 & 0 & 0 & 0 \end{bmatrix} \begin{Bmatrix} 0 \\ 0 \\ 0.3391 \\ -0.9319 \end{Bmatrix} = \begin{Bmatrix} -21.631 \\ 0 \\ 21.631 \\ 0 \end{Bmatrix} \text{ kN}$$

The forces on the bar element are represented graphically in Figure 2.7. It can be seen that the member is under a tensile force of 21.631 kN. This result can be checked using the method of joints.

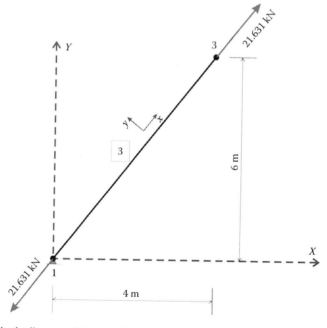

FIGURE 2.7   Free body diagram of element 3.

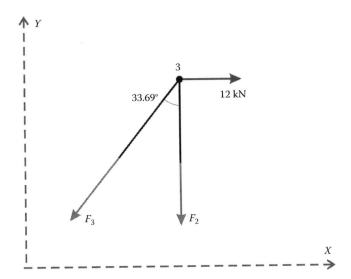

FIGURE 2.8   Equilibrium of node 3.

Consider the free body diagram of node (joint) 3 as shown in Figure 2.8. The equilibrium of the joint in the x direction requires

$$\Sigma_X = 12 - F_3 \times \sin(33.69) = 0 \Longrightarrow F_3 = 21.633 \text{ kN}$$

This confirms the obtained result with the finite element method.

Remark: The preceding sections illustrate the steps required in a finite element analysis. As can be noticed, even for a small number of elements (in this case 3), the calculations are rather involved. For very large structures, with a large number of elements, the calculation effort is so intensive that it is virtually impossible to carry out without the help of a digital computer. However, it can be also noticed that the calculations involve matrix algebra and the steps are quite repetitive, which makes them ideally suited for programming on a digital computer.

2.8   COMPUTER CODE: truss.m

The programming style and the syntax used by Smith and Griffiths [3] are adopted herein, except that the coding is done in MATLAB®. MATLAB is a high-level language specially designed for dealing with matrices. This makes it particularly suited for programming the finite element method. In addition, MATLAB will allow the reader to focus on the finite element method rather than on the programming details.

Programming the finite element method involves the following steps:

- **Step 1**: Data preparation and input.
- **Step 2**: Computation of element matrices.
- **Step 3**: Assembly of elements' stiffness matrices and elements' force vectors in the global stiffness matrix and global force vector.
- **Step 4**: Imposition of boundary conditions such as prescribed displacements. This is usually carried out simultaneously with the assembly of the global matrix in **Step 3**.
- **Step 5**: Solution of the global system of equations for the nodal unknowns.
- **Step 6**: Computation of secondary variables such as stresses and strains from displacements.
- **Step 7**: Print and/or plot desired results.

These steps are best illustrated by means of an example. Let us consider the truss represented in Figure 2.5. The main program is given in the M-file **truss.m**.

2.8.1 Data Preparation

Since the basic building block in MATLAB is a matrix, the data will be prepared in the form of tables whenever possible as they are very easily translated into matrices. Although there are many ways of reading data in MATLAB, in what follows we will use an M-file, **truss_1_data.m**, to read the data relevant to the truss. Note that a consistent set of units is required in any finite element analysis. In this case, *mm* are used for length and *N* for forces.

The input data for this structure consist of

- **nnd = 3**; number of nodes
- **nel = 3**; number of elements
- **nne = 2**; number of nodes per element
- **nodof = 2**; number of degrees of freedom per node

2.8.1.1 Nodes Coordinates

The coordinates x and y of the nodes are given in the form of a matrix **geom(nnd, 2)**:

$$\mathbf{geom} = \begin{bmatrix} 0 & 0 \\ 4000 & 0 \\ 4000 & 6000 \end{bmatrix}$$

2.8.1.2 Element Connectivity

The table of connectivity describes how the elements are connected to each other. The nodal coordinates are given in the matrix **connec(nel, 2)**:

$$\mathbf{connec} = \begin{bmatrix} 1 & 2 \\ 2 & 3 \\ 1 & 3 \end{bmatrix}$$

2.8.1.3 Material and Geometrical Properties

The material and geometrical properties are given in the matrix **prop(nel, 2)**:

$$\mathbf{prop} = \begin{bmatrix} 200000 & 2300 \\ 200000 & 2300 \\ 200000 & 2300 \end{bmatrix}$$

2.8.1.4 Boundary Conditions

Boundary conditions give information on how the structure is restrained in space against any rigid body movement. Without the introduction of boundary conditions, the global stiffness matrix is singular. To solve the equilibrium equations, we need to know how the nodes are restrained in space.

In what follows, we adopt the following convention:

- A restrained degree of freedom is assigned the digit 0
- A free degree of freedom is assigned the digit 1

The information on the boundary conditions is given in the matrix **nf(nnd, nodof)**. This matrix is called the matrix of nodal freedom matrix. It is first initialized to 1, then the degrees of freedom are read:

$$\mathbf{nf} = \begin{bmatrix} 0 & 0 \\ 1 & 0 \\ 1 & 1 \end{bmatrix}$$

The free degrees of freedom (different from zero) are then counted and their rank assigned back into the matrix $\mathbf{nf}(\mathbf{nnd}, \mathbf{nodof})$:

$$\mathbf{nf} = \begin{bmatrix} 0 & 0 \\ 1 & 0 \\ 2 & 3 \end{bmatrix}$$

In this case, the total number of active degrees of freedom is obtained as $\mathbf{n} = \mathbf{3}$. At this stage, it is possible to initialize the global matrix $\mathbf{KK}(\mathbf{n}, \mathbf{n}) = \mathbf{0}$ and the global force vector $\mathbf{F}(\mathbf{n}) = \mathbf{0}$:

$$\mathbf{KK} = \begin{bmatrix} 0 & 0 & 0 \\ 0 & 0 & 0 \\ 0 & 0 & 0 \end{bmatrix} \quad \text{and} \quad \mathbf{F} = \begin{Bmatrix} 0 \\ 0 \\ 0 \end{Bmatrix}$$

Note that we will only assemble the quantities corresponding to the active degrees of freedom; that is, the lines and the columns in the matrix $\mathbf{KK}$ corresponding respectively to the active degrees of freedom $\mathbf{1}$, $\mathbf{2}$, and $\mathbf{3}$. As to the restrained degrees of freedom, with a number equal to $\mathbf{0}$, they will be simply eliminated.

2.8.1.5 Loading

Finally, to be able to solve for the unknown displacements, we need to know how the structure is loaded. The information about the loading is also given in the form of a matrix, $\mathbf{load}(\mathbf{nnd}, \mathbf{2})$:

$$\mathbf{load} = \begin{bmatrix} 0 & 0 \\ 0 & 0 \\ 1200 & 0 \end{bmatrix}$$

The data are stored in the M-file: **truss_1_data.m**.

At this level in the main program **truss.m**, the model data are written to the file **truss_1_results.txt** using the M-file: **print_truss_model.m**. This is not necessary; however, it is always helpful to write the data to a file because it is easier to check for errors.

2.8.2 ELEMENT MATRICES

2.8.2.1 Stiffness Matrix in Local Coordinates

For each element, from 1 to *nel*, we set up the local stiffness and transformation matrices. Once the stiffness matrix **kl** is set up in local coordinates, it is transformed into global coordinates **kg** through the transformation matrix **C** and then assembled to the global stiffness matrix **KK**.

For any element **i**, we retrieve its first and second node from the connectivity matrix:

$$\mathbf{node\_1} = \mathbf{connec}(\mathbf{i}, \mathbf{1})$$

$$\mathbf{node\_2} = \mathbf{connec}(\mathbf{i}, \mathbf{2})$$

Then using the values of the nodes, we retrieve their x and y coordinates from the geometry matrix:

$$\mathbf{x1} = \mathbf{geom}(\mathbf{node\_1}, \mathbf{1}); \qquad \mathbf{y1} = \mathbf{geom}(\mathbf{node\_1}, \mathbf{2})$$

$$\mathbf{x2} = \mathbf{geom}(\mathbf{node\_2}, \mathbf{1}); \qquad \mathbf{y2} = \mathbf{geom}(\mathbf{node\_2}, \mathbf{2})$$

Next, using Pythagoras theorem, we evaluate the length of the element:

$$L = \sqrt{(x2 - x1)^2 + (y2 - y1)^2}$$

Finally, we retrieve the material and geometrical property of the section

$$E = prop(i, 1); \qquad A = prop(i, 2)$$

before evaluating the matrix **kl** using Equation (2.10).

The MATLAB script for evaluating the matrix **kl** is given in Appendix A in the M-file **truss_kl.m**.

2.8.2.2 Transformation Matrix

Once the nodal coordinates are retrieved, it is also possible to evaluate the angle θ that the element makes with the global X axis:

$$\theta = \tan^{-1}\left(\frac{y2 - y1}{x2 - x1}\right)$$

However, care should be taken when the element is at right angle with the global axis X as **x2−x1** = 0. The matrix **C** is evaluated using Equation (2.25). The MATLAB script is given in Appendix A in the M-file **truss_C.m**.

2.8.2.3 Stiffness Matrix in Global Coordinates

The element stiffness matrix **kg** is obtained using Equation (2.21):

$$\mathbf{kg} = \mathbf{C} \times \mathbf{kl} \times \mathbf{C^T}$$

2.8.2.4 "Steering" Vector

Once the matrix **kg** is formed, we retrieve the "steering vector" **g** containing the number of degrees of freedom of the nodes of the element:

$$\mathbf{g} = \begin{Bmatrix} \mathbf{nf(node\_1, 1)} \\ \mathbf{nf(node\_1, 2)} \\ \mathbf{nf(node\_2, 1)} \\ \mathbf{nf(node\_2, 2)} \end{Bmatrix}$$

For example, for element 1, the vector **g** will look like

$$\mathbf{g} = \begin{Bmatrix} 0 \\ 0 \\ 1 \\ 0 \end{Bmatrix}$$

The only nonzero component in the vector **g** is located in the **third** position, and its value is equal to 1. That is, only the element corresponding to the **third** line and **third** column in the matrix [**kg**(3, 3)] will be assembled, and it will occupy the position [**KK**(1, 1] in the global matrix. The MATLAB script for constructing the steering vector **g** is given in Appendix A in the M-file **truss_g.m**.

2.8.3 ASSEMBLY OF THE GLOBAL STIFFNESS MATRIX

The global stiffness matrix [**KK**] is assembled using a double loop over the components of the vector **g**:

loop **i**: for any $\mathbf{g(i) \neq 0}$
      loop **j**: for any $\mathbf{g(j) \neq 0}$
      add [**kg(i, j)**] to [**KK(g(i), g(j))**]
      end loop **j**
end loop **i**

The script is given in Appendix A in the M-file **form_KK.m**.

2.8.4 ASSEMBLY OF THE GLOBAL FORCE VECTOR

A loop is carried over all the nodes. If a degree of freedom j of a node i is free, that is, $\mathbf{nf(i, j) \neq 0}$, then it is susceptible of carrying an external force

$$\mathbf{F(nf(i, j)) = load(i, j)}$$

The global force vector is formed in Appendix A in the M-file **form_truss_F.m**.

2.8.5 SOLUTION OF THE GLOBAL SYSTEM OF EQUATIONS

In MATLAB, it is very easy to solve a system of linear equations: one statement does it all. In this case, the global displacements vector **delta** is obtained as

$$\mathbf{delta = KK\backslash F}$$

The backslash symbol \ is used to "divide" a matrix by a vector.

2.8.6 NODAL DISPLACEMENTS

Once the global displacements vector **delta** is obtained, it is possible to retrieve any nodal displacements. A loop is carried over all the nodes. If a degree of freedom j of a node i is free, that is, $\mathbf{nf(i, j) \neq 0}$, then it could have a displacement different from zero. The value of the displacement is extracted from the global displacements vector **delta**:

$$\mathbf{node\_disp(i, j) = delta(nf(i, j))}$$

2.8.7 ELEMENT FORCES

To obtain the member forces, a loop is carried over all the elements:

1. Form element stiffness matrix [**kl**] in local xy.
2. Form element transformation matrix [**C**].
3. Transform the element matrix from local to global coordinates $[\mathbf{kg}] = [\mathbf{C}] * [\mathbf{kl}] * [\mathbf{C}]^{\mathrm{T}}$.
4. Form element "steering" vector {**g**}.
   a. Loop over the degrees of freedom of the element to obtain element displacements vector **edg** in global coordinates.
   b. If $\mathbf{g(j) = 0}$, then the degree of freedom is restrained; $\mathbf{edg(j) = 0}$.
   c. Otherwise $\mathbf{edg(j) = delta(g(j))}$.

5. Obtain element force vector in global XY coordinates
   $\{\mathbf{fg}\} = [\mathbf{kg}] * \{\mathbf{edg}\}$.
6. Transform element force vector to local coordinates $\{\mathbf{fl}\} = [\mathbf{C}]^T * \{\mathbf{fg}\}$.
7. For each element, store the third component of $\{\mathbf{fl}\}$. If the component is positive, the element is under "tension," otherwise it is under "compression."

The results of the analysis are written to the file **truss_results.txt** using the M-file **print_1_results.m** given in Appendix A. A copy of the file **truss_results.txt** is included within Section 2.8.8.

2.8.8   PROGRAM SCRIPTS

File:truss.m

```
%                              truss.m
%
%  LINEAR STATIC ANALYSIS OF A TRUSS STRUCTURE
%
clc                % Clear screen
clear              % Clear all variables in memory
%
% Make these variables global so they can be shared
% by other functions
%
global nnd nel nne nodof eldof n global geom connec prop nf load
%
disp('Executing truss.m');


%
%%
%  ALTER THE NEXT LINES TO CHOOSE AN OUTPUT FILE FOR THE RESULTS
%  Open file for output of results
%
fid = fopen('truss_1_results.txt','w'); disp('Results printed in
file : truss_1_results.txt ');
%
% ALTER THE NEXT LINE TO CHOOSE AN INPUT FILE
%
truss_1_data              % Load the input file
%
print_truss_model         % Print model data
%
KK =zeros(n) ;            % Initialize global stiffness
                          % matrix to zero
%
F=zeros(n,1);             % Initialize global force
                          % vector to zero
%
for i=1:nel
    kl=truss_kl(i);       % Form element matrix in local xy
%
    C = truss_C(i);       % Form transformation matrix
%
    kg=C*kl*C' ;          % Transform the element matrix from
                          % local to global coordinates
%
    g=truss_g(i) ;        % Retrieve the element steering
                          % vector
%
    KK =form_KK(KK, kg, g);    % assemble global stiffness
                               % matrix
%
end
%
```

```
%
F = form_truss_F(F);                  % Form global force vector
%
%
%%%%%%%%%%%%  End of assembly  %%%%%%%%%%%
%
%
delta = KK\F ;          % solve for unknown displacements
%
% Extract nodal displacements
%
for i=1:nnd
    for j=1:nodof
        node_disp(i,j) = 0;
        if nf(i,j)~= 0;
        node_disp(i,j) = delta(nf(i,j)) ;
        end
    end
end
%
% Calculate the forces acting on each element
% in local coordinates, and store them in the
% vector force().
%
 for i=1:nel
    kl=truss_kl(i);        % Form element matrix in local xy
    C = truss_C(i);        % Form transformation matrix
    kg=C*kl*C' ;           % Transform the element matrix from
                           % local to global coordinates
    g=truss_g(i) ;         % Retrieve the element steering vector
    for j=1:eldof
        if g(j)== 0
            edg(j)=0.;  % displacement = 0. for restrained freedom
        else
            edg(j) = delta(g(j));
        end
    end
    fg = kg*edg';          % Element force vector in global XY
    fl=C'*fg ;             % Element force vector in local  xy
    force(i) = fl(3);
end
%
print_truss_results;
%
fclose(fid);
```

File:truss_1_data.m

```
%                   File:    truss_1_data.m
%
% The following variables are declared as global in order
% to be used by all the functions (M-files) constituting
% the program
%
global nnd nel nne nodof eldof n global geom connec prop nf load
%
format short e
%
%%%%%%%%%%%%%%% Beginning of data input %%%%%%%%%%%%%%%%%%
%
nnd = 3; % Number of nodes:
nel = 3; % Number of elements:
nne = 2 ; % Number of nodes per element:
nodof =2 ; % Number of degrees of freedom per node
eldof = nne*nodof; % Number of degrees of freedom
                  % per element
%
```

```
% Nodes coordinates X and Y
geom=zeros(nnd,2);
geom = [0.      0.    ; ... % X and Y coord. node 1
        4000.   0.    ; ... % X and Y coord. node 2
        4000.   6000.];     % X and X coord. node 3
%
% Element connectivity
%
connec=zeros(nel,2);
connec = [1   2 ; ... % 1st and 2nd node of element 1
          2   3 ; ... % 1st and 2nd node of element 2
          1   3];     % 1st and 2nd node of element 3
%
% Geometrical properties
%
% prop(1,1) = E; prop(1,2)= A
%
prop=zeros(nel,2);
prop = [200000    2300; ...  % E and A of element 1
        200000    2300; ...  % E and A of element 2
        200000    2300];     % E and A of element 3
%
% Boundary conditions
%
nf = ones(nnd, nodof); % Initialize the matrix nf to 1
nf(1,1) = 0; nf(1,2) =0 ; % Prescribed nodal freedom of node 1
nf(2,2) = 0 ;             % Prescribed nodal freedom of node 3
%
% Counting of the free degrees of freedom
%
n=0; for i=1:nnd
    for j=1:nodof
        if nf(i,j) ~= 0
        n=n+1;
        nf(i,j)=n;
        end
    end
end
%
% loading
%
load = zeros(nnd, 2);
load(3,:)=[1200. 0]; %forces in X and Y directions at node 3
%
%%%%%%%%%%%%%%%%%%%%%%%%% End of input %%%%%%%%%%%%%%%%%%%%%%%%
```

File:truss_1_results.txt

```
******* PRINTING MODEL DATA **************


-------------------------------------------------------
Number of nodes:                                    3
Number of elements:                                 3
Number of nodes per element:                        2
Number of degrees of freedom per node:              2
Number of degrees of freedom per element:           4

-------------------------------------------------------
Node        X           Y
 1,      0000.00,    0000.00
 2,      4000.00,    0000.00
 3,      4000.00,    6000.00

-------------------------------------------------------
Element     Node_1      Node_2
   1,         1,           2
```

```
     2,          2,          3
     3,          1,          3

------------------------------------------------------

Element        E              A
    1,       200000,        2300
    2,       200000,        2300
    3,       200000,        2300

------------------------------------------------------

Node      disp_U      disp_V
   1,        0,          0
   2,        1,          0
   3,        2,          3

------------------------------------------------------

Node      load_X          load_Y
   1,     0000.00,        0000.00
   2,     0000.00,        0000.00
   3,     1200.00,        0000.00
------------------------------------------------------


Total number of active degrees of freedom, n = 3

--------------------------------------------------------


  ****** PRINTING ANALYSIS RESULTS **************

------------------------------------------------------
Global force vector  F
    0
  1200
    0

------------------------------------------------------
Displacement solution vector:  delta
 -0.00000
  0.09635
 -0.02348

------------------------------------------------------
Nodal displacements
Node      disp_X          disp_Y
   1,     0.00000,        0.00000
   2,    -0.00000,        0.00000
   3,     0.09635,       -0.02348


------------------------------------------------------
Members actions
element      force         action
   1,        -0.00,      Compression
   2,     -1800.00,      Compression
   3,      2163.33,      Tension
```

2.9   PROBLEMS

Prepare a data file for the trusses shown next and carry out the analysis using the code *truss.m*.

2.9.1   PROBLEM 2.1 (FIGURE 2.9)

Input file

```
%               File:   truss_problem_1_data.m
%
% The following variables are declared as global in order
```

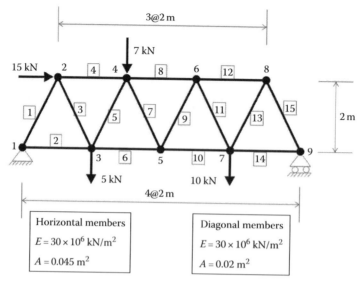

FIGURE 2.9   Model of Problem 2.1.

```
% to be used by all the functions (M-files) constituting
% the program
%
global nnd nel nne nodof eldof n global geom connec prop nf load
%
format short e
%
%%%%%%%%%%%%%% Beginning of data input %%%%%%%%%%%%%%%%%
%
nnd = 9; % Number of nodes:
nel = 15; % Number of elements:
nne = 2 ; % Number of nodes per element:
nodof =2 ; % Number of degrees of freedom per node
eldof = nne*nodof; % Number of degrees of freedom
                % per element
%
% Nodes coordinates X and Y
geom=zeros(nnd,2);
geom = [0.       0.; ...    % X and Y coord. node 1
        1.       2.; ...    % X and Y coord. node 2
        2.       0.; ...    % X and Y coord. node 3
        3.       2.; ...    % X and Y coord. node 4
        4.       0.; ...    % X and Y coord. node 5
        5.       2.; ...    % X and Y coord. node 6
        6.       0.; ...    % X and Y coord. node 7
        7.       2.; ...    % X and Y coord. node 8
        8.       0.] ;      % X and Y coord. node 9

%
% Element connectivity
%
connec=zeros(nel,2);
connec = [1    2 ; ...     % 1st and 2nd node of element 1
          1    3 ; ...     % 1st and 2nd node of element 2
          2    3 ; ...     % 1st and 2nd node of element 3
          2    4 ; ...     % 1st and 2nd node of element 4
          3    4 ; ...     % 1st and 2nd node of element 5
          3    5 ; ...     % 1st and 2nd node of element 6
          4    5 ; ...     % 1st and 2nd node of element 7
          4    6 ; ...     % 1st and 2nd node of element 8
```

```
          5      6 ; ...    % 1st and 2nd node of element 9
          5      7 ; ...    % 1st and 2nd node of element 10
          6      7 ; ...    % 1st and 2nd node of element 11
          6      8 ; ...    % 1st and 2nd node of element 12
          7      8 ; ...    % 1st and 2nd node of element 13
          7      9 ; ...    % 1st and 2nd node of element 14
          8      9 ]  ;     % 1st and 2nd node of element 15
%
% Geometrical properties
%
% prop(1,1) = E; prop(1,2)= A
%
prop=zeros(nel,2);
prop  = [30.e6        0.02  ; ...    % E and A of element 1
          30.e6       0.045 ; ...    % E and A of element 2
          30.e6       0.02  ; ...    % E and A of element 3
          30.e6       0.045 ; ...    % E and A of element 4
          30.e6       0.02  ; ...    % E and A of element 5
          30.e6       0.045 ; ...    % E and A of element 6
          30.e6       0.02  ; ...    % E and A of element 7
          30.e6       0.045 ; ...    % E and A of element 8
          30.e6       0.02  ; ...    % E and A of element 9
          30.e6       0.045 ; ...    % E and A of element 10
          30.e6       0.02  ; ...    % E and A of element 11
          30.e6       0.045 ; ...    % E and A of element 12
          30.e6       0.02  ; ...    % E and A of element 13
          30.e6       0.045 ; ...    % E and A of element 14
          30.e6       0.02  ];       % E and A of element 15
%
% Boundary conditions
%
nf = ones(nnd, nodof); % Initialize the matrix nf to 1
nf(1,1) = 0; nf(1,2) =0 ; % Prescribed nodal freedom of node 1
nf(9,2)= 0 ;               % Prescribed nodal freedom of node 3
%
% Counting of the free degrees of freedom
%
n=0; for i=1:nnd
     for j=1:nodof
          if nf(i,j) ~= 0
          n=n+1;
          nf(i,j)=n;
          end
     end
end
%
% loading
%
load = zeros(nnd, 2);
load(2,:)=[15.    0.]; %forces in X and Y directions at node 2
load(3,:)=[0.    -5.]; %forces in X and Y directions at node 3
load(4,:)=[0.    -7.]; %forces in X and Y directions at node 4
load(7,:)=[0.   -10.]; %forces in X and Y directions at node 7

%
%%%%%%%%%%%%%%%%%%%%%%%%%% End of input %%%%%%%%%%%%%%%%%%%%%%%%%
```

Results file

```
******* PRINTING MODEL DATA **************


---------------------------------------------------------
Number of nodes:                              9
Number of elements:                          15
Number of nodes per element:                  2
```

```
Number of degrees of freedom per node:          2
Number of degrees of freedom per element:       4
```

```
--------------------------------------------------------
Node         X               Y
 1,      0000.00,         0000.00
 2,      0001.00,         0002.00
 3,      0002.00,         0000.00
 4,      0003.00,         0002.00
 5,      0004.00,         0000.00
 6,      0005.00,         0002.00
 7,      0006.00,         0000.00
 8,      0007.00,         0002.00
 9,      0008.00,         0000.00
```

```
--------------------------------------------------------
Element       Node_1         Node_2
   1,           1,             2
   2,           1,             3
   3,           2,             3
   4,           2,             4
   5,           3,             4
   6,           3,             5
   7,           4,             5
   8,           4,             6
   9,           5,             6
  10,           5,             7
  11,           6,             7
  12,           6,             8
  13,           7,             8
  14,           7,             9
  15,           8,             9
```

```
--------------------------------------------------------
Element       E               A
   1,      3e+007,          0.02
   2,      3e+007,          0.045
   3,      3e+007,          0.02
   4,      3e+007,          0.045
   5,      3e+007,          0.02
   6,      3e+007,          0.045
   7,      3e+007,          0.02
   8,      3e+007,          0.045
   9,      3e+007,          0.02
  10,      3e+007,          0.045
  11,      3e+007,          0.02
  12,      3e+007,          0.045
  13,      3e+007,          0.02
  14,      3e+007,          0.045
  15,      3e+007,          0.02
```

```
--------------------------------------------------------
Node        disp_U         disp_V
 1,           0,             0
 2,           1,             2
 3,           3,             4
 4,           5,             6
 5,           7,             8
 6,           9,            10
 7,          11,            12
 8,          13,            14
 9,          15,             0
```

```
--------------------------------------------------------
Node        load_X         load_Y
 1,      0000.00,         0000.00
 2,      0015.00,         0000.00
```

```
3,      0000.00,       -005.00
4,      0000.00,       -007.00
5,      0000.00,       0000.00
6,      0000.00,       0000.00
7,      0000.00,       -010.00
8,      0000.00,       0000.00
9,      0000.00,       0000.00
------------------------------------------------------------
```

Total number of active degrees of freedom, n = 15

```
------------------------------------------------------------
```

   ****** PRINTING ANALYSIS RESULTS *************

```
------------------------------------------------------------
```
Global force vector   F
```
    15
     0
     0
    -5
     0
    -7
     0
     0
     0
     0
     0
   -10
     0
     0
     0
```

```
------------------------------------------------------------
```
Displacement solution vector: delta
```
   0.00014
  -0.00010
   0.00003
  -0.00019
   0.00010
  -0.00023
   0.00006
  -0.00023
   0.00007
  -0.00021
   0.00009
  -0.00018
   0.00005
  -0.00009
   0.00010
```

```
------------------------------------------------------------
```
Nodal displacements

Node	disp_X	disp_Y
1,	0.00000,	0.00000
2,	0.00014,	-0.00010
3,	0.00003,	-0.00019
4,	0.00010,	-0.00023
5,	0.00006,	-0.00023
6,	0.00007,	-0.00021
7,	0.00009,	-0.00018
8,	0.00005,	-0.00009
9,	0.00010,	0.00000

```
------------------------------------------------------------
```

```
Members actions
element       force        action
 1,          -7.69,       Compression
 2,          18.44,       Tension
 3,           7.69,       Tension
 4,         -21.87,       Compression
 5,          -2.10,       Compression
 6,          22.81,       Tension
 7,          -5.73,       Compression
 8,         -20.25,       Compression
 9,           5.73,       Tension
10,          17.69,        Tension
11,          -5.73,        Compression
12,         -15.12,        Compression
13,          16.91,       Tension
14,           7.56,       Tension
15,         -16.91,       Compression
```

2.9.2 PROBLEM 2.2 (FIGURE 2.10)

Input file

```
%                   File:    truss_problem_2_data.m
%
% The following variables are declared as global in order
% to be used by all the functions (M-files) constituting
% the program
%
global nnd nel nne nodof eldof n global geom connec prop nf load
%
format short e
%
```

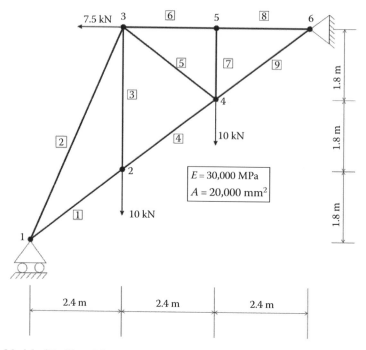

FIGURE 2.10   Model of Problem 2.2.

```
%%%%%%%%%%%%% Beginning of data input %%%%%%%%%%%%%%%%%
%
nnd = 6; % Number of nodes:
nel = 9; % Number of elements:
nne = 2 ; % Number of nodes per element:
nodof =2 ; % Number of degrees of freedom per node
eldof = nne*nodof; % Number of degrees of freedom
                   % per element
%
% Nodes coordinates X and Y
geom=zeros(nnd,2);
geom = [   0.        0.; ...   % X and Y coord. node 1
        2400.     1800.; ...   % X and Y coord. node 2
        2400.     5400.; ...   % X and Y coord. node 3
        4800.     3600.; ...   % X and Y coord. node 4
        4800.     5400.; ...   % X and Y coord. node 5
        7200.     5400.];      % X and Y coord. node 6
%
% Element connectivity
%
connec=zeros(nel,2);
connec = [1    2 ; ...   % 1st and 2nd node of element 1
          1    3 ; ...   % 1st and 2nd node of element 2
          2    3 ; ...   % 1st and 2nd node of element 3
          2    4 ; ...   % 1st and 2nd node of element 4
          3    4 ; ...   % 1st and 2nd node of element 5
          3    5 ; ...   % 1st and 2nd node of element 6
          4    5 ; ...   % 1st and 2nd node of element 7
          5    6 ; ...   % 1st and 2nd node of element 8
          4    6 ]  ;    % 1st and 2nd node of element 9
%
% Geometrical properties
%
% prop(1,1) = E; prop(1,2)= A
%
prop=zeros(nel,2);
prop  = [30000.      20000.; ...   % E and A of element 1
         30000.      20000.; ...   % E and A of element 2
         30000.      20000.; ...   % E and A of element 3
         30000.      20000.; ...   % E and A of element 4
         30000.      20000.; ...   % E and A of element 5
         30000.      20000.; ...   % E and A of element 6
         30000.      20000.; ...   % E and A of element 7
         30000.      20000.; ...   % E and A of element 8
         30000.      20000.];      % E and A of element 9
%
% Boundary conditions
%
nf = ones(nnd, nodof); % Initialize the matrix nf to 1
nf(1,2) =0 ;               % Prescribed nodal freedom of node 1
nf(6,1)= 0 ; nf(6,2)= 0 ;  % Prescribed nodal freedom of node 6
%
% Counting of the free degrees of freedom
%
n=0; for i=1:nnd
    for j=1:nodof
        if nf(i,j) ~= 0
        n=n+1;
        nf(i,j)=n;
        end
    end
end
%
% loading
%
load = zeros(nnd, 2);
load(2,:)=[0.      -10000.];  %forces in X and Y directions at node 2
```

```
load(3,:)=[-7500      0.];  %forces in X and Y directions at node 3
load(4,:)=[0.     -10000.]; %forces in X and Y directions at node 4

%
%%%%%%%%%%%%%%%%%%%%%%%%%% End of input %%%%%%%%%%%%%%%%%%%%%%%%%%
```

Results file

```
******* PRINTING MODEL DATA **************
```

```
--------------------------------------------------------
Number of nodes:                                6
Number of elements:                             9
Number of nodes per element:                    2
Number of degrees of freedom per node:          2
Number of degrees of freedom per element:       4

--------------------------------------------------------
Node       X            Y
  1,    0000.00,     0000.00
  2,    2400.00,     1800.00
  3,    2400.00,     5400.00
  4,    4800.00,     4600.00
  5,    4800.00,     5400.00
  6,    7200.00,     5400.00

--------------------------------------------------------
Element     Node_1      Node_2
    1,         1,          2
    2,         1,          3
    3,         2,          3
    4,         2,          4
    5,         3,          4
    6,         3,          5
    7,         4,          5
    8,         5,          6
    9,         4,          6

--------------------------------------------------------
Element       E           A
    1,      30000,      20000
    2,      30000,      20000
    3,      30000,      20000
    4,      30000,      20000
    5,      30000,      20000
    6,      30000,      20000
    7,      30000,      20000
    8,      30000,      20000
    9,      30000,      20000

--------------------------------------------------------
Node      disp_U      disp_V
  1,        1,          0
  2,        2,          3
  3,        4,          5
  4,        6,          7
  5,        8,          9
  6,        0,          0

--------------------------------------------------------
Node      load_X       load_Y
  1,     0000.00,      0000.00
  2,     0000.00,    -10000.00
  3,    -7500.00,      0000.00
  4,     0000.00,    -10000.00
```

```
5,       0000.00,       0000.00
6,       0000.00,       0000.00
-------------------------------------------------------

Total number of active degrees of freedom, n = 9

-------------------------------------------------------

******* PRINTING ANALYSIS RESULTS **************

-------------------------------------------------------
Global force vector   F
   0
   0
   -10000
   -7500
   0
   0
   -10000
   0
   0


-------------------------------------------------------
Displacement solution vector:   delta
 -0.80865
 -0.26183
 -0.65965
  0.18000
 -0.61631
  0.17710
 -0.95294
  0.09000
 -0.95294


-------------------------------------------------------
Nodal displacements
Node      disp_X         disp_Y
  1,     -0.80865,       0.00000
  2,     -0.26183,      -0.65965
  3,      0.18000,      -0.61631
  4,      0.17710,      -0.95294
  5,      0.09000,      -0.95294
  6,      0.00000,       0.00000


-------------------------------------------------------
Members actions
element       force          action
  1,        8333.33,       Tension
  2,      -16414.76,       Compression
  3,        7222.22,       Tension
  4,       10243.94,       Tension
  5,       24595.49,       Tension
  6,      -22500.00,       Compression
  7,          -0.00,       Compression
  8,      -22500.00,       Compression
  9,       31622.78,       Tension
```

2.10 ANALYSIS OF A SIMPLE TRUSS WITH ABAQUS

2.10.1 OVERVIEW OF ABAQUS

Abaqus is a suite of commercial finite element software. It consists of Abaqus Standard, which is a general purpose finite element software, and Abaqus Explicit for dynamic analysis. It is now owned by Dassault Systèms and is part of the SIMULIA range of products, http://www.simulia.com/products/unified_fea.html.

Modeling and Visualization	Reference
Abaqus/CAE User's Manual	Abaqus Keywords Reference Manual
	Abaqus Theory Manual
Analysis	Abaqus Verification Manual
Abaqus Analysis User's Manual	Abaqus User Subroutines Reference Manual
	Abaqus Glossary Manual
Examples	Programming
Abaqus Example Problems Manual	Abaqus Scripting User's Manual
Abaqus Benchmarks Manual	Abaqus Scripting Reference Manual
	Abaqus GUI Toolkit User's Manual
Tutorials	Abaqus GUI Toolkit Reference Manual
Getting Started with Abaqus: Interactive Edition	
Getting Started with Abaqus: Keywords Edition	Interfaces
	Abaqus Interface for MSC.ADAMS User's Manual
Information	Abaqus Interface for Moldflow User's Manual
Using Abaqus Online Documentation	
	Version 6.8 Update Information
Installation and Licensing	Abaqus Release Notes
Abaqus Installation and Licensing Guide	

FIGURE 2.11   Abaqus documentation.

Data input for a finite element analysis with Abaqus can be done either through Abaqus/CAE, which is an intuitive graphic user interface, that also allows monitoring and viewing the results, or through an input file prepared with a text editor and executed through the command line, or finally using a script prepared with Python, which is an object-oriented programming language. Python is included in Abaqus as Abaqus Python. The latter is an advanced option reserved for experienced users and will not be covered in this book. Note that Python is free to use, even for commercial products, because of its OSI-approved open-source license (http://www.python.org/).

Abaqus also comes with an integrated user manual, *Abaqus Documentation*, that can be opened in a browser; see Figure 2.11. New users usually prefer using the graphic interface, and they can start with the tutorial provided in the documentation: "Getting started with Abaqus: Interactive edition." This tutorial takes the user through all the steps required to build a finite element model, analyze it, and visualize the results. There are also many tutorials available on the web.

Students can join the SIMULIA Learning Community and they may be eligible for a free copy of Abaqus Student Version (http://www.simulia.com/academics/purchase.html).

2.10.2   ANALYSIS OF A TRUSS WITH ABAQUS INTERACTIVE EDITION

2.10.2.1   Modeling

In this section, we will analyze the truss shown in Figure 2.9 with the Abaqus interactive edition.

Click **Start**, **All Programs** and locate **Abaqus** as shown in Figure 2.12.

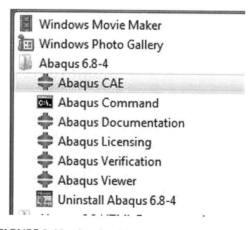

FIGURE 2.12   Starting Abaqus.

Double click on **Abaqus CAE** to reveal the main user interface. Click on **Create Model Database** to start a new analysis. On the main menu, click on **File** and **Set Work Directory** to choose your working directory. Click on **Save As** and name the file **Truss.cae** (Figure 2.13).

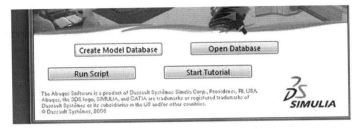

FIGURE 2.13   Abaqus CAE main user interface.

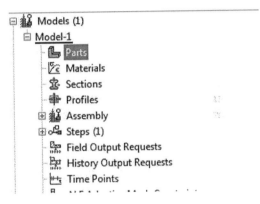

On the left-hand-side menu, click on **Part** to begin creating the model (Figure 2.14).

FIGURE 2.14   Creating a part.

The creating part window shown in Figure 2.15 appears on the screen. Name the part **Truss_part**, and check **2D Planar** as this is a planar truss, check on **Deformable** in the type. Choose **Wire** as the base feature. Enter an approximate size of 10 m and click on **Continue**. *WARNING: There are no predefined system of units within Abaqus, so the user is responsible for ensuring that the correct units are specified.*

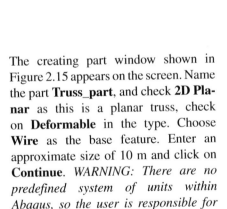

FIGURE 2.15   Choosing the geometry of the part.

Click on **Auto-fit View** to fit the view of the sketcher to the screen. You can also place the cursor on the center of the sketcher and zoom in and out using the middle mouse button (Figure 2.16).

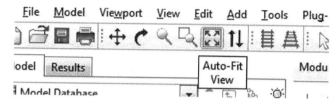

FIGURE 2.16   Fitting the sketcher to the screen.

In the sketcher menu, choose the **Create-Lines Connected** button to begin drawing the geometry of the truss (Figure 2.17).

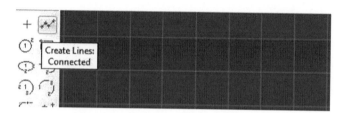

FIGURE 2.17   Drawing using the connected line button.

Begin drawing the truss. The coordinates of the cursor are given in the top-left corner. You could also enter them using the **Pick a point or enter X-Y coordinates** in the box situated in the bottom-left corner. Once finished, click on **Done** in the bottom-left corner to exit the sketcher (Figure 2.18).

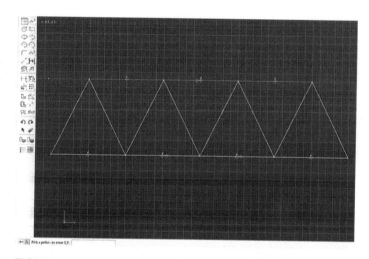

FIGURE 2.18   Drawing the truss geometry.

The finished part should appear as shown in Figure 2.19.

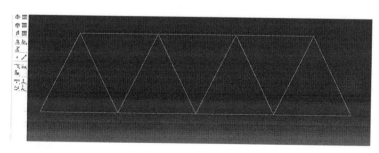

FIGURE 2.19   Finished part.

Next, under the model tree, click on **Materials** to create a material for the truss. Since all the members of the truss are made of the same material, we will only define one material, which we will name **Truss_material**. Then click on **Mechanical**, then **Elasticity**, and **Elastic** (Figure 2.20).

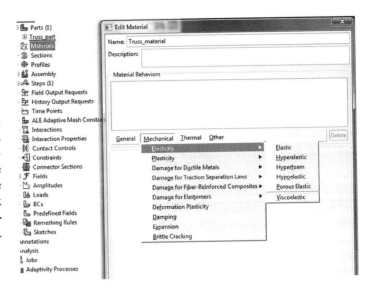

FIGURE 2.20   Material definition.

Enter $30.e6\,\text{kN/m}^2$ for the elastic modulus, and 0.3 for Poisson's ratio even though it is not applicable for a truss (Figure 2.21).

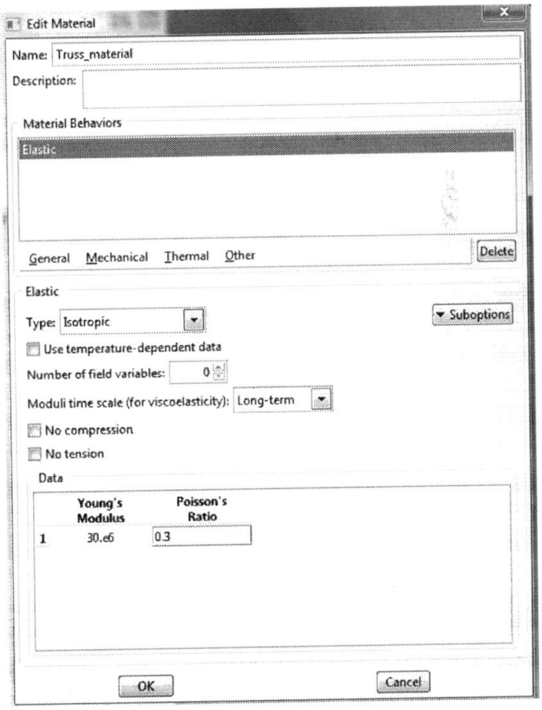

FIGURE 2.21   Material properties.

The longitudinal members of the truss have a cross area of 0.045 m² and the diagonal members have a cross area of 0.02 m². To input this data, we need to define two sections (Figure 2.22).

Under the Model tree, click on **Sections** and the **Create Section** window appears. Name the section **Longitudinal**. In the **Category** check **Beam**, and in the **Type**, choose **Truss**. Click on **Continue**.

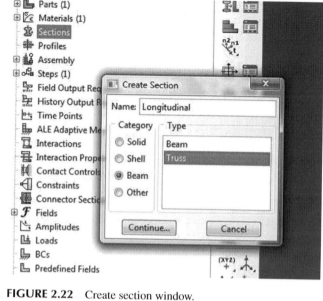

FIGURE 2.22   Create section window.

Next the **Edit Section** window appears. Scroll through **Material** and choose the already created material **Truss_material** to assign it to the section. In **Cross sectional area** enter 0.045 m² and click **OK** (Figure 2.23).

Follow exactly the same procedure to create another section named **Diagonal** and enter 0.02 m² for the cross area.

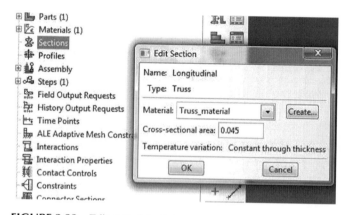

FIGURE 2.23   Edit material window.

Next we assign the defined sections to the corresponding members. Expand the menu under **Truss_part** and click on **Section assignment**. The message **Select the regions to be assigned a section** should appear on the bottom-left corner of the main window (Figure 2.24).

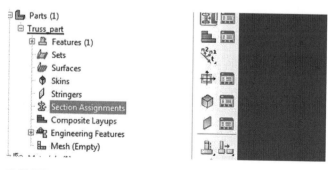

FIGURE 2.24   Section assignment.

Keep the Shift key down, and with the mouse select the horizontal members. Once a member is selected it changes color. Click on **done** in the bottom-left corner next to the message **Select the regions to be assigned a section**. The **Edit Section Assignment** window appears (Figure 2.25).

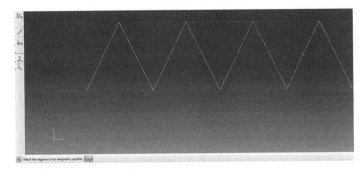

FIGURE 2.25   Regions to be assigned a section.

In **Section**, scroll to **Longitudinal** and click on **OK** (Figure 2.26).

Repeat the same thing for the diagonal members. Keep the Shift key down, and with the mouse select the diagonal members. Click on **done** in the bottom-left corner next to the message **Select the regions to be assigned a section**. The **Edit Section Assignment** window appears. In **Section**, scroll to **Diagonal** and click on **OK**.

FIGURE 2.26   Edit section assignment.

In the next step, we will define the elements. Expand the menu under **Truss_part** and click on **Mesh(empty)** to load the meshing menu (Figure 2.27).

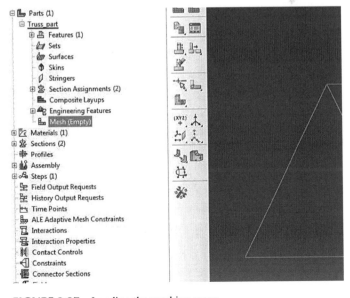

FIGURE 2.27   Loading the meshing menu.

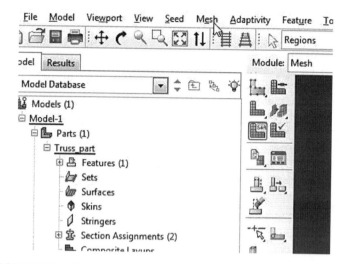

On the main menu, click on **Mesh** and then on **Element Type**, and with the mouse select the whole truss. Click on **Done** in the bottom-left corner of the main window (Figure 2.28).

FIGURE 2.28 Selecting regions to be assigned element type.

The element type dialog box appears. In **Element Library** click on **Standard**. In **Element family** scroll down and choose **Truss**. In **Geometric order**, choose **Linear**. The message **T2D2: A 2: node linear 2-D truss** should appear in the dialog box (Figure 2.29).

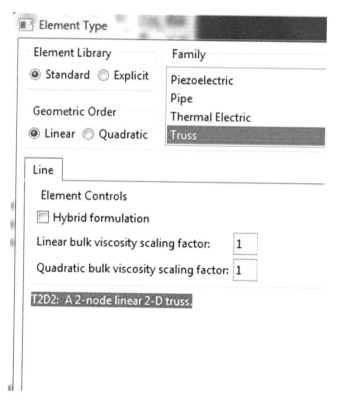

FIGURE 2.29 Selecting element type.

On the main menu click on **Seed**, then on **Edge by number**, and select the whole truss. Enter 1 in the bottom-left corner of the main window, and press **Enter**. The seeding on the truss should look like Figure 2.30.

On the main menu, click on **Mesh** again, and then on **Part** to mesh the truss. Once meshed, the truss changes color to blue.

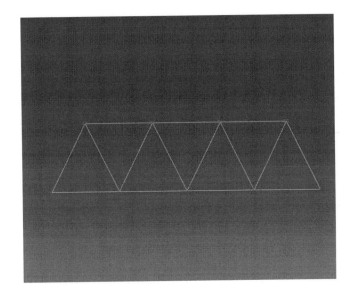

FIGURE 2.30   Mesh.

Expand the menu under **Assembly** and double click on **instances**.

In Abaqus you can create many parts and assemble them together to form a model. You can also create many instances from one part. For example, in a bridge, you do not have to draw all the girders. If they are similar, drawing one is enough. The others are created as instances of the first one (Figure 2.31).

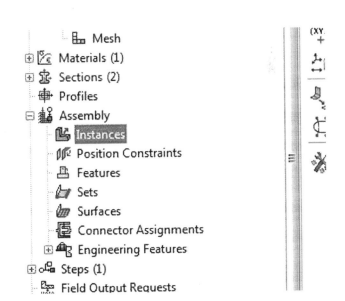

FIGURE 2.31   Assembling the model.

The **create instance** dialog box appears. In this case, we have only one part: **Truss_part**. Select it and click **OK** (Figure 2.32).

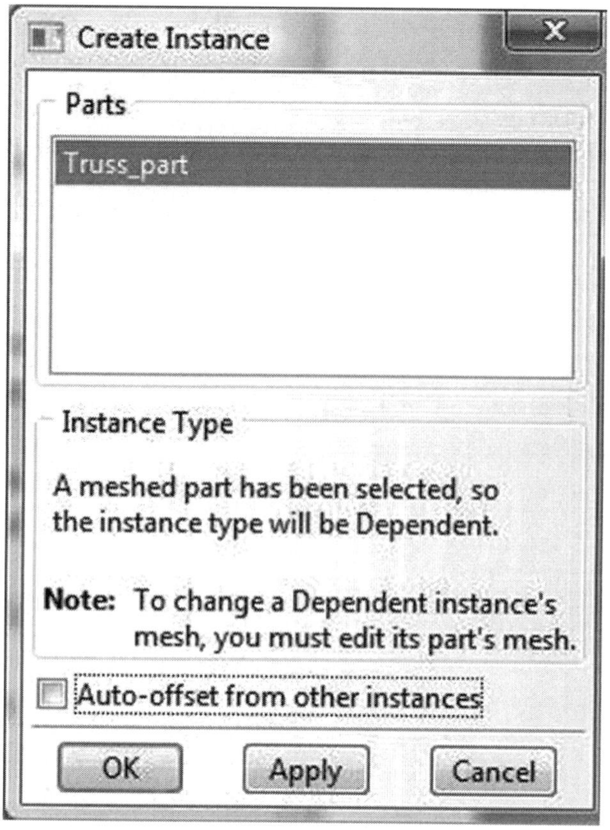

FIGURE 2.32   Creating instances.

Before introducing the boundary conditions, we need to understand how the degrees of freedom are numbered. The translations along the axes x, y, and z are respectively numbered 1, 2, and 3. The rotations around these axes are respectively numbered 4, 5, and 6 (Figure 2.33).

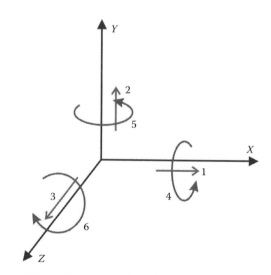

FIGURE 2.33   Numbering of the degrees of freedom.

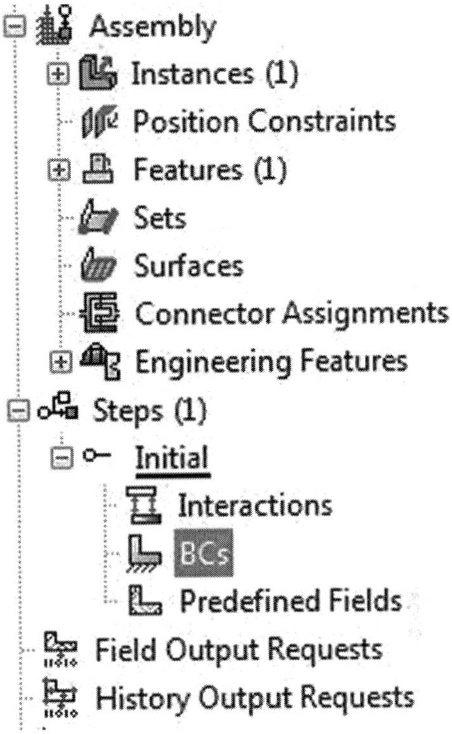

FIGURE 2.34 Creating boundary conditions.

Expand the menu under **Steps** and **Initial**, click on **BC** to introduce the boundary conditions (Figure 2.34).

The **Create Boundary Condition** dialog box appears. Name the boundary condition **Pinned_support**. Choose **Symmetry/Antisymmetry/Encastré** and click on **Continue** (Figure 2.35).

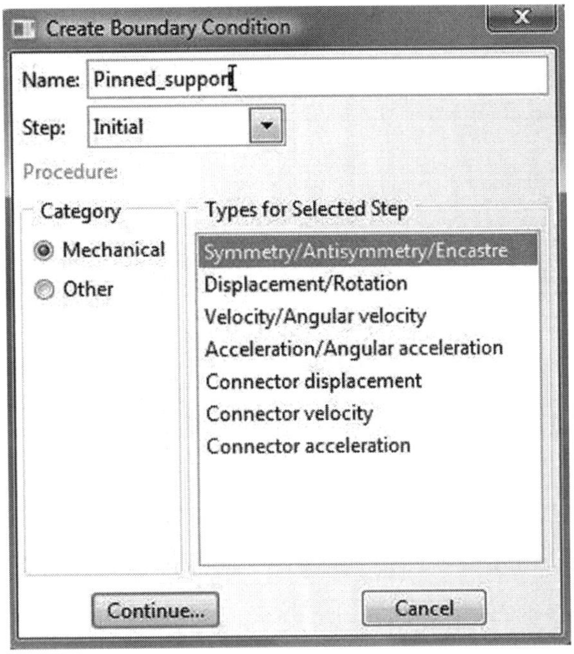

FIGURE 2.35 Type of boundary conditions.

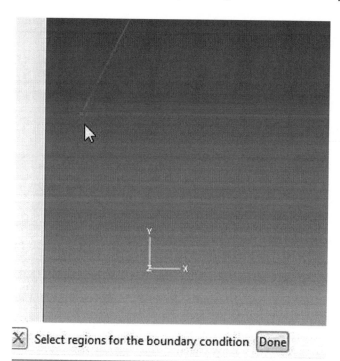

Select the left-side support and click on **Done** (Figure 2.36).

FIGURE 2.36   Selecting a region to be assigned boundary conditions.

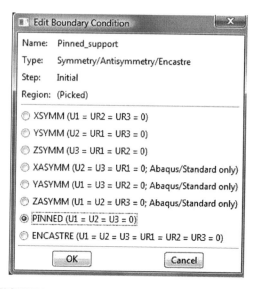

The **Edit Boundary Condition** dialog box appears. Select **PINNED(U1 = U2 = U3 = 0)** and click on **OK** (Figure 2.37).

FIGURE 2.37   Edit boundary condition dialog box for pinned support.

Under **Steps** and **Initial**, click on **BC** to create the boundary conditions for the roller. In the **Create Boundary Condition** dialog box, name the boundary condition **Roller_Support**. Choose **Symmetry/Antisymmetry/Encastré** and click on **Continue**. Select the right support and click on **Done**. In the **Edit Boundary Condition** dialog box, select **XASYMM(U2 = U3 = UR1 = 0)** and click on **OK** (Figure 2.38).

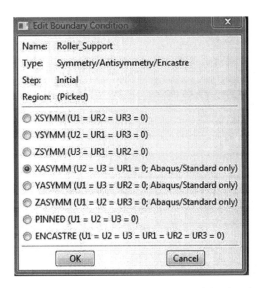

FIGURE 2.38 Edit boundary condition dialog box for roller support.

In the left-hand-side menu, right click on **Steps** to crate another step for applying the loads. Click on **Continue** (Figure 2.39).

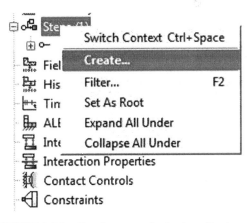

FIGURE 2.39 Creating a step for load application.

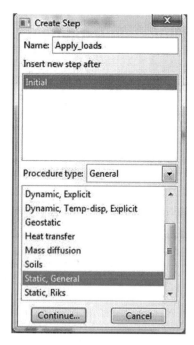

In the **Create Step** dialog box, name the step **Apply_Loads**, select **Static, General**, and click on **Continue** (Figure 2.40).

FIGURE 2.40   Create step dialog box.

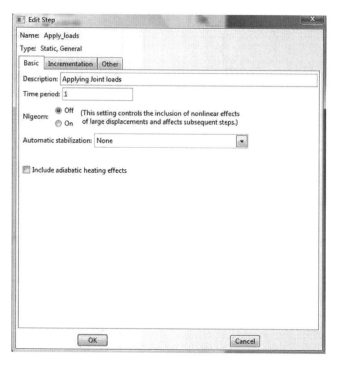

In the **Edit step** dialog box, although it is not necessary, you can still provide a description such as applying joint loads. Leave all the other details as they are, and click on **OK** (Figure 2.41).

FIGURE 2.41   Edit step dialog box.

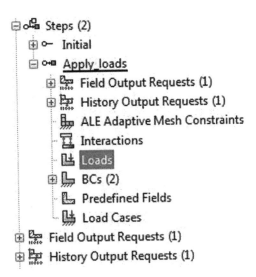

FIGURE 2.42 Creating a load.

In the left-hand-side menu, under **Steps** and **Apply_Loads**, click on **Loads** as shown in Figure 2.42.

In the **Create load** dialog box, name the load *Horizontal* 15 kN *force*. In **Step** scroll to **Apply_Loads**, which means that the load will be applied in this step. In **Category** choose **Mechanical**, and in **Type** choose **Concentrated Force**. Click on **Continue** (Figure 2.43).

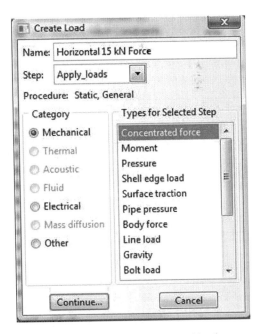

FIGURE 2.43 Creating a concentrated load.

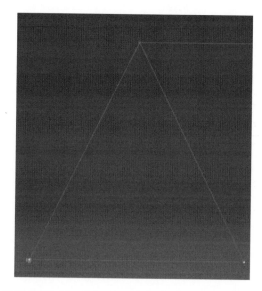

FIGURE 2.44　Selecting a joint for load application.

With the mouse, select the top-left joint as shown in Figure 2.44, and click on done in the bottom-left corner of the same window.

In the **Edit Load** dialog box, enter 15. for **CF1**, and click on **OK** (Figure 2.45).

Edit Load	
Name:	Horizontal 15 kN Force
Type:	Concentrated force
Step:	Apply_loads (Static, General)
Region:	(Picked)

CSYS: (Global) [Edit...] [Create...]

Distribution: Uniform ▾ [Create...]

CF1: 15.

CF2:

Amplitude: (Ramp) ▾ [Create...]

☐ Follow nodal rotation

Note: Force will be applied per node.

[OK]　　　　[Cancel]

FIGURE 2.45　Entering the magnitude of a joint force.

Repeat the same procedure for the other joint loads. Since they are vertical loads pointing in opposite direction to the axis y, their magnitude should be entered in **CF2** as negative. Once finished, the loaded truss should look like the one shown in Figure 2.46.

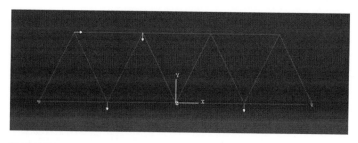

FIGURE 2.46　Loaded truss.

2.10.2.2 Analysis

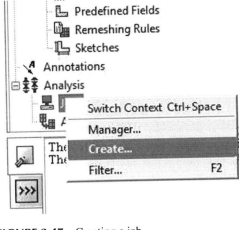

Under **Analysis**, right click on **Jobs** and then click on **Create** (Figure 2.47).

FIGURE 2.47 Creating a job.

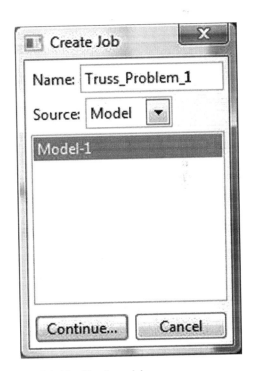

The **Create Job** dialog box appears. Name the job **Truss_Problem_1**, and click on **Continue** (Figure 2.48).

FIGURE 2.48 Naming a job.

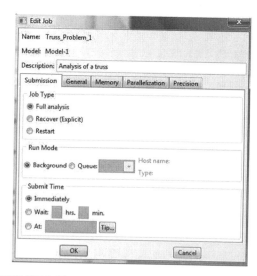

The **Edit Job** dialog box appears. Enter a description for the job. Check **Full analysis** and choose to run the job in **Background** and check to start it **immediately**. Click **OK** (Figure 2.49).

FIGURE 2.49   Editing a job.

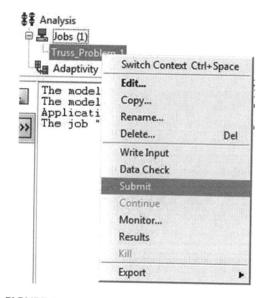

Expand the tree under **Jobs**, right click on **Truss_Problem_1**. Then, click on **Submit** (Figure 2.50).

FIGURE 2.50   Submitting a job.

If you get the following message **Job Truss_Problem_1 completed successfully** in the bottom window, then your job is free of errors and was executed properly. Now, it is time to view the analysis results (Figure 2.51).

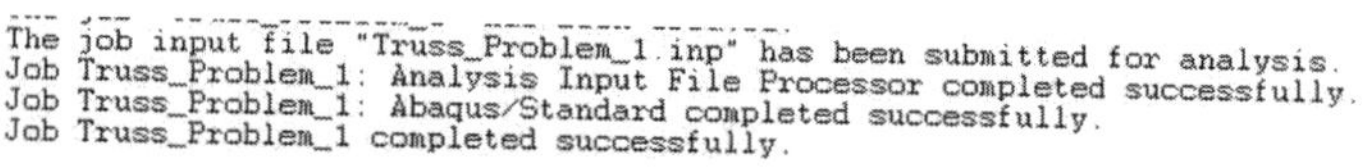

```
The job input file "Truss_Problem_1.inp" has been submitted for analysis.
Job Truss_Problem_1: Analysis Input File Processor completed successfully.
Job Truss_Problem_1: Abaqus/Standard completed successfully.
Job Truss_Problem_1 completed successfully.
```

FIGURE 2.51   Monitoring of a job.

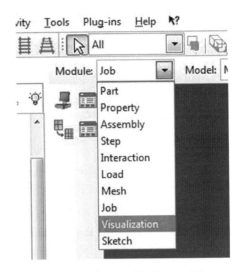

Under the top menu, in the **Module** scroll to **Visualization**, and click to load **Abaqus Viewer**. On the main menu, under **File**, click **Open**, navigate to your working directory, and open the file **Truss_Problem_1.odb**. It should have the same name as the job you submitted (Figure 2.52).

FIGURE 2.52   Opening the visualization module.

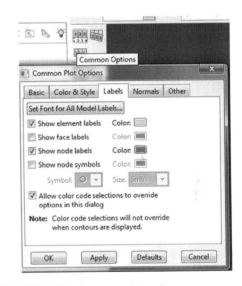

Click on the **Common options** icon to display the **Common Plot options** dialog box. Under **labels**, check **Show Element labels** and **Show Node labels** to display elements and nodes' numbering (Figure 2.53).

FIGURE 2.53   Common plot options.

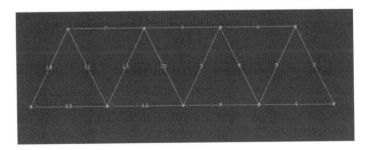

You may obtain a different nodes and elements numbering to the one shown in Figure 2.54. However, you must ensure that there are 15 elements and 9 nodes only.

FIGURE 2.54   Elements and nodes' numbering.

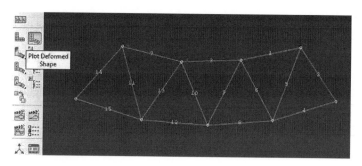

Click on the icon **Plot Deformed Shape** to display the deformed shape of the truss (Figure 2.55).

FIGURE 2.55   Deformed shape.

On the main menu, click on **Results** then on **Field Output** to open the **Field Output** dialog box. Choose **U Spatial displacements at nodes**. For component, choose $U2$ to plot the vertical displacement (Figure 2.56).

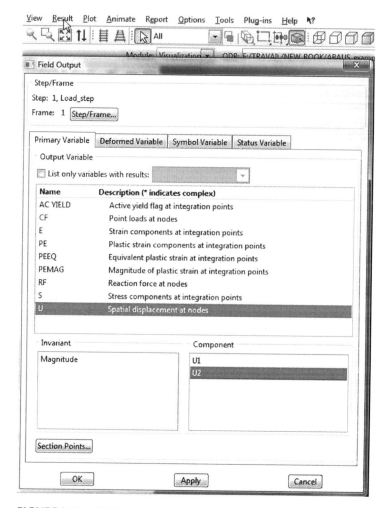

FIGURE 2.56   Field output dialog box.

Figure 2.57 shows the contour plot of the vertical displacement $U2$ as well as the legend block.

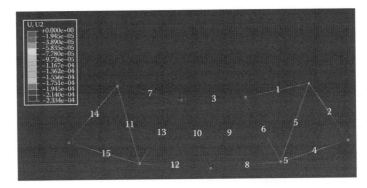

FIGURE 2.57   Contour plot of the vertical displacement $U2$.

If you cannot read the displacements values in the legend block, on the main menu click on **Viewport Annotation Options**. Under **Legend**, click on **Set font** and enter a bigger font (Figure 2.58).

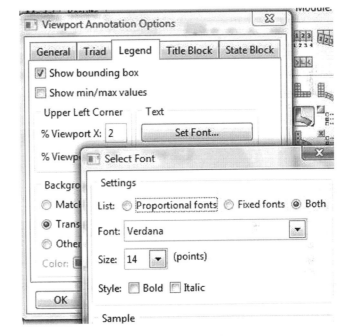

FIGURE 2.58   Viewport annotations options.

On the main menu, click on **Results**, then on **Field Output** to open the **Field Output** dialog box. Choose **S Stress components at integration points**. For component, choose $S11$ to plot the stresses in the bars.

Note that Abaqus does not plot the normal forces in the bars (Figure 2.59).

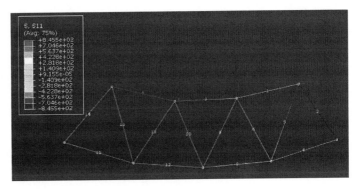

FIGURE 2.59   Normal stresses in the bars.

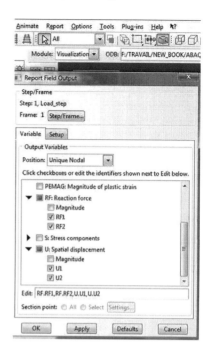

To create a text file containing the reaction forces and nodal displacements, in the menu bar click on **Report** and **Field Output**. In the **Report Field Output** dialog box, for **Position** select **Unique nodal**, check **RF1** and **RF2** for **RF: Reaction force**, and check **U1** and **U2** for **U: Spatial displacement**. Then click on click on **Set up** (Figure 2.60).

FIGURE 2.60   Selecting variables to print to a report.

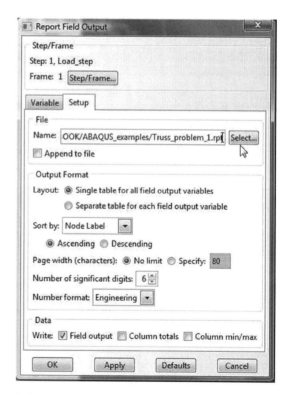

Click on **Select** to navigate to your working directory. Name the file **Truss_Problem_1.rpt**.
Uncheck **Append to file** and click **OK** (Figure 2.61).

FIGURE 2.61   Choosing a directory and the file name to which to write the report.

Open your working directory and take a look at the files generated by Abaqus. You can happily ignore most of them. However, you should keep the ***.inp** file as it contains all the information about the model. We will write a similar file in the next section. The ***.odb** is binary and contains all the information about the model and the results of the analysis. It is used by the visualization module to view the results. The ***.dat** contains written output such as results. Most importantly, it contains any errors made during the setting up of the model. The ***.msg** file that appears as an outlook item in Windows contains any error that arises during the analysis. It is particularly useful in nonlinear problems. The rest of the files, you can ignore them for the time being.

Use your favorite text editor, and open the file **Truss_Problem_1.rpt**

```
*************************************************************************
Field Output Report, written Fri Apr 01 09:17:09 2011

Source 1
---------

   ODB: F:/TRAVAIL/NEW_BOOK/Abaqus_examples/truss_problem_1.odb
   Step: Load_step
   Frame: Increment       1: Step Time =      1.000

Loc 1 : Nodal values from source 1

Output sorted by column "Node Label".

Field Output reported at nodes for part: TRUSS-1
```

Node Label	RF.RF1 @Loc 1	RF.RF2 @Loc 1	U.U1 @Loc 1	U.U2 @Loc 1
1	0.	0.	73.6886E-06	-213.242E-06
2	0.	0.	51.2812E-06	-94.0781E-06
3	0.	15.125	98.5185E-06	-15.125E-36
4	0.	0.	103.689E-06	-233.414E-06
5	0.	0.	87.3148E-06	-182.554E-06
6	0.	0.	136.096E-06	-100.075E-06
7	0.	0.	61.1111E-06	-230.828E-06
8	0.	0.	27.3148E-06	-186.493E-06
9	-15.	6.875	15.E-36	-6.875E-36

Note that at node 9, the horizontal reaction is equal to -15 kN, and the vertical reaction is equal to 6.875. The horizontal and vertical displacements at node 7 are respectively equal to $61.1111e - 06, -230.828e - 06$ m, which are the same as previously obtained with the MATLAB code **truss.m**, *node* 5, 0.00006, -0.00023 m.

2.10.3 Analysis of a Truss with Abaqus Keyword Edition

In Abaqus you can create a complete finite element model by simply using a text editor. The input file must have the extension **.inp**. It contains Abaqus commands in the format of **Keywords**. A keyword starts with a *. In the **Abaqus Documentation**, click on the **Abaqus Keywords Reference Manual** to find the meaning and usage of all the Abaqus keywords. They are organized in an alphabetical order.

In this section, we will prepare an input file for the truss shown in Figure 2.9. We will keep the same node and element numbering. The problem at hand is very simple; therefore, the file should be very easy to understand.

Using a text editor, create a file and save it as **truss_problem_1_keyword.inp**. Before creating the model, make sure you adhere to the following rules:

- Any line that starts with two stars ** represents a comment that will be ignored by Abaqus.
- Any line that starts with only one * represents a command, and Abaqus will attempt to execute it. If it is not a proper keyword, an error will result.
- Any line that does not start with (*) or (**) represents data.
- Do not leave blank lines, instead use two stars **.

```
*HEADING
Example Truss_Problem_1_Keyword_Edition
**
** the HEADING (Example Truss_Problem_1_Keyword_Edition) will appear on any output files
** created by Abaqus
**
****************************************************
**
**   Geometry definition
**
**   Enter the nodal coordinates of the nodes
**
**
*Node, Nset = all_nodes
1,        0.,           0.
2,        1.,           2.
3,        2.,           0.
4,        3.,           2.
5,        4.,           0.
6,        5.,           2.
7,        6.,           0.
8,        7.,           2.
9,        8.,           0.
**
**Define node sets to be used for BC and applying loads
**
**
*Nset, nset=Pinned_support
1
*Nset, nset=Roller_support
9
**
*Nset, nset=HF15
2
*Nset, nset=VF5
3
*Nset, nset=VF7
4
*Nset, nset=VF10
7
**
** Select element type as T2D2 (planar truss element)
** and define element connectivity
**
*Element, type=T2D2
1,      1,        2
2,      1,        3
3,      2,        3
4,      2,        4
5,      3,        4
6,      3,        5
7,      4,        5
8,      4,        6
9,      5,        6
10,     5,        7
11,     6,        7
12,     6,        8
13,     7,        8
14,     7,        9
15,     8,        9
```

```
**
** Create two element sets one for the horizontal elements named "Horizontal"
** and one for the diagonal elements named "Diagonal"
**
*elset, elset = Horizontal
2, 4, 6, 8, 10, 12, 14
*elset, elset = Diagonal
1, 3, 5, 7, 9, 11, 13, 15
**
** Define material, and name it "My_material"
**
**
*Material, name=My_material
*Elastic
 3e+07,
**
**
** Define a section for the horizontal members
**
*Solid Section, elset= Horizontal, material=My_material
0.045,
**
**
** Define a section for the diagonal members
**
*Solid Section, elset=Diagonal, material=My_material
0.02,
**
**
** Define Boundary Conditions
**
*Boundary
Roller_support, XASYMM Pinned_support, PINNED
** -------------------------------------------------------------
**
** Define step and name it "Load_step"
**
*Step, name=Load_step
*Static
1., 1., 1e-05, 1.
**
** Apply the loads as concentrated forces
**
*Cload
HF15, 1, 15.
VF5, 2, -5.
VF10, 2, -10.
VF7, 2, -7.
**
** OUTPUT REQUESTS
**
** FIELD OUTPUT
** Only request the default field output
**
*Output, field, variable=PRESELECT
**
** HISTORY OUTPUT
** Only request the default History output
**
*Output, history, variable=PRESELECT
*End Step
```

The file starts with the keyword ***HEADING**. Below in the data line put any text you want to describe the model. The text will appear on any output files created by Abaqus.

Next define the geometry of the nodes using the keyword ***node**. You can group all the nodes in a node set named **all_nodes**. In the data line, below the keyword, enter the node number, followed by its x and y coordinates. Use one line per node and make sure you separate the entered values by commas ",". Otherwise, you will get an error.

Once all the nodes are defined, create node sets that will be used later for imposing the boundary conditions and applying the loads:

- ***Nset, nset=Pinned_support** creates a node set named **Pinned_support** that contains the node 1 entered in the data line.
- ***Nset, nset=Roller_support** creates a node set named **Roller_support** containing node 9.
- ***Nset, nset=HF15** creates a node set named **HF15** containing node 2.
- ***Nset, nset=VF5** creates a node set named **VF5** containing node 3.
- ***Nset, nset=VF7** creates a node set named **VF7** containing node 4.
- ***Nset, nset=VF10** creates a node set named **VF10** containing node 7.

Next using the keyword ***elset** create two elements sets, one for the horizontal members named **Horizontal** and one for the diagonal members named **Diagonal**.

Using the keyword ***Material** create a material named **My_material**. The created material is elastic and has a Young's modulus of $3e + 07$ given in the data line of the keyword ***Elastic**.

Using the keyword ***Solid Section** create a section for the horizontal members with the element set **Horizontal** and **My_material** for material. Enter the cross section of 0.045 in the data line. Create another one for the diagonal members using the element set **Diagonal** and the same material. This time enter 0.02 for the cross section.

Using the keyword ***Boundary** apply the boundary condition. We assign **YSYMM** (symmetry about a plane $Y = constant$) to node set **Roller_support**. It means the degrees of freedom 2, 4, and 6 are suppressed. In the next data line, we assign **PINNED** to node set **Pinned_support**. It means the degrees of freedom 1, 3, and 3 are suppressed.

Next using the keyword ***step** create a step and name it **Load_step**. The keyword ***static** indicates that it will be a general static analysis. It is important to note that there are four values in the data line of the keyword ***static**. These values represents pseudo-time in Abaqus Standard; that is, a mapping between time and load. The first value equal to 1 represents the initial time increment. In other words, Abaqus will initially try to apply the total load as one increment. The second value also equal to 1 is the total time period of the step. The third value corresponds to the minimum time increment. This particularly happens in nonlinear analysis. If Abaqus cannot apply the load as a whole, it keeps reducing the increment until it reaches this minimum value. The fourth and last value is the maximum time increment allowed.

The keyword ***cload** indicates that the loads will be applied as concentrated loads. In the data lines,

- **HF15, 1, 15.** indicates that a positive 15 kN load is applied in the direction 1 (x direction) to node set HF15 defined previously
- **VF5, 2, −5.** indicates that a negative 5 kN load is applied in the direction 2 (y direction) to node set VF5 defined previously
- **VF10, 2, −10.** indicates that a negative 10 kN load is applied in the direction 2 (Y direction) to node set VF10 defined previously
- **VF7, 2, −7.** indicates that a negative 7 kN load is applied in the direction 2 (Y direction) to node set VF7 defined previously

You can request outputs that will be written to the database file (***.odb**) using the keyword ***output**. There are two types of outputs: **field** and **history**. When the **variable** is set equal to **PRESELECT**, only the default variables will be printed. Field output is intended for infrequent requests for a large portion of the model and can be used to generate contour plots, animations, and so on. History output, on the other hand, is intended for relatively frequent output requests for small

FIGURE 2.62 Running Abaqus from the command line.

portions of the model and is displayed in *XY* data plots. For example, if we want to monitor the displacement of a node with load, this is the type of output that needs to be requested.

You can create many steps in Abaqus, but each one of them must end with the keyword ***end step**.

If your operating system is **Windows**, in **Start Menu**, click on **Accessories** and then on **Command prompt** to open a DOS shell. Using DOS commands, navigate to your working directory. At the command line type **Abaqus job=truss_problem_1_keyword inter** followed by **Return**. The outcome should be similar to the one shown in Figure 2.62. If you get an error, open the file with extension ***.dat** to see what type of error. To load the visualization model, type **Abaqus Viewer** at the command line.

3 Beam Element

3.1 INTRODUCTION

A beam constitutes the simplest way of spanning a gap between two objects. As structural elements, beams are prominent in both civil and mechanical engineering. They are used as supports for floors in buildings, decks in bridges, wings in aircraft, or axles for cars.

A beam is generally slender and carries loadings applied perpendicular to its longitudinal axis. In matrix structural analysis, or finite element for that matter, a beam is regarded as an element with a node at each end. When the element is loaded as shown in Figure 3.1a, each node will undergo a vertical displacement w and a rotation θ as shown in Figure 3.1b. The end nodes 1 and 2 are subject to shear forces and moments, which result in vertical translations and rotations. Each node, therefore, has two degrees of freedom. In total, the element has four degrees of freedom. The nodal forces and displacements can be expressed in vector form as

$$\{F_e\} = \{F_1, M_1, F_2, M_2\}^T \tag{3.1}$$

$$\{d_e\} = \{w_1, \theta_1, w_2, \theta_2\}^T \tag{3.2}$$

The differential equations describing the behavior of a beam element are well known. They are referred to as the Euler–Bernoulli theory of bending or simply known as the engineering beam theory. For a differential element dx of the beam as shown in Figure 3.2, the relationships between deflection, slope, load, shear, and moment are given in the form of differential equations as

$$\frac{d^2w}{dx^2} = \frac{M}{EI} \tag{3.3}$$

$$\frac{d^3w}{dx^3} = \frac{1}{EI}\frac{dM}{dx} = \frac{S}{EI} \tag{3.4}$$

$$\frac{d^4w}{dx^4} = \frac{1}{EI}\frac{dS}{dx} = \frac{q(x)}{EI} \tag{3.5}$$

where w, M, S, EI, and $q(x)$ represent respectively the deflection, moment, shear force, stiffness, and uniformly distributed load.

3.2 STIFFNESS MATRIX

It is possible to develop the matrix relationship between the nodal forces, $\{F_1, M_1, F_2, M_2\}^T$, and the nodal displacements, $\{w_1, \theta_1, w_2, \theta_2\}^T$, by integrating the differential equations (3.3) through (3.5). The integration produces constants of integration that can be identified by considering the boundary conditions of the element. A simpler way of establishing the matrix relationship is to operate as for the bar element (see Section 2.2.1). It consists in placing simple supports at each end of the beam, then set the degrees of freedom to unity one at a time, and calculate the nodal forces needed to produce the deformed state. The reactions at the supports resulting from the imposition of unit displacements/rotations at the nodes are called stiffness influence coefficients. To obtain these coefficients, we will use the theorem of Castigliano.

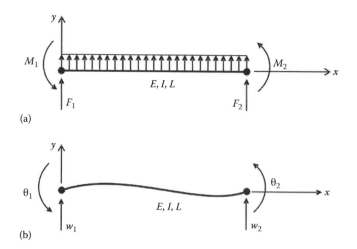

FIGURE 3.1   Beam element. (a) Forces and (b) displacements.

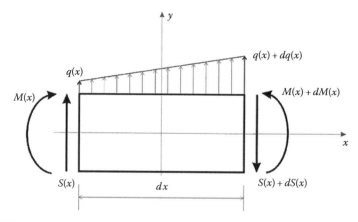

FIGURE 3.2   Differential element of a beam.

ROTATION θ_2

Consider the beam element shown in Figure 3.3a. The member is initially straight. If we try to rotate node 2 by an amount θ_2, then reaction forces will be developed at nodes 1 and 2. Considering vertical equilibrium yields

$$F_{y1} + F_{y2} = 0 \tag{3.6}$$

Taking moments around z with respect to node 2 gives

$$M_1 + M_2 - F_{y1}L = 0 \tag{3.7}$$

Taking moments around z with respect to x as shown yields

$$M(x) = -M_1 + F_{y1}x \tag{3.8}$$

The moment $M(x)$ may also be written as a function of M_2:

$$M(x) = -F_{y1}(L - x) + M_2 \tag{3.9}$$

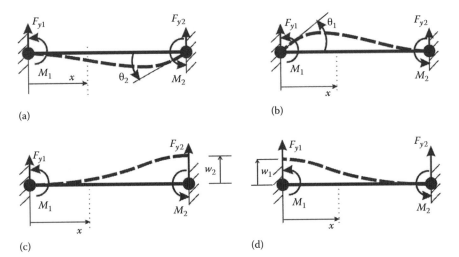

FIGURE 3.3 Nodal degrees of freedom. (a) Rotation θ_2, (b) rotation θ_1, (c) displacement w_2, and (d) displacement w_1.

The strain energy of a beam in bending is given as

$$\Pi = \int_0^L \frac{M(x)^2}{2EI}dx = \frac{1}{2EI}\int_0^L (-F_{y1}(L-x) + M_2)^2 dx$$

$$= \frac{1}{2EI}\left(F_{y1}^2 \frac{L^3}{3} + M_2^2 L - F_{y1}M_2 L^2\right) \tag{3.10}$$

Using the theorem of Castigliano and taking the derivative with respect to F_{y1} yields

$$\frac{\partial \Pi}{\partial F_{y1}} = \frac{2}{3}L^3 F_{y1} - M_2 L^2 = w_1 = 0 \tag{3.11}$$

and

$$\frac{\partial \Pi}{\partial M_2} = \frac{1}{2EI}(2M_2 L - F_{y1}L^2) = \theta_2 \tag{3.12}$$

Solving for M_2 and F_{y1} using Equations (3.11) and (3.12) yields

$$M_2 = \frac{4EI}{L}\theta_2 \tag{3.13}$$

$$F_{y1} = \frac{6EI}{L^2}\theta_2 \tag{3.14}$$

Since $M_1 + M_2 - F_{y1} = 0$, we also have

$$M_1 = \frac{2EI}{L}\theta_2 \tag{3.15}$$

ROTATION θ_1

By simply transposing the suffices, similar expressions can be obtained for M_1, M_2, and F_{y2} when considering a rotation θ_1 (Figure 3.3b); that is,

$$M_1 = \frac{4EI}{L}\theta_1 \tag{3.16}$$

$$M_2 = \frac{2EI}{L}\theta_2 \tag{3.17}$$

$$F_{y2} = \frac{6EI}{L^2}\theta_1 \tag{3.18}$$

DISPLACEMENT w_2

The initially straight member is now given a vertical displacement w_2 as represented in Figure 3.3c. The bending moment at a distance x is obtained as

$$M(x) = -M_1 + F_{y1}x \tag{3.19}$$

or as

$$M(x) = -M_1 - F_{y2}x \tag{3.20}$$

Substituting in the expression of the bending energy yields

$$\Pi = \int_0^L \frac{M(x)^2}{2EI}dx = \frac{1}{2EI}\int_0^L (-M_1 - F_{y2}x)^2 dx$$
$$= \frac{1}{2EI}\left(M_1^2 L + F_{y2}^2 \frac{L^3}{3} + F_{y2}M_1L^2\right) \tag{3.21}$$

Using the theorem of Castigliano, we obtain

$$\frac{\partial \Pi}{\partial M_1} = \frac{1}{2EI}(2M_1 L + F_{y2}L^2) = \theta_1 = 0 \tag{3.22}$$

and

$$\frac{\partial \Pi}{\partial F_{y2}} = \frac{1}{2EI}\left(2F_{y2}\frac{L^3}{3} + M_1L^2\right) = w_2 \tag{3.23}$$

Solving for M_1 and F_{y2} using Equations (3.22) and (3.23) yields

$$M_1 = -\frac{6EI}{L^2}w_2 \tag{3.24}$$

$$F_{y2} = \frac{12EI}{L^3}w_2 \tag{3.25}$$

From equilibrium of the moments, we obtain M_2 as

$$M_2 = -\frac{6EI}{L^2}w_2 \tag{3.26}$$

DISPLACEMENT W_1

Again, by simply transposing the suffices, similar expressions can be obtained for M_1, M_2, and F_{y1}; when considering a displacement w_1 (Figure 3.3d); that is,

$$M_1 = -\frac{6EI}{L^2}w_1 \tag{3.27}$$

$$M_2 = -\frac{6EI}{L^2}w_1 \tag{3.28}$$

$$F_{y1} = \frac{12EI}{L^3}w_1 \tag{3.29}$$

The preceding results can be grouped in a matrix form:

$$\begin{Bmatrix} F_{y1} \\ M_1 \\ F_{y2} \\ M_2 \end{Bmatrix} = \begin{bmatrix} 12EI/L^3 & 6EI/L^2 & -12EI/L^3 & 6EI/L^2 \\ 6EI/L^2 & 4EI/L & -6EI/L^2 & 2EI/L \\ -12EI/L^3 & -6EI/L^2 & 12EI/L^3 & -6EI/L^2 \\ 6EI/L^2 & 2EI/L & -6EI/L^2 & 4EI/L \end{bmatrix} \begin{Bmatrix} w_1 \\ \theta_1 \\ w_2 \\ \theta_2 \end{Bmatrix} \tag{3.30}$$

or simply as

$$\{f_e\} = [K_e]\{\delta_e\} \tag{3.31}$$

where $[K_e]$ is the stiffness matrix that relates the nodal displacements to the nodal forces.

3.3   UNIFORMLY DISTRIBUTED LOADING

The stiffness matrix for a beam element was developed for loadings applied only at its nodes. Quite often, however, beams support uniformly distributed loading along (or part of) their length. This requires modification in order to be used in an analysis. The distributed loading is replaced by a system of statically equivalent nodal forces that are always of opposite sign from the fixed end reactions, as shown in Figure 3.4. Figure B.1 in Appendix B shows the equivalent nodal loads for the most common loadings on beams.

The displacements computed using equivalent nodal loads are exact in a finite element sense; however, the internal reactions computed in individual elements using the relation $\{F_e\} = [K_e]\{d_e\}$ are not. Instead, to obtain the correct internal reactions, the following relation must be used:

$$\{F_e\} = [K_e]\{d_e\} - \{F_0\} \tag{3.32}$$

where $\{F_0\}$ represents the vector of equivalent nodal forces at element level.

To illustrate the computation of the reaction forces, let us consider a beam for which a solution can be easily obtained. Such a beam is presented in Figure 3.5 together with the bending moment and shear force diagrams, which have been obtained with the method of moment distribution.

From the shear force diagram, the support reactions at A, B, and C are, respectively, given as

$$R_A = 1.6 \text{ kN} \downarrow \quad R_B = 11.8 \text{ kN} \uparrow \quad R_C = 1.8 \text{ kN} \uparrow \tag{3.33}$$

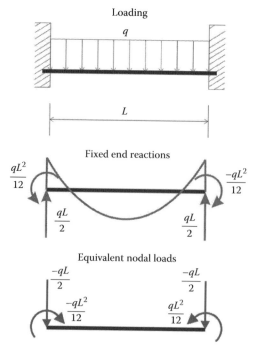

FIGURE 3.4   Statically equivalent nodal loads.

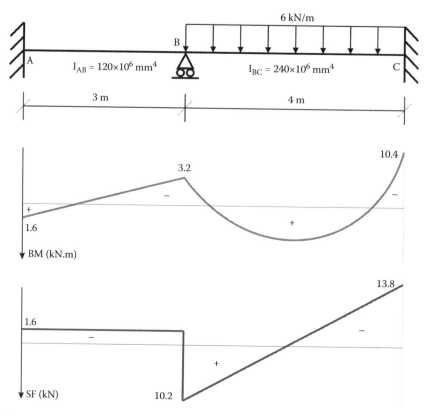

FIGURE 3.5   Loading, bending moment, and shear force diagrams.

From the bending moment diagrams, the support moments are obtained as

$$M_A = 1.6 \text{ kN.m} \curvearrowright \quad M_B = 3.2 \text{ kN.m} \curvearrowright\curvearrowright \quad M_C = 10.4 \text{ kN.m} \curvearrowright \quad (3.34)$$

Using the finite element method, let us calculate these support reactions.

Element AB:
Considering that the beam is made of steel with an elastic modulus of 200×10^6 kN/m², and using a consistent set of units, kN and m, from Equation (3.30) the stiffness matrix of element AB is obtained as

$$[K_{AB}] = \begin{bmatrix} 10667 & 16000 & 10667 & 16000 \\ 16000 & 32000 & 16000 & 16000 \\ -10667 & -16000 & 10667 & -16000 \\ 16000 & 16000 & -16000 & 32000 \end{bmatrix} \quad (3.35)$$

Element AB is not subjected to any external loading

$$\{F_{AB}\} = \begin{Bmatrix} 0 \\ 0 \\ 0 \\ 0 \end{Bmatrix} \quad (3.36)$$

Element BC:

$$[K_{BC}] = \begin{bmatrix} 9000 & 18000 & -9000 & 18000 \\ 18000 & 48000 & -18000 & 24000 \\ -9000 & -18000 & 9000 & -18000 \\ 18000 & 24000 & -18000 & 48000 \end{bmatrix} \quad (3.37)$$

The applied uniformly distributed load is transformed into equivalent static loads as shown in Figure 3.4:

$$\{F_{AB}\} = \begin{Bmatrix} -qL/2 = -12 \text{ kN} \\ -qL^2/12 = -8 \text{ kN.m} \\ -qL/2 = -12 \text{ kN} \\ qL^2/12 = 8 \text{ kN.m} \end{Bmatrix} \quad (3.38)$$

Assembling the global stiffness matrix and force vector results in

$$\begin{bmatrix} 10667 & 16000 & 10667 & 16000 & 0 & 0 \\ 16000 & 32000 & 16000 & 16000 & 0 & 0 \\ -10667 & -16000 & \mathbf{19667} & \mathbf{2000} & -9000 & 18000 \\ 16000 & 16000 & \mathbf{2000} & \mathbf{80000} & -18000 & 24000 \\ 0 & 0 & -9000 & -18000 & 9000 & -18000 \\ 0 & 0 & 18000 & 24000 & -18000 & 48000 \end{bmatrix} \begin{Bmatrix} w_A \\ \theta_A \\ w_B \\ \theta_B \\ w_C \\ \theta_C \end{Bmatrix} = \begin{Bmatrix} 0 \\ 0 \\ -12 \\ -8 \\ -12 \\ 8 \end{Bmatrix} \quad (3.39)$$

The boundary conditions for the beam are given as

$$w_A = \theta_A = w_B = w_C = \theta_C = 0 \tag{3.40}$$

Eliminating the lines and columns corresponding to these degrees of freedom results in one single equation:

$$80000 \times \theta_B = -8 \implies \theta_B = -0.0001 rd \tag{3.41}$$

The results for each span will be computed individually.
The nodal displacements of element AB are obtained as

$$\{d_{AB}\} = \begin{Bmatrix} w_A = 0 \\ \theta_A = 0 \\ w_B = 0 \\ \theta_B = -0.0001 \end{Bmatrix} \tag{3.42}$$

The final reactions for element AB are caused by the rotation of joint B

$$\begin{Bmatrix} V_A \\ M_A \\ V_{B1} \\ M_B \end{Bmatrix} = \begin{bmatrix} 10667 & 16000 & 10667 & 16000 \\ 16000 & 32000 & 16000 & 16000 \\ -10667 & -16000 & 10667 & -16000 \\ 16000 & 16000 & -16000 & 32000 \end{bmatrix} \begin{Bmatrix} 0 \\ 0 \\ 0 \\ -0.0001 \end{Bmatrix} = \begin{Bmatrix} -1.6 \\ -1.6 \\ 1.6 \\ -3.2 \end{Bmatrix} \tag{3.43}$$

It can be noticed that

$$V_A = R_A = 1.6 \text{ kN} \downarrow$$

$$M_A = 1.6 \text{ kN.m} \curvearrowright$$

$$M_B = 3.2 \text{ kN.m} \curvearrowright$$

As to the notation V_{B1}, it means that only the end shear at point B is considered. The total reaction at B is equal to the end shear from element AB plus the end shear at point B from element BC, that is

$$R_B = V_{B1} + V_{B2}$$

Similarly, the final reactions for element BC are caused by joint B rotation minus the equivalent nodal loads that replaced the uniformly distributed load, that is

$$\begin{Bmatrix} V_{B2} \\ M_B \\ V_C \\ M_C \end{Bmatrix} = \begin{bmatrix} 9000 & 18000 & -9000 & 18000 \\ 18000 & 48000 & -18000 & 24000 \\ -9000 & -18000 & 9000 & -18000 \\ 18000 & 24000 & -18000 & 48000 \end{bmatrix} \begin{Bmatrix} 0 \\ -0.0001 \\ 0 \\ 0 \end{Bmatrix} - \begin{Bmatrix} -12 \\ -8 \\ -12 \\ 8 \end{Bmatrix} = \begin{Bmatrix} 10.2 \\ 3.2 \\ 13.8 \\ -10.4 \end{Bmatrix} \tag{3.44}$$

Finally, we obtain

$$R_B = V_{B1} + V_{B2} = 1.6 + 10.2 = 11.8 \text{ kN}$$

The final results shown in Figure 3.6 are exactly the same as the ones shown in Figure 3.5.

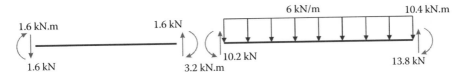

6 kN/m   10.4 kN.m

1.6 kN.m   1.6 kN

1.6 kN   10.2 kN   13.8 kN

1.6 kN   3.2 kN.m

FIGURE 3.6 Support reactions for individual members.

3.4 INTERNAL HINGE

In some cases, a beam may contain an internal hinge, which results in a discontinuity in the slope of the deflection curve as well as a zero value of the bending moment. If we are to analyze the beam shown in Figure 3.7 using the finite element method, we will discretize the beam using two elements. The hinge should be accounted for only once; either associated with element 1 or with element 2. If the beam is discretized with two elements, one with a hinge at its right end and the other with a hinge at its left, the result will be a singular stiffness matrix. Using Equation (3.30), the force–displacement relationship for element 1 is written as

$$
\begin{bmatrix}
12EI/L^3 & 6EI/L^2 & -12EI/L^3 & 6EI/L^2 \\
6EI/L^2 & 4EI/L & -6EI/L^2 & 2EI/L \\
-12EI/L^3 & -6EI/L^2 & 12EI/L^3 & -6EI/L^2 \\
6EI/L^2 & 2EI/L & -6EI/L^2 & 4EI/L
\end{bmatrix}
\begin{Bmatrix}
w_{11} \\ \theta_{11} \\ w_{12} \\ \theta_{12}
\end{Bmatrix}
=
\begin{Bmatrix}
F_{11} \\ M_{11} \\ F_{12} \\ M_{12}=0
\end{Bmatrix}
\tag{3.45}
$$

To eliminate the moment M_{12}, which is equal to zero, we partition the system of equations as follows:

$$
\begin{bmatrix}
12EI/L^3 & 6EI/L^2 & -12EI/L^3 & \vdots & 6EI/L^2 \\
6EI/L^2 & 4EI/L & -6EI/L^2 & \vdots & 2EI/L \\
-12EI/L^3 & -6EI/L^2 & 12EI/L^3 & \vdots & -6EI/L^2 \\
\cdots & \cdots & \cdots & \cdots & \cdots \\
6EI/L^2 & 2EI/L & -6EI/L^2 & \vdots & 4EI/L
\end{bmatrix}
\begin{Bmatrix}
w_{11} \\ \theta_{11} \\ w_{12} \\ \cdots \\ \theta_{12}
\end{Bmatrix}
=
\begin{Bmatrix}
F_{11} \\ M_{11} \\ F_{12} \\ \cdots \\ M_{12}=0
\end{Bmatrix}
\tag{3.46}
$$

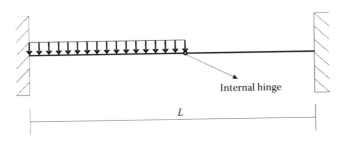

Internal hinge

L

FIGURE 3.7 Beam with an internal hinge.

or in a more compact form as

$$
\begin{bmatrix} k_{11} & \vdots & k_{12} \\ \cdots & \cdots & \cdots \\ k_{21} & \vdots & k_{22} \end{bmatrix} \begin{Bmatrix} d \\ \cdots \\ \theta_{12} \end{Bmatrix} = \begin{Bmatrix} F \\ \cdots \\ M_{12} = 0 \end{Bmatrix} \tag{3.47}
$$

Expanding Equation (3.47) yields

$$
[k_{11}]\{d\} + [k_{12}]\{\theta_{12}\} = \{F\} \tag{3.48}
$$
$$
[k_{21}]\{d\} + [k_{22}]\{\theta_{12}\} = \{M_{12}\}
$$

Solving for $\{\theta_{12}\}$ using the second equation of (3.48) yields

$$
\{\theta_{12}\} = [k_{22}]^{-1}(\{M_{12}\} - [k_{21}]\{d\}) \tag{3.49}
$$

Substituting for $\{\theta_{12}\}$ in the first equation of (3.48) and rearranging yields

$$
([k_{11}] - [k_{12}][k_{22}]^{-1}[k_{21}])\{d\} = (\{F\} - [k_{12}][k_{22}]^{-1}\{M_{12}\}) \tag{3.50}
$$

or in a more compact form as

$$
[K_C]\{d\} = \{F_C\} \tag{3.51}
$$

where $[K_C]$ is a condensed matrix. When the partitioned parts of Equation (3.48) are substituted in Equation (3.51), the condensed matrix becomes

$$
[K_C] = \begin{bmatrix} 3EI/L^3 & 3EI/L^2 & -3EI/L^3 \\ EI/L^2 & 3EI/L & -3EI/L^2 \\ -3EI/L^3 & -3EI/L^2 & 3EI/L^3 \end{bmatrix} \tag{3.52}
$$

It is true that moment M_{12} is equal to zero at the hinge, but not the rotation θ_{12}, and, as such, it should not have been eliminated from Equation (3.51). To include the rotation θ_{12}, we expand Equation (3.51) as follows:

$$
\begin{bmatrix} 3EI/L^3 & 3EI/L^2 & -3EI/L^3 & 0 \\ 3EI/L^2 & 3EI/L & -3EI/L^2 & 0 \\ -3EI/L^3 & -3EI/L^2 & 3EI/L^3 & 0 \\ 0 & 0 & 0 & 0 \end{bmatrix} \begin{Bmatrix} w_{11} \\ \theta_{11} \\ w_{12} \\ \theta_{12} \end{Bmatrix} = \begin{Bmatrix} F_{11} \\ M_{11} \\ F_{12} \\ M_{12} \end{Bmatrix} \tag{3.53}
$$

For element 2 with a hinge at its left end, Equation (3.53) is rewritten as

$$
\begin{bmatrix} 3EI/L^3 & 0 & -3EI/L^3 & 3EI/L^2 \\ 0 & 0 & 0 & 0 \\ -3EI/L^3 & 0 & 3EI/L^3 & -3EI/L^2 \\ 3EI/L^2 & 0 & -3EI/L^2 & 3EI/L \end{bmatrix} \begin{Bmatrix} w_{21} \\ \theta_{21} \\ w_{22} \\ \theta_{22} \end{Bmatrix} = \begin{Bmatrix} F_{21} \\ M_{21} \\ F_{22} \\ M_{22} \end{Bmatrix} \tag{3.54}
$$

It is very important that the hinge should be accounted for only once. Otherwise, the result will be a singular stiffness matrix (Figure 3.8).

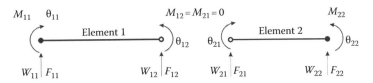

FIGURE 3.8   Beam elements with a hinge.

3.5   COMPUTER CODE: BEAM.m

Except for slight differences that need to be taken into account, writing a MATLAB® code for the analysis of slender beams is not much different from that for a truss structure. First, the elements' stiffness do not need to be transformed from local to global coordinates. Second, each element will have two types of loading: one that consists of the external forces directly applied to the nodes and another that only consists of the statically equivalent nodal loads. Therefore, in the development of the program BEAM.m, we will follow the same style as that used in the program TRUSS.m.

Let us consider the beam shown in Figure 3.9.

3.5.1   DATA PREPARATION

To read the data, we will use the M-file **beam_1_data.m**. Again, we will use a consistent set of units: mm for length and N for force.

The input data for this beam consist of the following:

- **nnd** $= 4$; number of nodes
- **nel** $= 3$; number of elements
- **nne** $= 2$; number of nodes per element
- **nodof** $= 2$; number of degrees of freedom per node

3.5.1.1   Nodes Coordinates

The abscissae x of the nodes are given in the form of a vector **geom(nnd, 1)**:

$$\mathbf{geom} = \begin{bmatrix} 0 \\ 4000 \\ 9000 \\ 16000 \end{bmatrix}$$

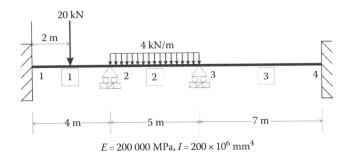

FIGURE 3.9   Example of a continuous beam.

3.5.1.2 Element Connectivity

The table of connectivity describes how the elements are connected to each other. The element connectivity is given in the matrix **connec(nel, 2)**:

$$\mathbf{connec} = \begin{bmatrix} 1 & 2 \\ 2 & 3 \\ 3 & 4 \end{bmatrix}$$

3.5.1.3 Material and Geometrical Properties

The material and geometrical properties are given in the matrix **prop(nel, 2)**. The first column represents the Young's modulus while the second represents the second moment of inertia of the cross section:

$$\mathbf{prop} = \begin{bmatrix} 200000 & 200.e + 6 \\ 200000 & 200.e + 6 \\ 200000 & 200.e + 6 \end{bmatrix}$$

3.5.1.4 Boundary Conditions

In the same fashion as for the truss, a restrained degree of freedom is assigned the digit 0, while a free degree of freedom is assigned the digit 1. As previously explained, a node in a beam element has two degrees of freedom: a vertical translation along the axis y and a rotation around the axis z perpendicular to the plan xy. As shown in Figure 3.9, nodes 1 and 4 are fully fixed (encastré). Their degrees of freedom are all assigned the digit 0. Nodes 2 and 3 are simple supports. They are restrained vertically but are free to rotate. Therefore, their degrees of freedom w and θ are respectively assigned the digits 0 and 1. The information on the boundary conditions is given in the matrix **nf(nnd, nodof)**:

$$\mathbf{nf} = \begin{bmatrix} 0 & 0 \\ 0 & 1 \\ 0 & 1 \\ 0 & 0 \end{bmatrix}$$

The free degrees of freedom (different from zero) are then counted and their rank assigned back into the matrix **nf(nnd, nodof)**:

$$\mathbf{nf} = \begin{bmatrix} 0 & 0 \\ 0 & 1 \\ 0 & 2 \\ 0 & 0 \end{bmatrix}$$

In this case, the total number of active degrees of freedom is obtained as **n = 2**.

3.5.1.5 Internal Hinges

To account for internal hinges, we create a vector **Hinge(nel, 2)** that we initialize to 1. If a particular element k has a hinge at its left end, then we assign it the digit 0 at the position of its first node; that is,

$$\mathbf{Hinge(k, 1)} = 0$$

On the other hand, the hinge may be accounted for with the element j having it at its right. In that case, we assign it the digit 0 at the position of its second node; that is,

$$\textbf{Hinge}(j, 2) = 0$$

A hinge must be considered for one element only.

3.5.1.6   Loading

When it comes to loading, a beam element differs from a rod element. As previously explained, a beam element can have two types of loading: loads applied directly at the nodes and statically equivalent nodal loads. A good computer code should cater for both loadings. To distinguish between the two loading systems, we will use two matrices: **Joint_loads(nnd, 2)** and **Element_loads(nel, 4)**.

There are no loads applied directly at the nodes. Therefore, the matrix **Joint_loads(nnd, 2)** is empty:

$$\textbf{Joint\_loads} = \begin{bmatrix} 0 & 0 \\ 0 & 0 \\ 0 & 0 \\ 0 & 0 \end{bmatrix}$$

Elements 1 and 2 have loads applied along their length, which need to be transformed to statically equivalent nodal loads, as shown in Figure 3.9

Element	F_{y1}	M_1	F_{y2}	M_2
1	-10^4	-10^7	-10^4	10^7
2	-10^4	-8.333×10^6	-10^4	8.333×10^6
3	0	0	0	0

These data are stored in the M-file **beam_1_data.m** in the matrix **Element_loads**.

The two systems of loads are added to form the global force vector $\textbf{F}(n)$. This is carried out in the M-file **form_beam_F.m** as follows:

- *Joint loads*: To assemble the nodal loads, we create a loop over the nodes. If a degree of freedom $\textbf{nf}(i, j)$ is not restrained, then it is susceptible of carrying a load. That load is **Joint_loads(i, j)**, and it is assembled into the global force vector at the position $\textbf{F}(\textbf{nf}(i, j))$.
- *Element loads*: To assemble the statically equivalent nodal loads, we create a loop over the elements. Since the loads are element based, we need the "steering vector" $\textbf{g}$ containing the number of the degrees of freedom of the nodes of the element. It is formed in the same way as in the program **truss.m**. The script is given in the M-file **beam_g.m**. Then, we create a loop over the degrees of freedom of the element. If a degree of freedom $\textbf{nf}(i, j)$ is not restrained, then it is susceptible of carrying a load. That load is **Element_loads(i, j)**, and it is assembled into global force vector at the position $\textbf{F}(\textbf{g}(j))$.

The data preparation is now complete, and the model data are written to the file **beam_1_results.txt** using the M-file **print_beam_model.m**. At this stage, it is possible to initialize the global matrix $\textbf{KK}(n, n) = \textbf{0}$:

$$\textbf{KK} = \begin{bmatrix} 0 & 0 \\ 0 & 0 \end{bmatrix}$$

Again, we will only assemble the quantities corresponding to the active degrees of freedom; that is, the lines and the columns in the matrix **KK** corresponding respectively to the active degrees of freedom **1** and **2**. The restrained degrees of freedom, with a number equal to **0**, will be eliminated.

3.5.1.7 Stiffness Matrix

For a beam element, there is no need to transform the element stiffness matrix from local to global coordinates since both sets of axes are colinear. Therefore, for each element, from 1 to *nel*, we set up the local stiffness matrix and directly assemble it into the global stiffness matrix **KK**.

For any element **i**, we retrieve its first and second nodes from the connectivity matrix:

$$\textbf{node\_1} = \textbf{connec}(\textbf{i}, \textbf{1})$$

$$\textbf{node\_2} = \textbf{connec}(\textbf{i}, \textbf{2})$$

Then using the values of the nodes, we retrieve their *x* coordinates from the geometry matrix:

$$\textbf{x1} = \textbf{geom}(\textbf{node\_1});$$

$$\textbf{x2} = \textbf{geom}(\textbf{node\_2});$$

Next, we evaluate the length of the element as

$$\textbf{L} = |\textbf{x2} - \textbf{x1}|$$

Finally, we retrieve the material and geometrical property of the section

$$\textbf{E} = \textbf{prop}(\textbf{i}, \textbf{1}); \qquad \textbf{I} = \textbf{prop}(\textbf{i}, \textbf{2})$$

Depending on whether nodes 1 or 2 are internal hinges, we evaluate the stiffness matrix **kl** as follows:

- **if Hinge(i, node_1) = 0**, evaluate the matrix **kl** using Equation (3.53)
- **if Hinge(i, node_2) = 0**, evaluate the matrix **kl** using Equation (3.52)
- **else**, evaluate the matrix **kl** using Equation 3.30

The MATLAB script for evaluating the matrix **kl** is given in the M-file **beam_k.m**.

3.5.2 ASSEMBLY AND SOLUTION OF THE GLOBAL SYSTEM OF EQUATIONS

The global stiffness matrix [**KK**] is assembled using the same script **form_KK.m** as in the program **truss.m**. The global displacements vector **delta** is obtained as

$$\textbf{delta} = \textbf{KK}\backslash\textbf{F}$$

3.5.3 NODAL DISPLACEMENTS

To retrieve the nodal displacements, a loop is carried over all the nodes. If a degree of freedom *j* of a node *i* is free, that is, **nf(i, j)** ≠ **0**, then it could have a displacement different from zero. The value of the displacement is extracted from the global displacements vector **delta**:

$$\textbf{node\_disp}(\textbf{i}, \textbf{j}) = \textbf{delta}(\textbf{nf}(\textbf{i}, \textbf{j}))$$

3.5.4 ELEMENT FORCES

To obtain the member forces, a loop is carried over all the elements:

1. Form element stiffness matrix **[kl]**
2. Form element "steering" vector **{g}**
   a. Loop over the degrees of freedom of the element to obtain element displacements vector **edg**
   b. If **g(j) = 0**, then the degree of freedom is restrained; **ed(j) = 0**
   c. Otherwise **ed(j) = delta(g(j))**
3. Obtain element force vector due to joint loads as
   {fl} = [kl] ∗ {ed}
4. Obtain element equivalent nodal forces as
   {f0} = *Element_loads*(i, :).
5. Obtain element forces as
   force(i, :) = {fl} − {f0}

The results of the analysis are written to the file **beam_1_results.txt** using the M-file **print_beam_results.m**. A copy of the file **beam_1_results.txt** is included within the section **Program scripts**.

File:beam.m

```
%                              beam.m
%
%   LINEAR STATIC ANALYSIS OF A CONTINUOUS BEAM
%
clc                % Clear screen
clear              % Clear all variables in memory
%
% Make these variables global so they can be shared
% by other functions
%
global nnd nel nne nodof eldof n geom connec F ...
       prop nf Element_loads Joint_loads force Hinge
%
disp('Executing beam.m');
%
% Open file for output of results
%%
%   ALTER THE NEXT LINES TO CHOOSE AN OUTPUT FILE FOR THE RESULTS
%
disp('Results printed to file : beam_1_results.txt '); fid
=fopen('beam_1_results.txt','w');
%
%
% ALTER THE NEXT LINE TO CHOOSE AN INPUT FILE
%
beam_1_data                % Load the input file
%
%
KK =zeros(n) ;             % Initialize global stiffness
                           % matrix to zero

%
F=zeros(n,1);              % Initialize global force vector to zero
F = form_beam_F(F);        % Form global force vector
%
print_beam_model          % Print model data
%
for i=1:nel
```

```
    kl=beam_k(i);          % Form element matrix
%
    g=beam_g(i) ;          % Retrieve the element steering
                           % vector
%
    KK =form_KK(KK, kl, g);     % assemble global stiffness
                                % matrix
%
end
%
%%%%%%%%%%%%  End of assembly  %%%%%%%%%%%%
%
%
delta = KK\F ;         % solve for unknown displacements
%
% Extract nodal displacements
%
for i=1:nnd
    for j=1:nodof
        node_disp(i,j) = 0;
        if nf(i,j)~= 0;
        node_disp(i,j) = delta(nf(i,j)) ;
        end
    end
end
%
% Calculate the forces acting on each element
% in local coordinates, and store them in the
% vector force().
%
 for i=1:nel
    kl=beam_k(i);        % Form element matrix
%
    g=beam_g(i) ;          % Retrieve the element steering vector
    for j=1:eldof
        if g(j)== 0
            ed(j)=0.; % displacement = 0. for restrained freedom
        else
            ed(j) = delta(g(j));
        end
    end
    fl = kl*ed'         % Element force vector in global XY
    f0 = Element_loads(i,:)
    force(i,:) = fl-f0'
end
%
print_beam_results;
%
fclose(fid);
```

File:beam_1_data.m

```
%                File:     Beam_1_data.m
%
% The following variables are declared as global in order
% to be used by all the functions (M-files) constituting
% the program
%
global nnd nel nne nodof eldof n geom connec ...
    prop nf  Element_loads Joint_loads Hinge
%
format short e
%
%%%%%%%%%%%%%%% Beginning of data input %%%%%%%%%%%%%%%%%%
%
```

```
nnd = 4; % Number of nodes:
nel = 3; % Number of elements:
nne = 2 ; % Number of nodes per element:
nodof =2 ; % Number of degrees of freedom per node
eldof = nne*nodof; % Number of degrees of freedom
                   % per element
%
% Nodes coordinates X and Y
geom=zeros(nnd,1);
geom= [ 0.;   ... % X coord. node 1
        4000.;... % X  coord. node 2
        9000.;... % X coord. node 3
        16000. ] ; % X coord. node 4
%
% Element connectivity
%
connec=zeros(nel,2);
connec = [1    2 ; ... % 1st and 2nd node of element 1
          2    3 ; ... % 1st and 2nd node of element 2
          3    4 ]; % 1st and 2nd node of element 3
%
% Geometrical properties
%
% prop(1,1) = E; prop(1,2)= I
%
prop=zeros(nel,2);
prop = [200000    200.e+6; ... % E and I of element 1
        200000    200.e+6; ... % E and I of element 2
        200000    200.e+6]; % E and I of element 3
%
% Boundary conditions
%
nf = ones(nnd, nodof); % Initialize the matrix nf to 1
nf(1,1) = 0; nf(1,2) =0 ; % Prescribed nodal freedom of node 1
nf(2,1) = 0;              % Prescribed nodal freedom of node 2
nf(3,1) = 0;              % Prescribed nodal freedom of node 3
nf(4,1) = 0; nf(4,2)= 0 ; % Prescribed nodal freedom of node 4


%
% Counting of the free degrees of freedom
%
n=0;
for i=1:nnd
    for j=1:nodof
        if nf(i,j) ~= 0
        n=n+1;
        nf(i,j)=n;
        end
    end
end
%
%
% Internal Hinges
%
Hinge = ones(nel, 2);
%
% loading
%
Joint_loads= zeros(nnd, 2);
% Enter here the forces in X and Y directions at node i
%
Element_loads= zeros(nel, 4);
Element_loads(1,:)= [ -1.e4      -1.e7      -1.e4      1.e7];
Element_loads(2,:)= [ -1.e4      -8.333e6   -1.e4      8.333e6 ];
%
%%%%%%%%%%%%%%%%%%%%%%%% End of input %%%%%%%%%%%%%%%%%%%%%%%%
```

File:beam_1_results.m

```
******* PRINTING MODEL DATA **************


--------------------------------------------------------
Number of nodes:                                    4
Number of elements:                                 3
Number of nodes per element:                        2
Number of degrees of freedom per node:              2
Number of degrees of freedom per element:           4


--------------------------------------------------------
Node        X
 1,       0000.00
 2,       4000.00
 3,       9000.00
 4,       16000.00


--------------------------------------------------------
Element       Node_1        Node_2
  1,            1,            2
  2,            2,            3
  3,            3,            4


--------------------------------------------------------
Element        E              I
  1,        200000,         2e+008
  2,        200000,         2e+008
  3,        200000,         2e+008

--------------------------------------------------------
------------Nodal freedom---------------------------
Node       disp_w      Rotation
 1,          0,           0
 2,          0,           1
 3,          0,           2
 4,          0,           0


--------------------------------------------------------
-----------------Applied Nodal Loads------------------
Node       load_Y        Moment
 1,       0000.00,       0000.00
 2,       0000.00,       1667000.00
 3,       0000.00,       8333000.00
 4,       0000.00,       0000.00
--------------------------------------------------------

Total number of active degrees of freedom, n = 2

--------------------------------------------------------


******* PRINTING ANALYSIS RESULTS **************


--------------------------------------------------------
Global force vector  F
   1.667e+006
   8.333e+006


--------------------------------------------------------
Displacement solution vector:  delta
 -0.00001
  0.00016


--------------------------------------------------------
```

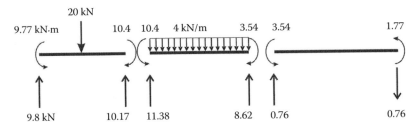

FIGURE 3.10   Example 1: Continuous beam results.

```
Nodal displacements
Node      disp_y        rotation
 1,      0.00000,       0.00000
 2,      0.00000,      -0.00001
 3,      0.00000,       0.00016
 4,      0.00000,       0.00000


----------------------------------------------------

Members actions
element      fy1           M1          Fy2            M2
 1,       9829.92,    9773230.20,    10170.08,    -10453539.60
 2,      11381.17,   10453539.60,     8618.83,     -3547673.27
 3,        760.22,    3547673.27,     -760.22,      1773836.63
```

The results are shown graphically for each element in Figure 3.10.

3.6   PROBLEMS

Prepare a data file for the beams shown next and carry out the analysis using the code *beam.m*.

3.6.1   PROBLEM 3.1 (FIGURE 3.11)

Input file

```
%              File:   Beam_problem1_data.m
%
% The following variables are declared as global in order
% to be used by all the functions (M-files) constituting
% the program
%
global nnd nel nne nodof eldof n geom connec ...
    prop nf  Element_loads Joint_loads Hinge
%
format short e
%
```

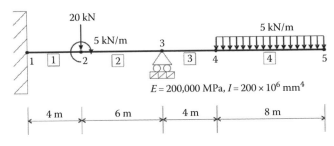

FIGURE 3.11   Problem 3.1.

```
%%%%%%%%%%%%%% Beginning of data input %%%%%%%%%%%%%%%%%
%
nnd = 5; % Number of nodes:
nel = 4; % Number of elements:
nne = 2 ; % Number of nodes per element:
nodof =2 ; % Number of degrees of freedom per node
eldof = nne*nodof; % Number of degrees of freedom
                   % per element
%
% Nodes coordinates X and Y
geom=zeros(nnd,1);
geom = [    0.; ...   % X coord. node 1
         4000.; ...   % X  coord. node 2
        10000.; ...   % X coord. node 3
        14000.; ...   % X coord. node 4
        22000.] ;     % X coord. node 5
%
% Element connectivity
%
connec=zeros(nel,2);
connec = [1   2 ; ...   % 1st and 2nd node of element 1
          2   3 ; ...   % 1st and 2nd node of element 2
          3   4 ; ...   % 1st and 2nd node of element 3
          4   5 ];      % 1st and 2nd node of element 4
%
% Geometrical properties
%
prop=zeros(nel,2);
prop   = [ 200000    200.e+6; ... % E and I of element 1
           200000    200.e+6; ... % E and I of element 2
           200000    200.e+6; ... % E and I of element 3
           200000    200.e+6 ];   % E and I of element 4
%
% Boundary conditions
%
nf = ones(nnd, nodof); % Initialize the matrix nf to 1
nf(1,1) = 0; nf(1,2) =0 ; % Prescribed nodal freedom of node 1
nf(3,1) = 0;              % Prescribed nodal freedom of node 3
%
% Counting of the free degrees of freedom
%
n=0;
for i=1:nnd
    for j=1:nodof
        if nf(i,j) ~= 0
        n=n+1;
        nf(i,j)=n;
        end
    end
end
%
%
% Internal Hinges
%
Hinge = ones(nel, 2);
%
% loading
%
Joint_loads= zeros(nnd, 2);
Joint_loads(2,:)=[-20000     -5e+6]
%
Element_loads= zeros(nel, 4);
Element_loads(4,:)= [ -2.e4    -2.66666e7    -2.e4   2.66666e7];

%
%%%%%%%%%%%%%%%%%%%%%%%%% End of input %%%%%%%%%%%%%%%%%%%%%%%%%
```

Results file

```
******* PRINTING MODEL DATA **************


--------------------------------------------------------
Number of nodes:                              5
Number of elements:                           4
Number of nodes per element:                  2
Number of degrees of freedom per node:        2
Number of degrees of freedom per element:     4


--------------------------------------------------------
Node        X
  1,      0000.00
  2,      4000.00
  3,     10000.00
  4,     14000.00
  5,     22000.00


--------------------------------------------------------
Element      Node_1       Node_2
   1,          1,           2
   2,          2,           3
   3,          3,           4
   4,          4,           5


--------------------------------------------------------
Element        E             I
   1,       200000,       2e+008
   2,       200000,       2e+008
   3,       200000,       2e+008
   4,       200000,       2e+008


--------------------------------------------------------
------------Nodal freedom-------------------------
Node       disp_w     Rotation
  1,          0,          0
  2,          1,          2
  3,          0,          3
  4,          4,          5
  5,          6,          7


--------------------------------------------------------
-----------------Applied Nodal Loads-----------------
Node       load_Y       Moment
  1,      0000.00,      0000.00
  2,     -20000.00,    -5000000.00
  3,      0000.00,      0000.00
  4,     -20000.00,   -26666600.00
  5,     -20000.00,    26666600.00
--------------------------------------------------------


Total number of active degrees of freedom, n = 7


--------------------------------------------------------

 ******* PRINTING ANALYSIS RESULTS **************


--------------------------------------------------------
Global force vector  F
   -20000
   -5e+006
   0
   -20000
   -2.66666e+007
   -20000
   2.66666e+007
```

```
------------------------------------------------------
Displacement solution vector:  delta
  15.57600
   0.00561
  -0.01870
 -128.13333
  -0.04270
 -533.73339
  -0.05337


------------------------------------------------------
Nodal displacements
Node       disp_y         rotation
  1,       0.00000,       0.00000
  2,      15.57600,       0.00561
  3,       0.00000,      -0.01870
  4,    -128.13333,      -0.04270
  5,    -533.73339,      -0.05337


------------------------------------------------------
Members actions
element     fy1            M1              Fy2            M2
  1,     -32640.00,   -121400000.00,     32640.00,     -9160000.00
  2,     -52640.00,      4160000.00,     52640.00,   -320000000.00
  3,      40000.00,    320000000.00,    -40000.00,   -160000000.00
  4,      40000.00,    160000000.00,        -0.00,           0.00
```

3.6.2 PROBLEM 3.2 (FIGURE 3.12)

Input file

```
%               File:    Beam_problem2_data.m
%
% The following variables are declared as global in order
% to be used by all the functions (M-files) constituting
% the program
%
global nnd nel nne nodof eldof n geom connec ...
    prop nf  Element_loads Joint_loads Hinge
%
```

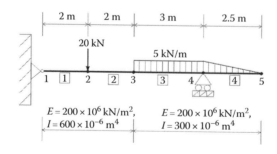

Equivalent nodal loads

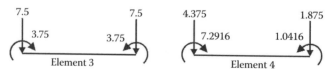

FIGURE 3.12   Problem 3.2 and equivalent nodal loads for elements 3 and 4.

```
format short e
%
%%%%%%%%%%%%% Beginning of data input %%%%%%%%%%%%%%%%%
%
nnd = 5; % Number of nodes:
nel = 4; % Number of elements:
nne = 2 ; % Number of nodes per element:
nodof =2 ; % Number of degrees of freedom per node
eldof = nne*nodof; % Number of degrees of freedom
                   % per element
%
% Nodes coordinates X and Y
geom=zeros(nnd,1);
geom = [0.; ... % X coord. node 1
        2.; ... % X  coord. node 2
        4.; ... % X coord. node 3
        7.; ... % X coord. node 4
        9.5] ;  % X coord. node 5
%
% Element connectivity
%
connec=zeros(nel,2);
connec = [1    2 ; ...   % 1st and 2nd node of element 1
          2    3 ; ...   % 1st and 2nd node of element 2
          3    4 ; ...   % 1st and 2nd node of element 3
          4    5];       % 1st and 2nd node of element 4
%
% Geometrical properties
%
% prop(1,1) = E; prop(1,2)= I
%
prop=zeros(nel,2);
prop  = [200e+6  600.e-6; ...  % E and I of element 1
         200e+6  600.e-6; ...  % E and I of element 2
         200e+6  300.e-6; ...  % E and I of element 3
         200e+6  300.e-6] ;    % E and I of element 4
%
% Boundary conditions
%
nf = ones(nnd, nodof); % Initialize the matrix nf to 1
nf(1,1) = 0;   ; % Prescribed nodal freedom of node 1
nf(4,1) = 0;             % Prescribed nodal freedom of node 3
%
% Counting of the free degrees of freedom
%
n=0;
for i=1:nnd
    for j=1:nodof
        if nf(i,j) ~= 0
        n=n+1;
        nf(i,j)=n;
        end
    end
end
%
%
% Internal Hinges
%
Hinge = ones(nel, 2);
%
% loading
%
Joint_loads= zeros(nnd, 2);
Joint_loads(2,:)=[-20       0]
%
Element_loads= zeros(nel, 4);
Element_loads(3,:)= [ -7.5      -3.75      -7.5       3.75];
```

```
Element_loads(4,:)= [ -4.375    -7.2916   -1.875   1.0416];
%
%%%%%%%%%%%%%%%%%%%%%%%% End of input %%%%%%%%%%%%%%%%%%%%%%%%
```

Results file

```
******* PRINTING MODEL DATA **************


-----------------------------------------------------
Number of nodes:                               5
Number of elements:                            4
Number of nodes per element:                   2
Number of degrees of freedom per node:         2
Number of degrees of freedom per element:      4

-----------------------------------------------------
Node       X
 1,      0000.00
 2,      0002.00
 3,      0004.00
 4,      0007.00
 5,      0009.50

-----------------------------------------------------
Element      Node_1       Node_2
  1,           1,            2
  2,           2,            3
  3,           3,            4
  4,           4,            5

-----------------------------------------------------
Element        E             I
  1,        2e+008,       0.0006
  2,        2e+008,       0.0006
  3,        2e+008,       0.0003
  4,        2e+008,       0.0003

-----------------------------------------------------
------------Nodal freedom-------------------------
Node      disp_w     Rotation
 1,         0,          1
 2,         2,          3
 3,         4,          5
 4,         0,          6
 5,         7,          8

-----------------------------------------------------
----------------Applied Nodal Loads------------------
Node      load_Y       Moment
 1,      0000.00,      0000.00
 2,      -020.00,      0000.00
 3,      -007.50,      -003.75
 4,      0000.00,      -003.54
 5,      -001.88,      0001.04
-----------------------------------------------------

Total number of active degrees of freedom, n = 8

-----------------------------------------------------


******* PRINTING ANALYSIS RESULTS **************

-----------------------------------------------------
```

```
Global force vector   F
    0
  -20
    0
  -7.5
  -3.75
  -3.5416
  -1.875
   1.0416


-------------------------------------------------------
Displacement solution vector:   delta
 -0.00065
 -0.00113
 -0.00039
 -0.00142
  0.00008
  0.00058
  0.00135
  0.00053


-------------------------------------------------------
Nodal displacements
Node      disp_y         rotation
 1,       0.00000,      -0.00065
 2,      -0.00113,      -0.00039
 3,      -0.00142,       0.00008
 4,       0.00000,       0.00058
 5,       0.00135,       0.00053


-------------------------------------------------------
Members actions
element      fy1          M1         Fy2          M2
 1,         15.94,      -0.00,     -15.94,       31.88
 2,         -4.06,     -31.87,       4.06,       23.75
 3,         -4.06,     -23.75,      19.06,      -10.94
 4,          6.25,      10.94,       0.00,        0.00
```

3.6.3   PROBLEM 3.3 (FIGURE 3.13)

Input file

```
%                 File:    beam_problem3_data.m
%
% The following variables are declared as global in order
% to be used by all the functions (M-files) constituting
% the program
%
global nnd nel nne nodof eldof n geom connec ...
    prop nf  Element_loads Joint_loads Hinge
%
```

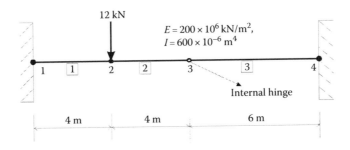

FIGURE 3.13   Problem 3.3.

```
format short e
%
%%%%%%%%%%%%%% Beginning of data input %%%%%%%%%%%%%%%%%
%
nnd = 4; % Number of nodes:
nel = 3; % Number of elements:
nne = 2 ; % Number of nodes per element:
nodof =2 ; % Number of degrees of freedom per node
eldof = nne*nodof; % Number of degrees of freedom
                   % per element
%
% Nodes coordinates X and Y
geom=zeros(nnd,1);
geom = [0.; ... % X coord. node 1
        4.; ... % X  coord. node 2
        8.; ... % X coord. node 3
        14.] ;  % X coord. node 4
%
% Element connectivity
%
connec=zeros(nel,2);
connec = [1   2 ; ...   % 1st and 2nd node of element 1
          2   3 ; ...   % 1st and 2nd node of element 2
          3   4] ;      % 1st and 2nd node of element 3
%
% Geometrical properties
%
% prop(1,1) = E; prop(1,2)= I
%
prop=zeros(nel,2);
prop  = [200e+6  600.e-6; ... % E and I of element 1
         200e+6  600.e-6; ... % E and I of element 2
         200e+6  600.e-6 ];   % E and I of element 3
%
% Boundary conditions
%
nf = ones(nnd, nodof); % Initialize the matrix nf to 1
nf(1,1) = 0; nf(1,2)=0 ; % Prescribed nodal freedom of node 1
nf(4,1) = 0; nf(4,2)=0 ; % Prescribed nodal freedom of node 4
%
% Counting of the free degrees of freedom
%
n=0;
for i=1:nnd
    for j=1:nodof
        if nf(i,j) ~= 0
        n=n+1;
        nf(i,j)=n;
        end
    end
end
%
%
% Internal Hinges
%
Hinge = ones(nel, 2);
Hinge(2,2) = 0;
%
% loading
%
Joint_loads= zeros(nnd, 2);
Joint_loads(2,:)=[-12      0]
%
Element_loads= zeros(nel, 4);
%
%%%%%%%%%%%%%%%%%%%%%%%%%%% End of input %%%%%%%%%%%%%%%%%%%%%%%%%
```

Results file

```
****** PRINTING MODEL DATA **************


-------------------------------------------------------
Number of nodes:                               4
Number of elements:                            3
Number of nodes per element:                   2
Number of degrees of freedom per node:         2
Number of degrees of freedom per element:      4


-------------------------------------------------------
Node        X
 1,      0000.00
 2,      0004.00
 3,      0008.00
 4,      0014.00

-------------------------------------------------------
Element      Node_1      Node_2
   1,          1,          2
   2,          2,          3
   3,          3,          4

-------------------------------------------------------
Element        E            I
   1,       2e+008,      0.0006
   2,       2e+008,      0.0006
   3,       2e+008,      0.0006


-------------------------------------------------------
------------Nodal freedom-------------------------
Node      disp_w     Rotation
 1,         0,          0
 2,         1,          2
 3,         3,          4
 4,         0,          0


-------------------------------------------------------
-----------------Applied Nodal Loads------------------
Node      load_Y        Moment
 1,      0000.00,      0000.00
 2,     -012.00,       0000.00
 3,      0000.00,      0000.00
 4,      0000.00,      0000.00
-------------------------------------------------------


Total number of active degrees of freedom, n = 4

-------------------------------------------------------


****** PRINTING ANALYSIS RESULTS **************

-------------------------------------------------------
Global force vector   F
   -12
    0
    0
    0

-------------------------------------------------------
Displacement solution vector:   delta
  -0.00096
  -0.00027
```

```
   -0.00158
    0.00040


---------------------------------------------------------
Nodal displacements
Node        disp_y         rotation
 1,        0.00000,        0.00000
 2,       -0.00096,       -0.00027
 3,       -0.00158,        0.00040
 4,        0.00000,        0.00000


---------------------------------------------------------
Members actions
element     fy1          M1          Fy2           M2
 1,        9.36,       26.90,       -9.36,        10.55
 2,       -2.64,      -10.55,        2.64,         0.00
 3,       -2.64,       -0.00,        2.64,       -15.82
```

3.7 ANALYSIS OF A SIMPLE BEAM WITH ABAQUS

3.7.1 INTERACTIVE EDITION

In this section, we will analyze the continuous beam shown in Figure 3.14 with the Abaqus interactive edition. The cross section of the beam is shown in Figure 3.15. The material is steel with an elastic modulus of 200 GPa.

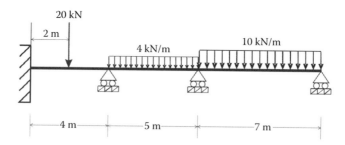

FIGURE 3.14 Continuous beam.

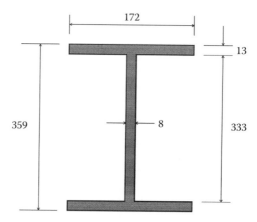

FIGURE 3.15 Beam cross section; dimensions are in mm.

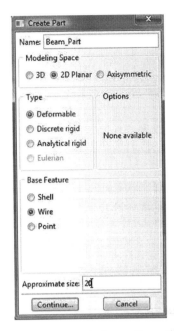

Start **Abaqus CAE**. Click on **Create Model Database**. On the main menu, click on **File** and **Set Work Directory** to choose your working directory. Click on **Save As** and name the file **Beam.cae**. On the left-hand-side menu, click on **Part** to begin creating the model. Name the part **Beam_Part**, check **2D Planar**, and check **Deformable** in the type. Choose **Wire** as the base feature. Enter an approximate size of 20 m and click on **Continue** (Figure 3.16).

FIGURE 3.16   Creating the Beam_Part.

In the sketcher menu, choose the **Create-Lines Connected** icon to begin drawing the geometry of the beam. Click on **Done** in the bottom-left corner of the viewport window (Figure 3.17).

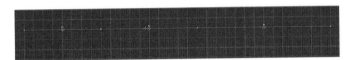

FIGURE 3.17   Drawing using the connected line icon.

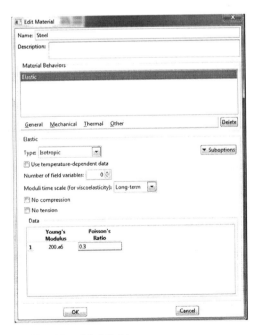

Under the model tree, click on **material** to create a material, and name it **Steel**. Click on **Mechanical**, then **Elasticity**, and **Elastic**. Enter $200.e6 \, \text{kN/m}^2$ for the elastic modulus, and 0.3 for Poisson's ratio (Figure 3.18).

FIGURE 3.18   Material definition.

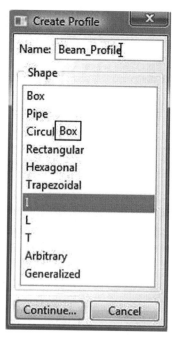

Under the model tree, click on **Profiles** to create a profile, and name it **Beam_Profile**. Click on **Continue** (Figure 3.19).

FIGURE 3.19   Creating a beam profile.

In the **Edit Profile** dialog box, enter the dimensions of the profile. Make sure you enter them in meters to keep a consistent set of units. Click on **OK** (Figure 3.20).

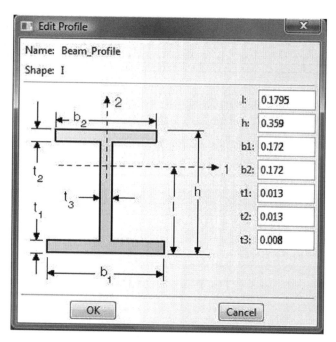

FIGURE 3.20   Entering the dimensions of a profile.

Under the model tree, click on **Sections** to create a section and name it **Beam_section**. In the **Category** check **Beam**, and in the **Type**, choose **Beam**. Click on **Continue** (Figure 3.21).

FIGURE 3.21   Creating a section.

In the **Edit Section** dialog box, in the **Profile name** choose **Beam_Profile**, and in **Material** choose **Steel**. Leave the Poisson's ratio as zero. Click on **OK** (Figure 3.22).

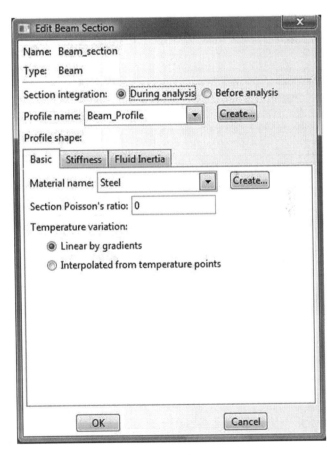

FIGURE 3.22   Editing a beam section.

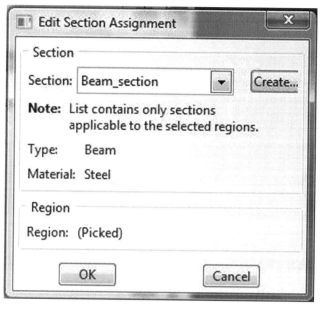

FIGURE 3.23   Editing section assignments.

Expand the menu under **Beam_Part** and double click on **Section Assignments**. With the mouse, select the whole beam in the drawing area, and click on **Done** in the left bottom corner. In the **Edit Section Assignments** dialog box, make sure that **Beam_section** appears in the section. Click on **OK** (Figure 3.23).

In Abaqus, a beam element must have an orientation in space. The default orientation is the one shown in Figure 3.24. The axis n_1 is in opposite direction to the global axis Z. For beams in a plane the n_1-direction is always $(0.0, 0.0, -1.0)$; that is, normal to the plane in which the motion occurs. Therefore, planar beams can bend only about the first beam-section axis.

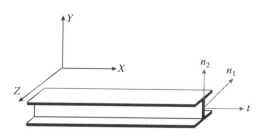

FIGURE 3.24   Beam orientation.

Change the Module to **Property**. Click on the **Assign Beam Orientation** icon and select the entire geometry from the viewport. In the prompt in the left-bottom corner of the viewport, accept $(0.0, 0.0, -1.0)$ as the direction for n_1, and click **Return**. Click **OK** to confirm (Figure 3.25).

FIGURE 3.25   Assigning beam orientation.

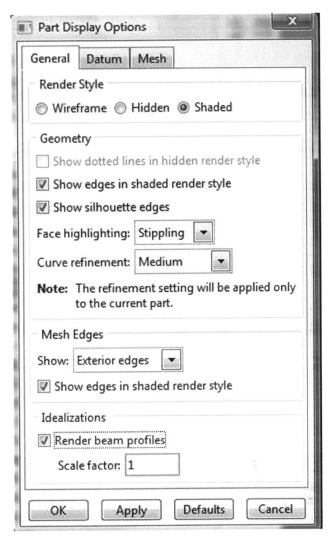

FIGURE 3.26 Rendering beam profile.

In the menu bar select **View**, then **Part Display Options**. In the **Part Display Options**, in **Idealizations**, check **Render beam profiles**. Click **Apply** (Figure 3.26).

Using the **Rotate View icon** you can rotate the beam to appear as shown in Figure 3.27. If you are happy with what you see, go back and uncheck **Render beam profiles**

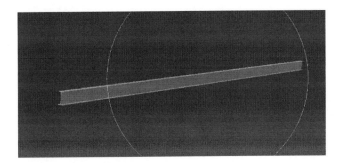

FIGURE 3.27 Rendered beam.

In the model tree, double click on **Mesh** under the **Beam_Part**, and in the main menu, under **Mesh**, click on **Element Type**. With the mouse highlight all members in the viewport and select **Done**. In the dialog box, select **Standard** for element type, **Linear** for geometric order, and **beam** for family. The name of the element **B21** and its description are given below the element controls. Click on **OK** (Figure 3.28).

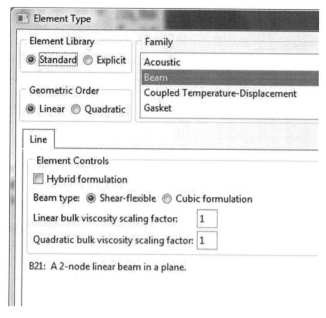

FIGURE 3.28   Selecting a beam element.

In the main menu, under **Seed**, click on **Edge by size**. With the mouse highlight all the beam in the viewport. In the prompt area of the viewport, enter **1.0**; that is, each element will have a length of 1 m. Click on **Return**, then click **Done** (Figure 3.29).

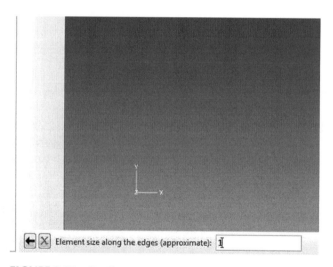

FIGURE 3.29   Seeding a mesh by size.

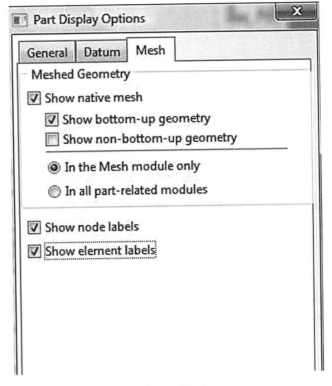

In the main menu, under **Mesh**, click on **Part**. In the prompt area of the viewport, click on **Yes**. In the menu bar select **View**, then **Part Display Options**. In the **Part Display Options**, under **Mesh**, check **Show node labels** and **Show element labels**. Click **Apply**. The element and node labels will appear in the viewport (Figure 3.30).

FIGURE 3.30   Node and element labels.

In the model tree under **Beam_Part**, double click on **Sets**. In the dialog box, name the set **Fixed_Support**, check **Node** in type, and click on **Continue**. With the mouse highlight node 1, which is the fixed support, and click on **Done** in the prompt area of the viewport (Figure 3.31).

FIGURE 3.31   Creating a node set.

FIGURE 3.32 Selecting multiple nodes.

FIGURE 3.33 Creating element sets.

Again double click on **Sets**. In the dialog box, name the set **Roller_Supports**, check **Node** in type, and click on **Continue**. While keeping the *SHIFT* key down, with the mouse highlight nodes 2, 3, and 4. When selected, they change color, as shown in Figure 3.32. Click on **Done** in the prompt area of the viewport. Again double click on **Sets**. Name the set **Loaded_Node**, check **Node** in type, and click on **Continue**. With the mouse highlight node 6. Click on **Done** in the prompt area of the viewport.

Next create two element sets: one for the elements subject to the 4 kN/m load and the other for the elements subject to 10 kN/m. Double click on **Sets**. In the dialog box, name the set **UDL4**, check **Element** in type, and click on **Continue**. While keeping the *SHIFT* key down, with the mouse highlight elements 5, 6, 7, 8, and 9. When selected, they change color as shown in Figure 3.33. Click on **Done** in the prompt area of the viewport. Create another element set named **UDL10** and select elements 10 to 16.

In the model tree, expand the **Assembly** and double click on **Instances**. Select **Dependent** for the instance type and click **OK**.

In the model tree, expand **Steps** and **Initial**, and double click on **BC**. Name the boundary condition **fixed**, select **Displacement/Rotation** for the type, and click on **Continue**. In the right-bottom corner of the viewport, you can see **Sets** (Figure 3.34) Double click on it.

FIGURE 3.34 Imposing BC using created sets.

In the dialog box that appears, select **Beam_Part-1. Fixed_Support** and check **Highlight selections in viewport**. Click on **Continue** (Figure 3.35).

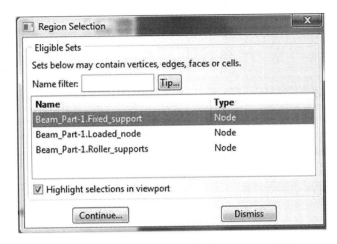

FIGURE 3.35 Selecting a node set for boundary conditions.

Fill up the **Edit Boundary Conditions** in the dialog box as shown by restricting all the degrees of freedom. Click on **OK** (Figure 3.36).

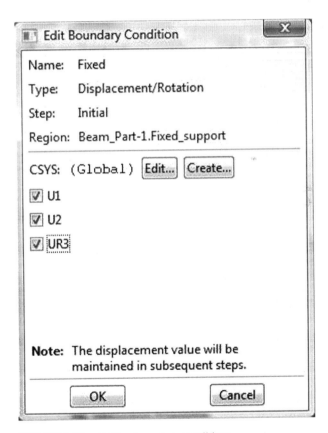

FIGURE 3.36 Editing boundary conditions.

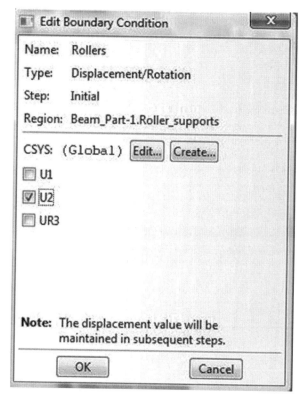

Click on **BC** again. Name the boundary condition **Rollers**, select **Displacement/ Rotation** for the type, and click on **Continue**. Double click on **Sets**. Select **Beam_Part-1.Roller_Supports**. Fill up the **Edit Boundary Conditions** by restricting only **U2**. Click on **OK** (Figure 3.37).

FIGURE 3.37   Imposing BC using created sets.

In the model tree, double click on **Steps**. Name the step **Apply_Loads**. Set the procedure to **General** and select **Static, General**. Click on **Continue**. Give the step a description and click **OK**. In the model tree, under **steps**, and under **Apply_Loads**, click on **Loads**. Name the load **Concentrated load** and select **Concentrated force** as the type. Click on **Continue**. In the **Region Selection** dialog box, select **Beam_Part-1.Loaded_node**. Click on **Continue**. In the **Edit Load** dialog box, enter −20 for **CF2**. Click **OK** (Figure 3.38).

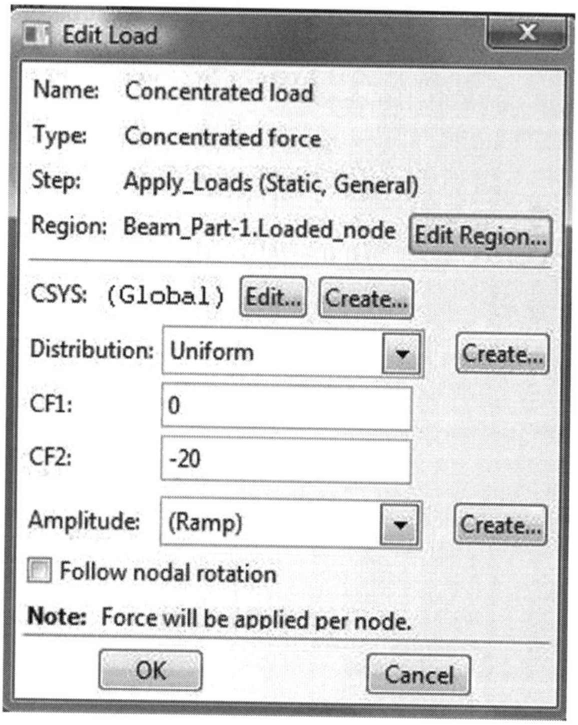

FIGURE 3.38   Imposing a concentrated load using a node set.

Click on **Loads** again. Name the load **UDL-4** and select **Line load** as the type. Click on **Continue**. In the **Region Selection** dialog box, select **Beam_Part-1.UDL4**. Click on **Continue**. In the **Edit Load** dialog box, enter −4 for **Component 2**. Click **OK**. Repeat the procedure again to create the 10 kN/m distributed load over element set **Beam_Part-1.UDL10** (Figure 3.39).

Edit Load

Name: UDL-4

Type: Line load

Step: Apply_Loads (Static, General)

Region: Beam_Part-1.UDL4

System: Global

Distribution: Uniform   Create...

Component 1:

Component 2: **-4**

Amplitude: (Ramp)   Create...

OK   Cancel

FIGURE 3.39   Imposing a line load on an element set.

In the model tree, expand the **Field Output Requests** and then double click on **F-Output-1**. **F-Output-1** is the default and is automatically generated when creating the step. Uncheck the variables **Contact** and select any other variable you wish to add to the field output. Click on **OK** (Figure 3.40).

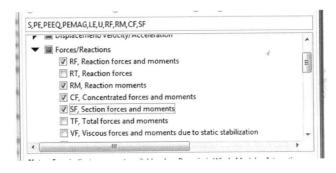

S,PE,PEEQ,PEMAG,LE,U,RF,RM,CF,SF

Displacement/Velocity/Acceleration

Forces/Reactions
☑ RF, Reaction forces and moments
☐ RT, Reaction forces
☑ RM, Reaction moments
☑ CF, Concentrated forces and moments
☑ SF, Section forces and moments
☐ TF, Total forces and moments
☐ VF, Viscous forces and moments due to static stabilization

FIGURE 3.40   Field output.

Under **Analysis**, right click on **Jobs** and then click on **Create**. In the **Create Job** dialog box, name the job **BEAM_Problem** and click on **Continue**. In the **Edit Job** dialog box, enter a description for the job. Check **Full analysis**, select to run the job in **Background**, and check to start it **immediately**. Click **OK**. Expand the tree under **Jobs**, right click on **BEAM_Problem**. Then, click on **Submit**. If you get the following message **BEAM_Problem completed successfully** in the bottom window, then your job is free of errors and was executed properly (Figure 3.41).

```
The job "Beam_Problem" has been created.
The job input file "Beam_Problem.inp" has been submitted for analysis.
Job Beam_Problem: Analysis Input File Processor completed successfully.
Job Beam_Problem: Abaqus/Standard completed successfully.
Job Beam_Problem completed successfully.
```

FIGURE 3.41   Submitting a job in Abaqus CAE.

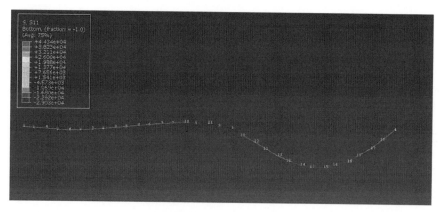

FIGURE 3.42 Plotting stresses in the bottom fiber.

Under the top menu, in the **Module** scroll to **Visualization**, and click to load **Abaqus Viewer**. On the main menu, under **File**, click **Open**, navigate to your working directory, and open the file **Beam_Problem.odb**. It should have the same name as the job you submitted. Click on the **Common options** icon to display the **Common Plot options** dialog box. Under **labels**, check **Show Element labels** and **Show Node labels** to display elements and nodes' numbering. Click on the icon **Plot Deformed Shape** to display the deformed shape of the beam. On the main menu, click on **Results**, then on **Field Output** to open the **Field Output** dialog box. Choose **S Stress components at integration points**. For component, choose $S11$ to plot the stresses in the bars (Figure 3.42). Click on **Section points** to open the section point dialog box. Check **bottom** to plot the stresses in the lower fiber or **Top** for the stresses in the top fiber. In the menu bar, click on **Report** and **Field Output**. In the **Report Field Output** dialog box, for **Position** select **Unique nodal**, check **RF2** and **RM3** for **RF: Reaction force**, and check **U2** and **UR3** for **U: Spatial displacement**. Then click on **Set up**. Click on **Select** to navigate to your working directory. Name the file **Beam_Problem.rpt**. Uncheck **Append to file** and click **OK**. Use your favorite text editor and open the file **Beam_Problem.rpt**, which should be the same as the one listed next.

```
********************************************************************************
Field Output Report, written Mon Apr 11 11:55:08 2011

Source 1
---------


   ODB: C:/Abaqus_Working Directory/Beam_Problem.odb
   Step: Apply_Loads
   Frame: Increment      1: Step Time =    1.000

Loc 1 : Nodal values from source 1

Output sorted by column "Node Label".

Field Output reported at nodes for part: BEAM_PART-1
```

Node Label	RF.RF2 @Loc 1	RM3 @Loc 1	U.U2 @Loc 1	UR3 @Loc 1
1	13.7126	14.6329	−13.7126E−36	−14.6329E−36
2	8.35087	0.	−8.35087E−36	352.684E−06
3	58.5745	0.	−56.5745E−36	−1.48255E−03
4	29.3621	0.	−24.3621E−36	2.99728E−03
5	0.	0.	−232.859E−06	−245.559E−06

6	0.	0.	-494.778E-06	-58.1202E-06
7	0.	0.	-350.09E-06	246.55E-06
8	0.	0.	356.109E-06	360.552E-06
9	0.	0.	721.62E-06	307.269E-06
10	0.	0.	972.231E-06	66.529E-06
11	0.	0.	857.33E-06	-487.975E-06
12	0.	0.	-2.1104E-03	-2.16607E-03
13	0.	0.	-4.34032E-03	-1.88215E-03
14	0.	0.	-5.88021E-03	-946.55E-06
15	0.	0.	-6.23626E-03	324.961E-06
16	0.	0.	-5.23046E-03	1.61662E-03
17	0.	0.	-3.00053E-03	2.61264E-03

Minimum	0.	0.	-6.23626E-03	-2.16607E-03
At Node	17	17	15	12
Maximum	58.5745	14.6329	972.231E-06	2.99728E-03
At Node	3	1	10	4
Total	110.000	14.6329	-24.9686E-03	1.61611E-03

3.7.2 ANALYSIS OF A BEAM WITH ABAQUS KEYWORD EDITION

In this section, we will prepare an input file for the beam shown in Figures 3.14 and 3.15. We will use the same number of elements and nodes as earlier.

The file is named **Beam_Problem_Keyword.inp** and is listed next:

```
*Heading
  Beam_Problem Model keyword edition
**
*Preprint, echo=No, model=NO, history=NO
**
**
**   Define the end nodes
**
*Node
     1,            -9.,           0.
    17,             6.,           0.
**
** Generate the remaining nodes
**
*Ngen
1,17,1
**
**   Define element 1
**
*Element, type=B21
1,1,2
**
** Generate the elements
**
*Elgen, elset = all_elements
1,16, 1, 1
**
**
**
*Nset, nset=Fixed_support
 1,
*Nset, nset=Roller_supports
 5,  10,  17
*Nset, nset=Loaded_node
 3
*Elset, elset=UDL4, generate
 5,  9,  1
```

```
*Elset, elset=UDL10, generate
 10,  16,   1
**
**
** Section: Beam_section  Profile: Beam_Profile
*Beam Section, elset=all_elements, material=Steel, section=I
0.1795, 0.359, 0.172, 0.172, 0.013, 0.013, 0.008
0.,0.,-1.
**
**
** MATERIALS
**
*Material, name=Steel
*Elastic
 2e+08, 0.3
**
** BOUNDARY CONDITIONS
**
**
*Boundary
Fixed_support, encastre
Roller_supports, 2, 2
** ------------------------------------------------------------
**
** STEP: Apply_Loads
**
*Step, name=Apply_Loads
*Static
1., 1., 1e-05, 1.
**
** LOADS
**
*Cload
Loaded_node, 2, -20.
**
*Dload
UDL4, PY, -4. UDL10, PY, -10.
**
** OUTPUT REQUESTS
**
**
*Output, field
*Node Output
CF, RF, RM, U
*Element Output
 S
**
*Output, history, variable=PRESELECT
*End Step
```

- The file starts with the keyword ***HEADING**, which in this case is entered as **Beam_Problem Model keyword edition**.
- Using the keyword ***node**, we define the two extreme nodes **1** and **17** and give their coordinates x and y.
- Using the keyword ***ngen**, which stands for node generate, we generate all the remaining nodes from **1** to **17** in an increment of **1**.
- Using the keyword ***Element, type = B21** representing a beam element in the plane. In the data line, we enter **1** as the element number with nodes **1** and **2** all separated by "**,**".
- Next, we generate the elements using the keyword ***elgen**. We group the elements in a set named **all_elements**. In the data line, we enter the master element that has been previously defined; that is element 1, then the number of elements to be generated, 16, followed by the increment in node numbers of corresponding nodes from element to element, which in this case is 1, then the increment in element numbers, which is again 1.

- Once all the elements and nodes are defined, using the keyword ***nset** we create the following node sets: **Fixed_support**, which contains node 1, **Roller_supports**, which contains nodes 5, 10, and 17, and **Loaded_node**, which contains node 3.
- Next, with the keyword ***elset**, and the parameter **generate**, we create element sets **UDL4** and **UDL10** containing respectively elements 5 to 9 and 10 to 16. When the parameter **generate** is included, each data line should give a first element, a last element, and the increment in element numbers between these elements. If it is not included, then all the elements forming the set must be listed in the data lines.
- With the keyword ***Beam Section** we define a section for the elements contained in the set **all_elements**, the material is **Steel**, and the section is the form of **I**. In the first data line we enter the dimensions of the section, and in the second its orientation with respect to the global coordinates.
- Using the keyword ***Material**, we create a material named **Steel**. The material is elastic and its properties are given in the data line of the keyword ***elastic**.
- Using the created node sets, we impose the boundary conditions with the keyword ***Boundary**. We fully fix the node set **Fixed_support** by using **encastre**. All the nodes in the node set **Roller_supports** are fixed in the direction 2.
- Next using the keyword ***step**, we create a step named **Apply_Loads**. The keyword ***static** indicates that it will be a general static analysis.
- Using the keyword ***cload**, we apply a concentrated load of $-20\,\text{kN}$ in the direction 2 to the node in node set **Loaded_node**.
- Using the keyword ***dload** for distributed load, we apply line loads of -4 and $-10\,\text{kN/m}$ to the elements contained respectively in element sets **UDL4** and **UDL10**.
- Using the keywords ***Output, field**, and ***Node Output**, we request the nodal variables **CF**: concentrated force, **RF**: reaction force, **RM**: reaction moment, and displacements **U** to be written to the database file ***.odb**. With ***Element Output**, we also add the stresses **S** to the database file.
- ***Output, history, variable = PRESELECT** requests the default variables for history output.
- Finally, we end the step and the file with ***End Step**.

At the command line type **Abaqus job=Beam_Problem_Keyword inter** followed by **Return**. If you get an error, open the file with extension ***.dat** to see what type of error. To load the visualization model, type **Abaqus Viewer** at the command line.

4 Rigid Jointed Frames

4.1 INTRODUCTION

Rigid jointed frames are often used in buildings. They resist the combined effects of horizontal and vertical loads. They derive their strength from the moment interactions between the beams and the columns at the rigid joints. As a result, the elements are subjected not only to bending but also to axial force. Such elements are referred to as *beam–column elements*. Their nodal displacements include both translations and rotation (u, v, θ), as shown in Figure 4.1. In total, there are six degrees of freedom

$$\{d_e\} = \{u_1, v_1, \theta_1, u_2, v_2, \theta_2\}^T \tag{4.1}$$

corresponding to six nodal loads

$$\{F_e\} = \{F_{x1}, F_{y1}, M_1, F_{x2}, F_{y2}, M_2\}^T \tag{4.2}$$

4.2 STIFFNESS MATRIX OF A BEAM–COLUMN ELEMENT

If we assume that the deformations are infinitesimally small, and the material is linear elastic, then the axial displacements of the beam–column element do not interact with the bending deformations. Consequently, the principle of superposition applies, and the displacements, forces, and stiffness matrix of the beam–column element can be obtained by simply adding the respective matrices of a truss element, Equation (2.10), and that of a beam element, Equation (3.30)

$$[K_e] = \begin{bmatrix} AE/L & 0 & 0 & -AE/L & 0 & 0 \\ 0 & 12EI/L^3 & 6EI/L^2 & 0 & -12EI/L^3 & 6EI/L^2 \\ 0 & 6EI/L^2 & 4EI/L & 0 & -6EI/L^2 & 2EI/L \\ -AE/L & 0 & 0 & AE/L & 0 & 0 \\ 0 & -12EI/L^3 & -6EI/L^2 & 0 & 12EI/L^3 & -6EI/L^2 \\ 0 & 6EI/L^2 & 2EI/L & 0 & -6EI/L^2 & 4EI/L \end{bmatrix} \tag{4.3}$$

4.3 STIFFNESS MATRIX OF A BEAM–COLUMN ELEMENT IN THE PRESENCE OF HINGED END

Sometimes a designer may specify an internal hinge in a frame, which results in a zero value for the bending moment. To account for the presence of a hinge, the stiffness matrix can be obtained by superimposing the respective matrices of a truss element, Equation (2.10), and that of a beam element with a hinge at its right end, Equation (3.52), or a hinge at its left end, Equation (3.53).

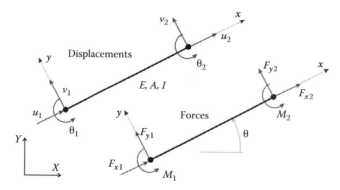

FIGURE 4.1 Beam column element with six degrees of freedom.

It follows that the stiffness matrix of a beam–column element with a hinge at its right end is given as

$$
[K_e] = \begin{bmatrix}
AE/L & 0 & 0 & -AE/L & 0 & 0 \\
0 & 3EI/L^3 & 3EI/L^2 & 0 & -3EI/L^3 & 0 \\
0 & 3EI/L^2 & 3EI/L & 0 & -3EI/L^2 & 0 \\
-AE/L & 0 & 0 & AE/L & 0 & 0 \\
0 & -3EI/L^3 & -3EI/L^2 & 0 & 3EI/L^3 & 0 \\
0 & 0 & 0 & 0 & 0 & 0
\end{bmatrix}
\tag{4.4}
$$

and with a hinge at its left end as

$$
[K_e] = \begin{bmatrix}
AE/L & 0 & 0 & -AE/L & 0 & 0 \\
0 & 3EI/L^3 & 0 & 0 & -3EI/L^3 & 3EI/L^2 \\
0 & 0 & 0 & 0 & 0 & 0 \\
-AE/L & 0 & 0 & AE/L & 0 & 0 \\
0 & -3EI/L^3 & 0 & 0 & 3EI/L^3 & -3EI/L^2 \\
0 & 3EI/L^2 & 0 & 0 & -3EI/L^2 & 3EI/L
\end{bmatrix}
\tag{4.5}
$$

As with a beam system, a hinge should be associated only with one element.

4.4 GLOBAL AND LOCAL COORDINATE SYSTEMS

Like for a truss member, beam–column (or frame) elements do not all have the same orientation in space. Similarly, when it comes to assembling the global stiffness, we need to have the element degrees of freedom (nodal displacements) given in terms of the common reference axes of the frame. The transformation is similar to that of a bar element except that the transformation matrix is given as

$$
\begin{bmatrix}
\cos\theta & -\sin\theta & 0 & 0 & 0 & 0 \\
\sin\theta & \cos\theta & 0 & 0 & 0 & 0 \\
0 & 0 & 1 & 0 & 0 & 0 \\
0 & 0 & 0 & \cos\theta & -\sin\theta & 0 \\
0 & 0 & 0 & \sin\theta & \cos\theta & 0 \\
0 & 0 & 0 & 0 & 0 & 1
\end{bmatrix}
\tag{4.6}
$$

The transformation is carried out as follows:

$$
[\overline{K_e}] = [C][K_e][C]^T
\tag{4.7}
$$

where $[\overline{K_e}]$ represents the element stiffness matrix in the global coordinate system.

4.5 GLOBAL STIFFNESS MATRIX ASSEMBLY AND SOLUTION FOR UNKNOWN DISPLACEMENTS

The assembly of the global stiffness matrix is similar to that of a truss detailed in Section 2.3.4, except that a beam column element has six degrees of freedom.

The introduction of the boundary conditions also follows the same principle. The only difference is that a node possesses three degrees of freedom: two translations and a rotation. Any of these degrees of freedom can be free or restrained. When all the degrees of freedom at a node are restrained, the node is sometimes referred to as encastré.

Distributed loads along a beam–column element are also treated in the same fashion as for a beam element, Section 3.3.

4.6 COMPUTER CODE: frame.m

Writing a MATLAB® code for the analysis of a frame is merely a combination of the codes previously written for a truss and a beam structure. The only differences reside in the matrices dimensions. Similar to a truss structure, the elements, stiffness matrices need to be transformed from local to global coordinates. Likewise to a beam structure, each beam–column element will have two types of loading: one that consists of the external forces directly applied to the nodes and another that only consists of the statically equivalent nodal loads. Therefore, in the development of the program **frame.m**, we will borrow the same style as that used in the programs **truss.m** and **beam.m**.

Let us consider the portal frame shown in Figure 4.2.

4.6.1 DATA PREPARATION

To read the data, we will use the M-file **frame_problem1_data.m**. Again, we will use a consistent set of units: *mm* for length and *N* for force.

The input data for this beam consist of

- **nnd** = **5**; number of nodes
- **nel** = **4**; number of elements
- **nne** = **2**; number of nodes per element
- **nodof** = **3**; number of degrees of freedom per node

Note that a beam–column element has three degrees of freedom per node:
nodof = **3**.

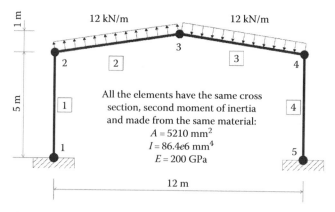

FIGURE 4.2   Example 1: Portal frame.

4.6.1.1   Nodes Coordinates

The coordinates x and y of the nodes are given in the form of a matrix **geom**(**nnd**, **2**):

$$\textbf{geom} = \begin{bmatrix} 0 & 0 \\ 0 & 5000 \\ 6000 & 6000 \\ 12000 & 5000 \\ 12000 & 0 \end{bmatrix}$$

4.6.1.2   Element Connectivity

The element connectivity is given in the matrix **connec**(**nel**, **2**):

$$\textbf{connec} = \begin{bmatrix} 1 & 2 \\ 2 & 3 \\ 3 & 4 \\ 4 & 5 \end{bmatrix}$$

4.6.1.3   Material and Geometrical Properties

The material and geometrical properties are given in the matrix **prop**(**nel**, **3**). The first column represents the Young's modulus, the second represents the cross-sectional area, and the third the second moment of inertia of the cross section:

$$\textbf{prop} = \begin{bmatrix} 200000 & 5310 & 86.4e + 6 \\ 200000 & 5210 & 86.4e + 6 \\ 200000 & 5210 & 86.4e + 6 \end{bmatrix}$$

4.6.1.4   Boundary Conditions

In the same fashion as for the truss and the beam, a restrained degree of freedom is assigned the digit 0, while a free degree of freedom is assigned the digit 1. As previously explained, a node in a beam–column element has three degrees of freedom: a horizontal translation along the axis X, a vertical translation along the axis Y, and a rotation around the axis Z perpendicular to the plan XY. As shown in Figure 4.2, nodes 1 and 5 are fully fixed (encastré). Their degrees of freedom are all assigned the digit 0. Nodes 2, 3, and 4 are free. Their degrees of freedom u, v, and θ are assigned the digit 1. The information on the boundary conditions is given in the matrix **nf(nnd, nodof)**:

$$\mathbf{nf} = \begin{bmatrix} 0 & 0 & 0 \\ 1 & 1 & 1 \\ 1 & 1 & 1 \\ 1 & 1 & 1 \\ 0 & 0 & 0 \end{bmatrix}$$

The free degrees of freedom (different from zero) are then counted, and their rank assigned back into the matrix **nf(nnd, nodof)**:

$$\mathbf{nf} = \begin{bmatrix} 0 & 0 & 0 \\ 1 & 2 & 3 \\ 4 & 5 & 6 \\ 7 & 8 & 9 \\ 0 & 0 & 0 \end{bmatrix}$$

In this case, the total number of active degrees of freedom is obtained as $\mathbf{n} = \mathbf{9}$.

4.6.1.5 Internal Hinges

To account for internal hinges, we create a vector **Hinge(nel, 2)** that we initialize to 1. If a particular element k has a hinge at its left end, then we assign it the digit 0 at the position of its first node; that is,

$$\mathbf{Hinge(k, 1)} = 0$$

On the other hand, the hinge may be accounted for with the element j having it at its right. In that case, we assign it the digit 0 at the position of its second node; that is,

$$\mathbf{Hinge(j, 2)} = 0$$

A hinge must be considered for one element only.

4.6.1.6 Loading

A beam–column element can have two types of loading: loads applied directly at the nodes and statically equivalent nodal loads. A good computer code should cater for both loadings. To distinguish between the two loading systems, we will use two matrices: **Joint_loads(nnd, 3)** and **Element_loads(nel, 4)**.

There are no loads applied directly at the nodes. Therefore, the matrix **Joint_loads(nnd, 3)** is empty:

$$\mathbf{Joint\_loads} = \begin{bmatrix} 0 & 0 & 0 \\ 0 & 0 & 0 \\ 0 & 0 & 0 \\ 0 & 0 & 0 \\ 0 & 0 & 0 \end{bmatrix}$$

Elements 2 and 3 have loads applied along their length which need to be transformed to statically equivalent nodal loads, as shown in Figure 3.4.

Element	F_{y1}	M_1	F_{y2}	M_2
2	$36.4965e3$	$37e6$	$36.4965e3$	$-37e6$
3	$-36.4965e3$	$-37e6$	$-36.4965e3$	$37e6$

These data are stored in the M-file: **beam_1_data.m** in the matrix **Element_loads**.

The two systems of loads are added to form the global force vector **F(n)**. This is carried out in the M-file **form_beam_F.m** as follows:

- *Joint loads*: The assembly of the joint loads is carried out using the script **Assem_Joint_Loads.m**. We create a loop over the nodes. If a degree of freedom **nf(i, j)** is not restrained, then it is susceptible of carrying a load. That load is **Joint_loads(i, j)**, and it is assembled into the global force vector at the position **F(nf(i, j))**.
- *Element loads*: The assembly of the statically equivalent nodal loads is carried out at the same time as the assembly of the global stiffness matrix in the same loop over all the elements.

The data preparation is now complete, and the model data are written to the file **frame_problem1_results.txt** using the M-file **print_frame_model.m** At this stage, it is possible to initialize the global matrix **KK(n, n) = 0**.

4.6.2 ELEMENT MATRICES

4.6.2.1 Stiffness Matrix in Local Coordinates

For each element, from 1 to *nel*, we set up the local stiffness and transformation matrices. Once the stiffness matrix **kl** is set up in local coordinates, it is transformed into global coordinates **kg** through the transformation matrix **C** and then assembled to the global stiffness matrix **KK**.

For any element **i**, we retrieve its first and second node from the connectivity matrix:

$$\textbf{node\_1} = \textbf{connec(i, 1)}$$
$$\textbf{node\_2} = \textbf{connec(i, 2)}$$

Then using the values of the nodes, we retrieve their *x* and *y* coordinates from the geometry matrix:

$$\textbf{x1} = \textbf{geom(node\_1, 1)}; \quad \textbf{y1} = \textbf{geom(node\_1, 2)}$$
$$\textbf{x2} = \textbf{geom(node\_2, 1)}; \quad \textbf{y2} = \textbf{geom(node\_2, 2)}$$

Next, using Pythagoras theorem, we evaluate the length of the element:

$$\textbf{L} = \sqrt{(\textbf{x2} - \textbf{x1})^2 + (\textbf{y2} - \textbf{y1})^2}$$

Finally, we retrieve the material and geometrical properties of the section

$$\textbf{E} = \textbf{prop(i, 1)}; \quad \textbf{A} = \textbf{prop(i, 2)}; \quad \textbf{I} = \textbf{prop(i, 3)}$$

before evaluating the matrix **kl** using Equation (4.3).

The MATLAB script for evaluating the matrix **kl** is given in the M-file **beam_column_k.m**.

4.6.2.2 Transformation Matrix

Once the nodal coordinates are retrieved, it is also possible to evaluate the angle θ that the element makes with the global X axis:

$$\theta = \tan^{-1}\left(\frac{\mathbf{y2} - \mathbf{y1}}{\mathbf{x2} - \mathbf{x1}}\right)$$

However, care should be taken when the element is at right angle with the global axis X as **x2** − **x1** = 0. The matrix **C** is evaluated using Equation (4.6). The MATLAB script is given in the M-file **beam_column_C.m**.

4.6.2.3 Stiffness Matrix in Global Coordinates

The element stiffness matrix **kg** is obtained as

$$\mathbf{kg} = \mathbf{C} \times \mathbf{kl} \times \mathbf{C}^{\mathbf{T}}$$

4.6.2.4 "Steering" Vector

Once the matrix **kg** is formed, we retrieve the "steering vector" **g** containing the number of the degrees of freedom of the nodes of the element:

$$\mathbf{g} = \begin{Bmatrix} \mathbf{nf}(\text{node\_1}, 1) \\ \mathbf{nf}(\text{node\_1}, 2) \\ \mathbf{nf}(\text{node\_1}, 3) \\ \mathbf{nf}(\text{node\_2}, 1) \\ \mathbf{nf}(\text{node\_2}, 2) \\ \mathbf{nf}(\text{node\_2}, 3) \end{Bmatrix}$$

The MATLAB script for constructing the steering vector **g** is given in the M-file **beam_column_g**.m.

4.6.2.5 Element Loads

For each element **i**, we also retrieve its statically equivalent nodal loads using the statement **fl = Element_loads(i, :)**. Since elements, loads are generally given in the element local coordinates system, we transform **fl** from local to global coordinates. To assemble the transformed vector of statically equivalent nodal loads into the global force vector F, we make use of the "steering vector" **g** containing the number of the degrees of freedom of the nodes of the element. The steering vector is built in the same way as in the program **truss**.m. The script is given in the M-file **beam_column_g**.m. Then, we create a loop over the degrees of freedom of the element. If a degree of freedom **nf(i, j)** is not restrained, then it is susceptible of carrying a load. That load is **Element_loads(i, j)**, and it is assembled in the global force vector at the position **F(g(j))**. The script is given in the M-file **Assem_Elem_loads.m**.

4.6.3 Assembly of the Global Stiffness Matrix

The global stiffness matrix [**KK**] is assembled using a double loop over the components of the vector **g**. The script is given in the M-file **form_KK.m**.

4.6.4 SOLUTION OF THE GLOBAL SYSTEM OF EQUATIONS

The solution of the global system of equations is obtained with one statement:

$$\textbf{delta} = \textbf{KK}\backslash\textbf{F}$$

The backslash symbol \ is used to "divide" a matrix by a vector.

4.6.5 NODAL DISPLACEMENTS

Once the global displacements vector **delta** is obtained, it is possible to retrieve any nodal displacements. A loop is carried over all the nodes. If a degree of freedom j of a node i is free, that is, $\textbf{nf}(\textbf{i},\textbf{j}) \neq \textbf{0}$, then it could have a displacement different from zero. The value of the displacement is extracted from the global displacements vector **delta**:

$$\textbf{node\_disp}(\textbf{i},\textbf{j}) = \textbf{delta}(\textbf{nf}(\textbf{i},\textbf{j}))$$

4.6.6 ELEMENT FORCES

To obtain the member forces, a loop is carried over all the elements:

1. Form element stiffness matrix [**kl**] in local xy
2. Form element transformation matrix [**C**]
3. Transform the element matrix from local to global coordinates
   [**kg**] = [**C**] ∗ [**kl**] ∗ [**C**]^{T}
4. Form element "steering" vector {**g**}
   a. Loop over the degrees of freedom of the element to obtain element displacements vector **edg** in global coordinates
   b. If **g(j)** = **0**, then the degree of freedom is restrained; **edg(j)** = **0**
   c. Otherwise **edg(j)** = **delta(g(j))**
5. Obtain element force vector in global XY coordinates
   {**fg**} = [**kg**] ∗ {**edg**}
6. Transform element force vector to local coordinates {**fl**} = [**C**]^{T} ∗ {**fg**}
7. Retrieve the element statically equivalents loads $\textbf{f}_0 = \textbf{Element\_loads}(\textbf{i}, :)$ if any
8. Obtain the elements internal forces as $\textbf{force}(\textbf{i}, :) = \textbf{fl} - \textbf{f}_0$

The results of the analysis are written to the file **frame_problem1_results.txt** using the M-file **print_frame_results.m**. A copy of the file **frame_problem1_results.txt** is included within.

File:frame.m

```
%   PROGRAM frame.m
%
%    LINEAR STATIC ANALYSIS OF A RIGID JOINTED FRAME
%
% Make these variables global so they can be shared by other functions
%
clc
clear all
%
global nnd nel nne nodof eldof n geom connec F
global  prop nf Element_loads Joint_loads force Hinge
%
format short e
%
disp('Executing frame.m');
%
% Open file for output of results
```

```
%
% ALTER NEXT LINES TO CHOOSE OUTPUT FILES
%
fid =fopen('frame_problem1_results.txt','w');
disp('Results printed to file : frame_problem1_results.txt ');
%
%%%%%%%%%%%  Beginning of data input     %%%%%%%%%%%%%%%%%%%%%%%
%
frame_problem1_data                      % Load the input file
%
F = zeros(n,1);    % Initialize global force vector to zero
%
F = Assem_Joint_Loads(F);  % Assemble joint loads to global force vector
%
print_frame_model           % Print model data
%
KK = zeros(n, n);  % Initialize the global stiffness matrix to zero
%
for i=1:nel
    kl=beam_column_k(i) ;    % Form element matrix in local xy
    C = beam_column_C(i);    % Form transformation matrix
    kg=C*kl*C' ;             % Transform the element matrix from local
                            % to global coordinates
    fl= Element_loads(i,:);  % Retrieve element equivalent nodal forces
                            % in local xy
    fg=C*fl'  ;             % Transform the element force vector from
                            % local to global coordinates
    g=beam_column_g(i) ;     % Retrieve the element degrees of freedom
    KK =form_kk(KK , kg, g);    % assemble global stiffness matrix
    F = Assem_Elem_loads(F , fg, g);   % assemble global force vector
end
%
%%%%%%%%%%%%%%%%%     End of assembly        %%%%%%%%%%%%%
%
%
delta = KK\F;               % solve for unknown displacements
%
% %
% Extract nodal displacements
%
for i=1:nnd
    for j=1:nodof
        node_disp(i,j) = 0;
        if nf(i,j)~= 0;
        node_disp(i,j) = delta(nf(i,j)) ;
        end
    end
end
%
%
 for i=1:nel
    kl=beam_column_k(i);      % Form element matrix in local xy
    C = beam_column_C(i);     % Form transformation matrix
    kg=C*kl*C' ;             % Transform the element matrix from local
                            % to global coordinates
    g=beam_column_g(i) ;     % Retrieve the element degrees of freedom
    for j=1:eldof
        if g(j)== 0
            edg(j)=0.;   % displacement = 0. for restrained freedom
        else
            edg(j) = delta(g(j));
        end
    end

    fg = kg*edg';       % Element force vector in global XY
    fl = C'*fg ;        % Element force vector in local  xy
    f0 = Element_loads(i,:) % Equivalent nodal loads
```

```
    force_l(i,:) = fl-f0';
    force_g(i,:) = C*(fl-f0');
end
%
print_frame_results;
%
fclose(fid);
```

File:frame_problem1_data.m

```
%                    File:    frame_problem1_data.m
%
% The following variables are declared as global in order
% to be used by all the functions (M-files) constituting
% the program
%
%
global nnd nel nne nodof eldof n geom connec F ...
        prop nf Element_loads Joint_loads force Hinge
%
format short e
%
nnd = 5 ;               % Number of nodes:
nel = 4 ;               % Number of elements:
nne = 2 ;               % Number of nodes per element:
nodof =3;               % Number of degrees of freedom per node
eldof = nne*nodof;      % Number of degrees of freedom per element
%
% Nodes coordinates x and y
%
geom=zeros(nnd,2);
geom(1,1)=0.      ; geom(1,2)= 0.;        % x and y coordinates of node 1
geom(2,1)=0.      ; geom(2,2)= 5000.;     % x and y coordinates of node 2
geom(3,1)=6000.   ; geom(3,2)= 6000.;     % x and y coordinates of node 3
geom(4,1)=12000.  ; geom(4,2)= 5000.;     % x and y coordinates of node 4
geom(5,1)=12000.  ; geom(5,2)= 0.;        % x and y coordinates of node 4
%
% Element connectivity
%
connec=zeros(nel,2);
connec(1,1) = 1;  connec(1,2) =2 ;   % First and second node of element 1
connec(2,1) = 2;  connec(2,2) =3 ;   % First and second node of element 2
connec(3,1) = 3;  connec(3,2) =4 ;   % First and second node of element 3
connec(4,1) = 4;  connec(4,2) =5 ;   % First and second node of element 4
%
% Geometrical properties
%
prop=zeros(nel,3);
prop(1,1)=2.0e+5; prop(1,2)=5210; prop(1,3)=86.4e+6; % E,A and I element 1
prop(2,1)=2.0e+5; prop(2,2)=5210; prop(2,3)=86.4e+6; % E,A and I element 2
prop(3,1)=2.0e+5; prop(3,2)=5210; prop(3,3)=86.4e+6; % E,A and I element 3
prop(4,1)=2.0e+5; prop(4,2)=5210; prop(4,3)=86.4e+6; % E,A and I element 4
%
% Boundary conditions
%
nf = ones(nnd, nodof);                    % Initialize the matrix nf to 1
nf(1,1) = 0; nf(1,2) =0; nf(1,3) = 0; % Prescribed nodal freedom of node 1
nf(5,1) = 0; nf(5,2)= 0; nf(5,3) = 0; % Prescribed nodal freedom of node 5
%
% Counting of the free degrees of freedom
%
n=0;
for i=1:nnd
    for j=1:nodof
        if nf(i,j) ~= 0
            n=n+1;
            nf(i,j)=n;
```

```
        end
      end
end
%
% Internal Hinges
%
Hinge = ones(nel,2);
%
% loading
%
Joint_loads= zeros(nnd, 3);
%
% Joint loads are usually entered in global coordinates
% Enter here the forces in X and Y directions and any
% concentrated moment at node i
%
% Staticaly equivalent loads are entered in local
% coordinates of the element
%
Element_loads= zeros(nel, 6);
Element_loads(2,:)= [0   36.4965e3   37e6   0   36.4965e3   -37e6];
Element_loads(3,:)= [0  -36.4965e3  -37e6   0  -36.4965e3    37e6];
%
%
%%%%%%%%%%%%   End of input   %%%%%%%%%%%%
```

File:frame_problem1_results.txt

```
******* PRINTING MODEL DATA **************
```

```
-------------------------------------------------------
Number of nodes:                                  5
Number of elements:                               4
Number of nodes per element:                      2
Number of degrees of freedom per node:            3
Number of degrees of freedom per element:         6
```

```
-------------------------------------------------------
Node        X              Y
 1,      0000.00        0000.00
 2,      0000.00        5000.00
 3,      6000.00        6000.00
 4,     12000.00        5000.00
 5,     12000.00        0000.00
```

```
-------------------------------------------------------
Element     Node_1      Node_2
   1,         1,           2
   2,         2,           3
   3,         3,           4
   4,         4,           5
```

```
-------------------------------------------------------
Element       E            A           I
   1,       200000,       5210      8.64e+007
   2,       200000,       5210      8.64e+007
   3,       200000,       5210      8.64e+007
   4,       200000,       5210      8.64e+007
```

```
-------------------------------------------------------
-------------Nodal freedom--------------------------
Node     disp_u        disp_u       Rotation
  1,       0,            0,            0
  2,       1,            2,            3
  3,       4,            5,            6
  4,       7,            8,            9
  5,       0,            0,            0
```

```
-------------------------------------------------------
----------------Applied joint Loads------------------
Node      load_X       load_Y         Moment
  1,      0000.00,     0000.00,        0000.00
  2,     -5999.99,     35999.93,        37000000.00
  3,    -11999.98,     0000.00,        -74000000.00
  4,     -5999.99,    -35999.93,        37000000.00
  5,      0000.00,     0000.00,        0000.00
-------------------------------------------------------


Total number of active degrees of freedom, n = 9

-------------------------------------------------------


    ******* PRINTING ANALYSIS RESULTS ************


-------------------------------------------------------
Global force vector   F
   -5999.99
   35999.9
   3.7e+007
   -12000
   0
   -7.4e+007
   -5999.99
   -35999.9
   3.7e+007


-------------------------------------------------------
Displacement solution vector:  delta
 -25.03159
  0.16363
  0.00712
 -25.04119
  0.00000
 -0.00686
 -25.03159
 -0.16363
  0.00712


-------------------------------------------------------
Nodal displacements
Node       disp_x          disp_y           rotation
  1,     0.00000e+000,    0.00000e+000,     0.00000e+000
  2,    -2.50316e+001,    1.63630e-001,     7.11912e-003
  3,    -2.50412e+001,    7.98515e-015,    -6.85508e-003
  4,    -2.50316e+001,   -1.63630e-001,     7.11912e-003
  5,     0.00000e+000,    0.00000e+000,     0.00000e+000


-------------------------------------------------------
Members actions in local coordinates
element    fx1        fy1          M1          fx2         Fy2          M2
1,    -34100.5529,   -11999.9753,    -54603626.4780,    34100.5529,   11999.9753,   -5396249.9239
2,     6230.6063,   -35609.3620,    5396249.9239,   -6230.6063,   -37383.6380,    -0.0000
3,    -6230.6063,    37383.6380,     0.0000,    6230.6063,   35609.3620,   5396249.9239
4,    34100.5529,   -11999.9753,    -5396249.9239,   -34100.5529,   11999.9753,   -54603626.4780
-------------------------------------------------------
Members actions in global coordinates
element    fx1        fy1          M1          fx2         Fy2          M2
1,    11999.9753,   -34100.5529,    -54603626.4780,   -11999.9753,   34100.5529,   -5396249.9239
2,    11999.9753,   -34100.5529,    5396249.9239,    0.0000,   -37899.2988,    -0.0000
3,     0.0000,    37899.2988,     0.0000,    11999.9753,   34100.5529,   5396249.9239
4,   -11999.9753,   -34100.5529,    -5396249.9239,   11999.9753,   34100.5529,   -54603626.4780
```

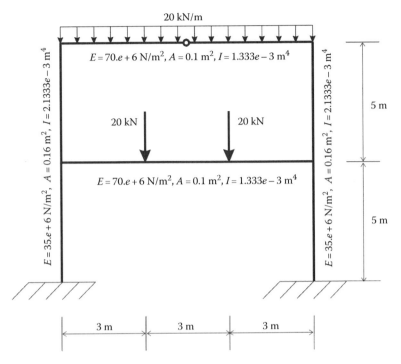

FIGURE 4.3   Frame with an internal hinge.

Figure 4.3 shows a two-level frame with an internal hinge in the top beam. The frame is made from two different materials. The columns have an elastic modulus of $35.e6\,\text{kN/m}^2$, a cross area of $0.16\,\text{m}^2$, and second moment of inertia of $2.1333e-3\,\text{m}^4$. The beams have an elastic modulus of $70e6\,\text{kN/m}^2$, a cross area of $0.1\,\text{m}^2$, and second moment of inertia of $1.333e-3\,\text{m}^4$. In addition, two concentrated loads are applied along the lower beam. Instead of considering the lower beam as one element with the concentrated loads transformed into statically equivalent loads, we will simply discretize the beam into three elements such that the two concentrated loads are applied at joints. The finite element discretization is shown in Figure 4.4.

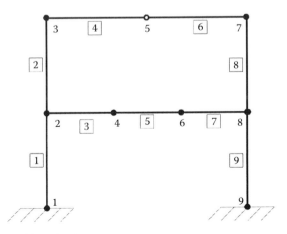

FIGURE 4.4   Finite element discretization.

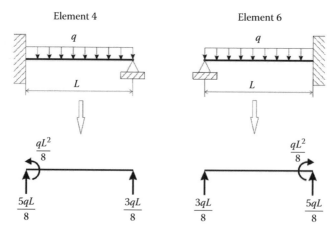

FIGURE 4.5 Statically equivalent nodal loads.

Elements 4 and 6 are both subject to a uniformly distributed load that needs to be transformed to statically equivalent nodal loads. However, the elements are joined by a hinge at node 5. In such a case, the statically equivalent nodal loads are the reactions of a propped cantilever (Figure 4.5).

Input File

```
%                File:    frame_problem2_data.m
%
% The following variables are declared as global in order
% to be used by all the functions (M-files) constituting
% the program
%
%
global nnd nel nne nodof eldof n geom connec F ...
      prop nf Element_loads Joint_loads force Hinge
%
format short e
%
nnd = 9 ;                   % Number of nodes:
nel = 9 ;                   % Number of elements:
nne = 2 ;                   % Number of nodes per element:
nodof =3;                   % Number of degrees of freedom per node
eldof = nne*nodof;          % Number of degrees of freedom per element
%
% Nodes coordinates x and y
%
geom=zeros(nnd,2);
geom(1,1)=0.     ; geom(1,2)= 0.;    %   x and y coordinates of node 1
geom(2,1)=0.     ; geom(2,2)= 5.;    %   x and y coordinates of node 2
geom(3,1)=0.     ; geom(3,2)= 10.;   %   x and y coordinates of node 3
geom(4,1)=3.     ; geom(4,2)= 5.;    %   x and y coordinates of node 4
geom(5,1)=4.5    ; geom(5,2)= 10.;   %   x and y coordinates of node 5
geom(6,1)=6.     ; geom(6,2)= 5.;    %   x and y coordinates of node 6
geom(7,1)=9.     ; geom(7,2)= 10.;   %   x and y coordinates of node 7
geom(8,1)=9.     ; geom(8,2)= 5.;    %   x and y coordinates of node 8
geom(9,1)=9.     ; geom(9,2)= 0.;    %   x and y coordinates of node 9
%
% Element connectivity
%
```

```
connec=zeros(nel,2);
connec(1,1) = 1;   connec(1,2) =2 ;        % First and second node of element 1
connec(2,1) = 2;   connec(2,2) =3 ;        % First and second node of element 2
connec(3,1) = 2;   connec(3,2) =4 ;        % First and second node of element 3
connec(4,1) = 3;   connec(4,2) =5 ;        % First and second node of element 4
connec(5,1) = 4;   connec(5,2) =6 ;        % First and second node of element 5
connec(6,1) = 5;   connec(6,2) =7 ;        % First and second node of element 6
connec(7,1) = 6;   connec(7,2) =8 ;        % First and second node of element 7
connec(8,1) = 7;   connec(8,2) =8 ;        % First and second node of element 8
connec(9,1) = 8;   connec(9,2) =9 ;        % First and second node of element 9
%
% Geometrical properties
%
prop=zeros(nel,3);
prop(1,1)=35e+6;   prop(1,2)=0.16;  prop(1,3)=2.1333e-3;  %E,A and I of element 1
prop(2,1)=35e+6;   prop(2,2)=0.16;  prop(2,3)=2.1333e-3;  %E,A and I of element 2
prop(3,1)=70e+6;   prop(3,2)=0.1 ;  prop(3,3)=1.3333e-3;  %E,A and I of element 3
prop(4,1)=70e+6;   prop(4,2)=0.1 ;  prop(4,3)=1.3333e-3;  %E,A and I of element 4
prop(5,1)=70e+6;   prop(5,2)=0.1 ;  prop(5,3)=1.3333e-3;  %E,A and I of element 5
prop(6,1)=70e+6;   prop(6,2)=0.1 ;  prop(6,3)=1.3333e-3;  %E,A and I of element 6
prop(7,1)=70e+6;   prop(7,2)=0.1 ;  prop(7,3)=1.3333e-3;  %E,A and I of element 7
prop(8,1)=35e+6;   prop(8,2)=0.16;  prop(8,3)=2.1333e-3;  %E,A and I of element 8
prop(9,1)=35e+6;   prop(9,2)=0.16;  prop(9,3)=2.1333e-3;  %E,A and I of element 9
%
% Boundary conditions
%
nf = ones(nnd, nodof);                      % Initialize the matrix nf to 1
nf(1,1) = 0; nf(1,2) =0; nf(1,3) = 0;       % Prescribed nodal freedom of node 1
nf(9,1) = 0; nf(9,2)= 0; nf(9,3) = 0;       % Prescribed nodal freedom of node 9
%
% Counting of the free degrees of freedom
%
n=0;
for i=1:nnd
    for j=1:nodof
        if nf(i,j) ~= 0
            n=n+1;
            nf(i,j)=n;
        end
    end
end
%
% Internal Hinges
%
Hinge = ones(nel, 2);
Hinge(4,2) = 0;           %Hinge accounted with element 4
%
% loading
%
% Joint loads are usually entered in global coordinates
% Enter here the forces in X and Y directions and any
% concentrated moment at node i
Joint_loads= zeros(nnd, 3);
Joint_loads(4,:)=[0    -20     0];
Joint_loads(6,:)=[0    -20     0];
%
% Staticaly equivalent loads are entered in local
% coordinates of the element
%
Element_loads= zeros(nel, 6);
Element_loads(4,:)= [ 0    -56.25    -50.625     0      -33.75     0];
Element_loads(6,:)= [ 0    -33.75      0         0      -56.25    50.625];
%
%
%%%%%%%%%%    End of input        %%%%%%%%%%
```

Results File

```
****** PRINTING MODEL DATA **************
```

```
-------------------------------------------------------
Number of nodes:                              9
Number of elements:                           9
Number of nodes per element:                  2
Number of degrees of freedom per node:        3
Number of degrees of freedom per element:     6
```

```
-------------------------------------------------------
Node       X          Y
  1,     0000.00     0000.00
  2,     0000.00     0005.00
  3,     0000.00     0010.00
  4,     0003.00     0005.00
  5,     0004.50     0010.00
  6,     0006.00     0005.00
  7,     0009.00     0010.00
  8,     0009.00     0005.00
  9,     0009.00     0000.00
```

```
-------------------------------------------------------
Element      Node_1      Node_2
    1,         1,          2
    2,         2,          3
    3,         2,          4
    4,         3,          5
    5,         4,          6
    6,         5,          7
    7,         6,          8
    8,         7,          8
    9,         8,          9
```

```
-------------------------------------------------------
Element       E           A            I
    1,     3.5e+007,     0.16        0.0021333
    2,     3.5e+007,     0.16        0.0021333
    3,     7e+007,       0.1         0.0013333
    4,     7e+007,       0.1         0.0013333
    5,     7e+007,       0.1         0.0013333
    6,     7e+007,       0.1         0.0013333
    7,     7e+007,       0.1         0.0013333
    8,     3.5e+007,     0.16        0.0021333
    9,     3.5e+007,     0.16        0.0021333
```

```
-------------------------------------------------------
------------Nodal freedom------------------------
Node     disp_u       disp_u       Rotation
  1,       0,           0,           0
  2,       1,           2,           3
  3,       4,           5,           6
  4,       7,           8,           9
  5,      10,          11,          12
  6,      13,          14,          15
  7,      16,          17,          18
  8,      19,          20,          21
  9,       0,           0,           0
```

```
-------------------------------------------------------
----------------Applied joint Loads------------------
Node     load_X       load_Y       Moment
  1,    0000.00,     0000.00,      0000.00
  2,    0000.00,     0000.00,      0000.00
  3,    0000.00,    -056.25,      -050.63
  4,    0000.00,    -020.00,       0000.00
```

```
5,     0000.00,    -067.50,     0000.00
6,     0000.00,    -020.00,     0000.00
7,     0000.00,    -056.25,     0050.63
8,     0000.00,    0000.00,     0000.00
9,     0000.00,    0000.00,     0000.00
-------------------------------------------------------

Total number of active degrees of freedom, n = 21

--------------------------------------------------------


  ******* PRINTING ANALYSIS RESULTS ************


--------------------------------------------------------
Global force vector   F
   0
   0
   0
   0
  -56.25
  -50.625
   0
  -20
   0
   0
  -67.5
   0
   0
  -20
   0
   0
  -56.25
   50.625
   0
   0
   0


----------------------------------------------------------
Displacement solution vector:  delta
  -0.00004
  -0.00010
   0.00049
   0.00004
  -0.00018
  -0.00366
  -0.00001
  -0.00008
  -0.00016
   0.00000
  -0.02762
   0.00732
   0.00001
  -0.00008
   0.00016
  -0.00004
  -0.00018
   0.00366
   0.00004
  -0.00010
  -0.00049


----------------------------------------------------------
Nodal displacements
Node      disp_x        disp_y        rotation
  1,    0.00000e+000,    0.00000e+000,    0.00000e+000
```

```
2,    -4.15908e-005,     -9.82143e-005,      4.89326e-004
3,     3.61455e-005,     -1.78571e-004,     -3.65810e-003
4,    -1.38636e-005,     -8.38724e-005,     -1.58328e-004
5,     8.98622e-018,     -2.76241e-002,      7.31947e-003
6,     1.38636e-005,     -8.38724e-005,      1.58328e-004
7,    -3.61455e-005,     -1.78571e-004,      3.65810e-003
8,     4.15908e-005,     -9.82143e-005,     -4.89326e-004
9,     0.00000e+000,      0.00000e+000,      0.00000e+000
```

```
-------------------------------------------------------
Members actions in local coordinates
element   fx1      fy1       M1    fx2      Fy2        M2
  1,    110.0000,    8.4705,    13.8690,  -110.0000,    -8.4705,    28.4833
  2,     90.0000,  -56.2264,   -78.6320,   -90.0000,    56.2264,  -202.5000
  3,    -64.6969,   20.0000,    50.1487,    64.6969,   -20.0000,     9.8513
  4,     56.2264,   90.0000,   202.5000,   -56.2264,    -0.0000,     0.0000
  5,    -64.6969,   -0.0000,    -9.8513,    64.6969,     0.0000,     9.8513
  6,     56.2264,    0.0000,    -0.0000,   -56.2264,    90.0000,  -202.5000
  7,    -64.6969,  -20.0000,    -9.8513,    64.6969,    20.0000,   -50.1487
  8,     90.0000,   56.2264,   202.5000,   -90.0000,   -56.2264,    78.6320
  9,    110.0000,   -8.4705,   -28.4833,  -110.0000,     8.4705,   -13.8690
-------------------------------------------------------
Members actions in global coordinates
element   fx1      fy1       M1    fx2      Fy2        M2
  1,     -8.4705,  110.0000,    13.8690,     8.4705,  -110.0000,    28.4833
  2,     56.2264,   90.0000,   -78.6320,   -56.2264,   -90.0000,  -202.5000
  3,    -64.6969,   20.0000,    50.1487,    64.6969,   -20.0000,     9.8513
  4,     56.2264,   90.0000,   202.5000,   -56.2264,    -0.0000,     0.0000
  5,    -64.6969,   -0.0000,    -9.8513,    64.6969,     0.0000,     9.8513
  6,     56.2264,    0.0000,    -0.0000,   -56.2264,    90.0000,  -202.5000
  7,    -64.6969,  -20.0000,    -9.8513,    64.6969,    20.0000,   -50.1487
  8,     56.2264,  -90.0000,   202.5000,   -56.2264,    90.0000,    78.6320
  9,     -8.4705, -110.0000,   -28.4833,     8.4705,   110.0000,   -13.8690
```

4.7 ANALYSIS OF A SIMPLE FRAME WITH ABAQUS

4.7.1 INTERACTIVE EDITION

In this section, we will analyze the portal frame shown in Figure 4.6 with the Abaqus interactive edition. The cross sections of the profiles used are shown in Figure 4.7. The material is steel with an elastic modulus of 200 GPa.

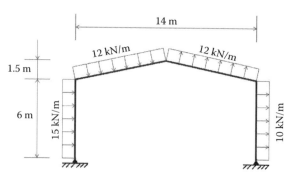

FIGURE 4.6　Portal frame.

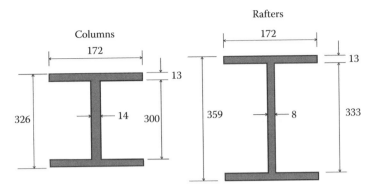

FIGURE 4.7 Profiles' sections; dimensions are in mm.

Start **Abaqus CAE**. Click on **Create Model Database**. On the main menu, click on **File** and **Set Work Directory** to choose your working directory. Click on **Save As** and name the file **Portal_frame.cae**. On the left-hand-side menu, click on **Part** to begin creating the model. Name the part **Portal_Frame**, check **2D Planar**, check **Deformable** in the type. Choose **Wire** as the base feature. Enter an approximate size of 20 m and click on **Continue**. In the sketcher menu, choose the **Create-Lines Connected** icon to begin drawing the geometry of the frame. Click on **Done** in the bottom-left corner of the viewport window (Figure 4.8).

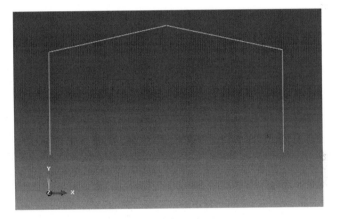

FIGURE 4.8 Creating the Portal_frame part.

Under the model tree, click on **material** to create a material, and name it **Steel**. Click on **Mechanical**, then **Elasticity**, and **Elastic**. Enter 200.*e*6 kN/m² for the elastic modulus, and 0.3 for Poisson's ratio. Next, click on **Profiles** to create a profile, and name it **Column_Profile**. Click on **Continue**. Enter the dimensions of the profile section. Repeat the procedure to create another profile for the rafters, which will be named **Rafter_Profile** (Figure 4.9).

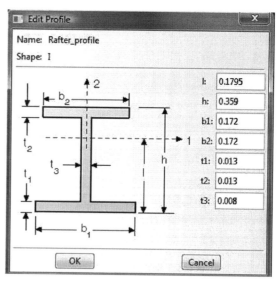

FIGURE 4.9   Material and profiles definitions.

Under the model tree, click on **Sections** to create a section and name it **Column_section**. In the **Category** check **Beam**, and in the **Type** choose **Beam**. Click on **Continue**. In the **Edit Section** dialog box, in the **Profile name** select **Column_Profile**, and in **Material** choose **Steel**. Leave the Poisson's ratio as zero. Click on **OK**. Repeat the procedure to create a section for the rafters named **Column_section** using the profile **Rafter_Profile** (Figure 4.10).

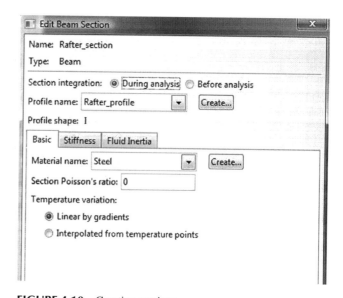

FIGURE 4.10   Creating sections.

Expand the menu under **Parts** and **Portal_Frame**, and double click on **Section Assignments**. By keeping the **Shift** key down, click on the columns in the viewport area. Click on **Done** in the left-bottom corner. In the **Edit Section Assignments** dialog box, select **Column_section**, and click on **OK**. Repeat the procedure by selecting this time the rafters, and in the **Edit Section Assignments** dialog box, select **Rafter_section**. Click on **OK** to finish (Figure 4.11).

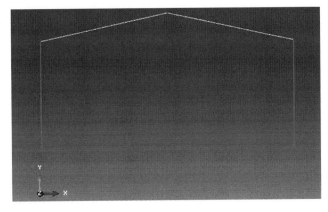

FIGURE 4.11   Editing section assignments.

To check the beam orientations, change the Module to **Property**. Click on the **Assign Beam Orientation** icon and select the entire geometry from the viewport. In the prompt in the left-bottom corner of the viewport, accept $(0.0, 0.0, -1.0)$ as the direction for n_1 and click **Return**. Click **OK** to confirm (Figure 4.12).

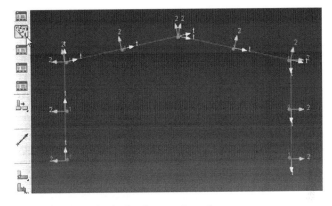

FIGURE 4.12   Assigning beam orientation.

In the menu bar select **View**, then **Part Display Options**. In the **Part Display Options**, in **Idealizations**, check **Render beam profiles**. Click **Apply** (Figure 4.13).

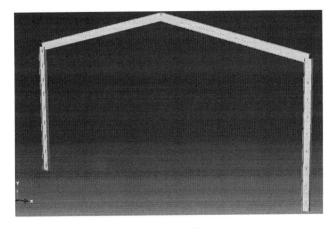

FIGURE 4.13   Rendering beam profile.

In the model tree, double click on **Mesh** under the **Portal_Frame**, and in the main menu, under **Mesh**, click on **Element Type**. With the mouse highlight all members in the viewport and select **Done**. In the dialog box, select **Standard** for element type, **Linear** for geometric order, and **beam** for family. Click on **OK**. In the main menu, under **Seed**, click on **Edges**. With the mouse highlight all the frame in the viewport. In the dialog box, select edge by number and enter **4**. Click on **Apply** and on **OK**. Under **mesh**, click on **Part**, and **Yes** in the prompt area (Figure 4.14).

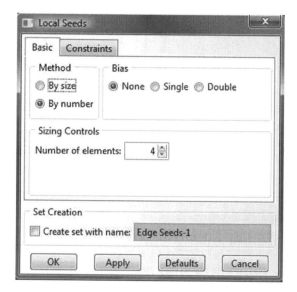

FIGURE 4.14   Seeding by number.

In the menu bar select **View**, then **Part Display Options**. In the **Part Display Options**, under **Mesh**, check **Show node labels** and **Show element labels**. Click **Apply**. The element and node labels will appear in the viewport (Figure 4.15).

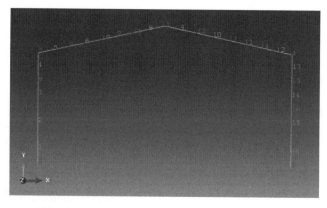

FIGURE 4.15   Mesh.

In the model tree under **Portal_Frame**, double click on **Sets**. In the dialog box, name the set **Pinned_Supports**, check **Node** in type, and click on **Continue**. With the mouse highlight the nodes forming the supports. Click on **Done** in the prompt area of the viewport. Make sure you select **Element** for **Type**, repeat the procedure to create the following element sets: **Left_column**, **Right_column**, **Columns**, **Left_Rafter**, **Right_Rafter**, and **Rafters** (Figure 4.16).

FIGURE 4.16   Creating the element set **Rafters**.

In the model tree, expand the **Assembly** and double click on **Instances**. Select **Portal_frame** for **Parts**, and click **OK**. In the model tree, expand **Steps** and **Initial**, and double click on **BC**. Name the boundary condition **Pinned**, select **Symmetry/Antisymmetry/Encastre** for the type, and click on **Continue**. In the right-bottom corner of the viewport, click on **Sets** and select **Portal_Frame-1.Pinned_supports**. In the **Edit Boundary Condition** select **PINNED**. Click **OK** (Figure 4.17).

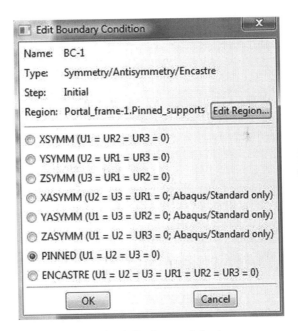

FIGURE 4.17   Imposing BC using created sets.

In the model tree, double click on **Steps**. Name the step **Apply_Loads**. Set the procedure to **General**, and select **Static, General**. Click on **Continue**. Give the step a description, and click **OK**. In the model tree, under **steps**, and under **Apply_Loads**, click on **Loads**. Name the load **W15** and select **Line load** as the type. Click on **Continue**. In the **Region Selection** dialog box, select **Portal_Frame-1.Left_Column**. Click on **Continue**. In the **Edit Load** dialog box, select **Global** for **System**, and enter 15 for **Component 1**. Click **OK**. Repeat exactly the same procedure for the right column, name the load **W10**, and enter 10 for the magnitude (Figure 4.18).

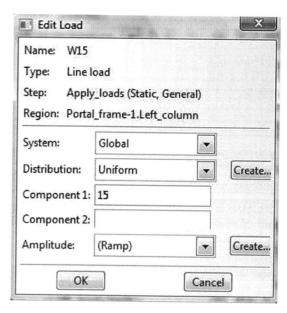

FIGURE 4.18 Imposing a line load in global coordinates.

In the model tree, under **steps**, and under **Apply_Loads**, click on **Loads**. Name the load **DOWN12** and select **Line load** as the type. Click on **Continue**. In the **Region Selection** dialog box, select **Portal_Frame-1.Left_Rafter**. Click on **Continue**. In the **Edit Load** dialog box, select **Local** for **System**, and enter −12 for **Component 2**. Click **OK**. Repeat exactly the same procedure for the right rafter, name the load **UP12**, and enter 12 for the magnitude (Figure 4.19).

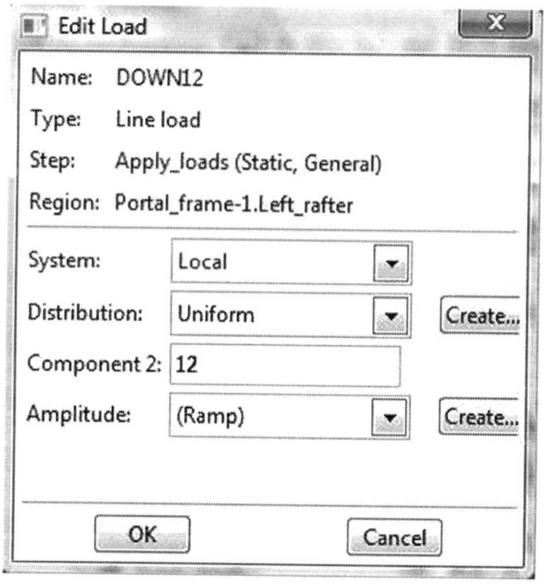

FIGURE 4.19 Imposing a line load in local coordinates.

In the model tree, expand the **Field Output Requests** and then double click on **F-Output-1**. **F-Output-1** is the default and is automatically generated when creating the step. Uncheck the variables **Contact** and select any other variable you wish to add to the field output. Click on **OK**. Under **Analysis**, right click on **Jobs** and then click on **Create**.

In the **Create Job** dialog box, name the job **Portal_frame** and click on **Continue**. In the **Edit Job** dialog box, enter a description for the job. Check **Full analysis**, select to run the job in

```
The model database "C:\ABAQUS_FILES\Portal_frame.cae" has been opened.
The job "Portal_frame" has been created.
The job input file "Portal_frame.inp" has been submitted for analysis.
Job Portal_frame: Analysis Input File Processor completed successfully.
Job Portal_frame: Abaqus/Standard completed successfully.
Job Portal_frame completed successfully.
```

FIGURE 4.20   Analyzing a job in Abaqus CAE.

Background, and check to start it **immediately**. Click **OK**. Expand the tree under **Jobs**, right click on **Portal_frame**. Then, click on **Submit**. If you get the following message **Portal_frame completed successfully** in the bottom window, then your job is free of errors and was executed properly. Notice that Abaqus has generated an input file for the job **Portal_frame.inp** (Figure 4.20). Open it with your preferred text editor and compare it with the one given in Section 4.7.2.

Under the top menu, in the **Module** scroll to **Visualization**, and click to load **Abaqus Viewer**. On the main menu, under **File**, click **Open**, navigate to your working directory, and open the file **Portal_frame.odb**. It should have the same name as the job you submitted. Click on the **Common options** icon to display the **Common Plot options** dialog box. Under **labels**, check **Show Element labels** and **Show Node labels** to display elements and nodes' numbering. Click on the icon **Plot Deformed Shape** to display the deformed shape of the beam. On the main menu, click on **Results** then on **Field Output** to open the **Field Output** dialog box. Choose **S Stress components at integration points**. For component, choose $S11$ to plot the stresses in the elements. Click on **Section points** to open the section point dialog box. Check **bottom** to plot the stresses in the lower fiber or **Top** for the stresses in the top fiber (Figure 4.21). In the menu bar, click on **Report** and **Field Output**. In the **Report Field Output** dialog box, for **Position** select **Unique nodal**, check **RF1**, **RF2**, and **RM3** for **RF: Reaction force**, and check **U1**, **U2**, and **UR3** for **U: Spatial displacement**. Then, click on **Set up**. Click on **Select** to navigate to your working directory. Name the file **Portal_Frame.rpt**. Uncheck **Append to file**, and click **OK**. Use your favorite text editor and open the file **Portal_Frame.rpt**, which should be the same as the one listed next.

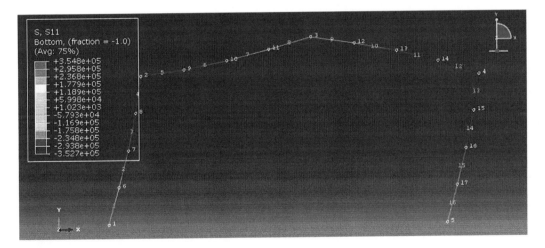

FIGURE 4.21   Plotting stresses in the bottom fiber (interactive edition).

```
****************************************************************************
Field Output Report, written Sun May 01 14:53:07 2011

Source 1
---------


    ODB: F:/TRAVAIL/NEW_BOOK/Abaqus_examples/Portal_frame.odb
    Step: Apply_loads
    Frame: Increment     1: Step Time =    1.000

Loc 1 : Nodal values from source 1

Output sorted by column ''Node Label''.

Field Output reported at nodes for part: PORTAL_FRAME-1
```

Node Label	RF.RF1 @Loc 1	RF.RF2 @Loc 1	RM3 @Loc 1	U.U1 @Loc 1	U.U2 @Loc 1	UR3 @Loc 1
1	-100.845	-7.5	0.	89.5948E-36	7.50000E-36	-74.8338E-03
2	0.	0.	0.	352.883E-03	25.9456E-06	-30.1636E-03
3	0.	0.	0.	352.071E-03	4.10271E-03	17.5904E-03
4	0.	0.	0.	351.052E-03	-25.9456E-06	-31.6545E-03
5	-85.1552	7.5	0.	77.6552E-36	-7.50000E-36	-72.8888E-03
6	0.	0.	0.	111.346E-03	6.48639E-06	-71.2556E-03
7	0.	0.	0.	212.186E-03	12.9728E-06	-61.4198E-03
8	0.	0.	0.	294.479E-03	19.4592E-06	-47.1234E-03
9	0.	0.	0.	360.710E-03	-36.4285E-03	-12.0266E-03
10	0.	0.	0.	362.105E-03	-42.8683E-03	3.42622E-03
11	0.	0.	0.	358.483E-03	-25.8914E-03	14.0226E-03
12	0.	0.	0.	358.292E-03	33.2892E-03	13.1950E-03
13	0.	0.	0.	361.470E-03	48.2755E-03	2.07421E-03
14	0.	0.	0.	359.490E-03	39.1929E-03	-13.5997E-03
15	0.	0.	0.	290.684E-03	-19.4592E-06	-47.9722E-03
16	0.	0.	0.	207.946E-03	-12.9728E-06	-61.0826E-03
17	0.	0.	0.	108.549E-03	-6.48639E-06	-69.7875E-03

4.7.2 KEYWORD EDITION

In this section, we will prepare an input file for the portal frame shown in Figures 4.6 and 4.7. The file named **Frame_Problem_Keyword.inp** is listed next:

```
*Heading
  Frame_Problem Model keyword edition
**
*Preprint, echo=No, model=NO, history=NO
**
**
**  Define the end nodes
**
*Node
      1,          0.,         0.
      5,          0.,         6.
      9,          7.,         7.5
     13,         14.,         6.
     17,         14.,          0.
**
** Generate the remaining nodes
**
*Ngen
1,5,1
5,9,1
9,13,1
13,17,1
**
```

```
**   Define element 1
**
*Element, type=B21
1,1,2
**
** Generate the elements
**
*Elgen, elset = all_elements
1,16, 1, 1
**
**
*Nset, nset=Pinned_supports
 1,  17
*Elset, elset=Left_Column, generate
 1,  4,  1
*Elset, elset=Right_Column, generate
 13,  16,   1
*Elset, elset=Columns
 Left_Column,  Right_Column
**
*Elset, elset=Left_Rafter, generate
5,  8,  1
*Elset, elset=Right_Rafter, generate
 9,  12,  1
*Elset, elset=Rafters
Left_Rafter,Right_Rafter,
**
**
** Section: Beam_section  Profile: Rafter_Profile
*Beam Section, elset=Rafters, material=Steel, section=I
0.1795, 0.359, 0.172, 0.172, 0.013, 0.013, 0.008
0.,0.,-1.
**
**
** Section: Beam_section  Profile: Column_Profile
*Beam Section, elset=Columns, material=Steel, section=I
0.163, 0.326, 0.172, 0.172, 0.013, 0.013, 0.014
0.,0.,-1.
**
** MATERIALS
**
*Material, name=Steel
*Elastic
 2e+08, 0.3
**
** BOUNDARY CONDITIONS
**
**
*Boundary
Pinned_supports,PINNED
** ----------------------------------------------------------------
**
** STEP: Apply_Loads
**
*Step, name=Apply_Loads
*Static
1., 1., 1e-05, 1.
**
** LOADS
**
**
*Dload
Right_column, PX, 10.
Left_column, PX, 15.
Left_rafter, P2, -12.
Right_rafter, P2, 12.
**
```

```
** OUTPUT REQUESTS
**
**
*Output, field
*Node Output
CF, RF, RM, U
*Element Output
 S
**
*Output, history, variable=PRESELECT
*End Step
```

At the command line, type **Abaqus job=Frame_Problem_Keyword inter** followed by **Return**. If you get an error, open the file with extension ***.dat** to see what type of error. To load the visualization model, type **Abaqus Viewer** at the command line.

On the main menu, under **File**, click **Open**, navigate to your working directory, and open the file **Frame_Problem_Keyword.odb**. It should have the same name as the job you submitted. Click on the **Common options** icon to display the **Common Plot options** dialog box. Under **labels**, check **Show Element labels** and **Show Node labels** to display elements and nodes' numbering. Click on the icon **Plot Deformed Shape** to display the deformed shape of the beam. On the main menu, click on **Results**, then on **Field Output** to open the **Field Output** dialog box. Choose **S Stress components at integration points**. For component, choose $S11$ to plot the stresses in the elements (Figure 4.22). Click on **Section points** to open the section point dialog box. Check **bottom** to plot the stresses in the lower fiber or **Top** for the stresses in the top fiber. Notice that the stress contour is exactly the same as obtained previously, except that the node and element numbering is different.

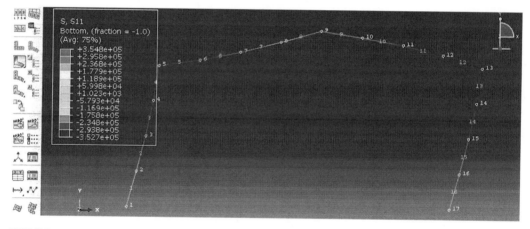

FIGURE 4.22   Plotting stresses in the bottom fiber (keyword edition).

5 Stress and Strain Analysis

5.1 INTRODUCTION

This chapter deals with the notions of stress–strain and strain–displacements relation, which are quite essential for understanding the remaining developments in the book. It marks the change of philosophy between matrix structural analysis and finite element analysis of a continuum. In the previous Chapters 2 through 4, we only considered structural elements whose behavior can be formulated as a function of a single variable x, which is the longitudinal direction of the element. This is of course possible because of the geometry, where two dimensions are insignificant compared to the third one. The only stress of interest therefore is the longitudinal stress σ_x along the dominant dimension. Yet, in a three-dimensional solid where all the dimensions are of the same size, this assumption is not valid anymore. When a three-dimensional solid is subjected to external forces and/or displacements, and at the same time is restrained against rigid body movement, internal forces are induced, and these result in more than one stress at a point. Additionally, these external forces result in material points within the body being displaced. When there is a change in distance between two points, straining has taken place. Again there is more than one strain at a point. As will be shown in the Sections 5.3.3 and 5.3.4, a segment of infinitesimal length not only experiences a change in length, but also a change in direction.

As stresses and strains are interrelated, we will also consider the relations between them. Such relations are called constitutive equations, since they describe the macroscopic behavior resulting from the internal constitution of the material. Materials, however, exhibit different behaviors over their entire range of deformations. As such, it is not possible to write one set of mathematical equations to describe these behaviors. Yet, for many engineering applications, the theory of linear elasticity offers a useful and reliable model for analysis.

5.2 STRESS TENSOR

5.2.1 DEFINITION

Let us consider a body in equilibrium under external forces as represented in Figure 5.1. Let us take a cut through the body, as represented by the plane Σ, and denote by dA an infinitesimal element of the internal cross section. A force $d\vec{F}$ is exerted on this small area. It represents the influence of the right section on the left section of the body.

The vector $d\vec{F}$ can be expressed in terms of its normal and tangential components, $d\vec{F}_n$ and $d\vec{F}_t$, to the surface dA. The stresses acting on the surface are then given as

$$\sigma_n = \lim_{dA \to 0} \frac{d\vec{F}_n}{dA} \tag{5.1}$$

$$\sigma_t = \lim_{dA \to 0} \frac{d\vec{F}_t}{dA} \tag{5.2}$$

It can be seen that $d\vec{F}_t$ has also two components on the plane of the surface dA. In total, therefore, there are three stress components: one normal and two tangential. However, as the infinitesimal element dA shrinks to a point, there will be an infinite number of planes passing through that point. It would be impossible therefore to consider all of them. However, if we choose three mutually perpendicular planes, as represented in Figure 5.2, the stresses can be written for all of

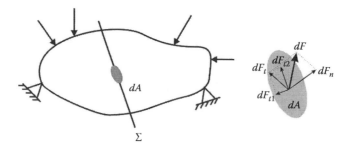

FIGURE 5.1   Internal force components.

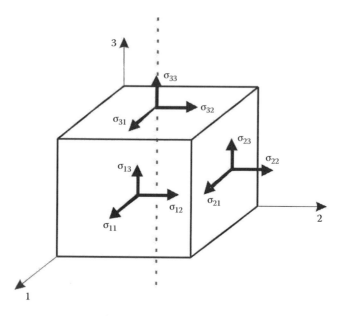

FIGURE 5.2   Stress components at a point.

them. It should be emphasized however that the parallelepiped represented in Figure 5.2 is not a block of material cut from the body, but a simple yet convenient schematic device to represent the stresses acting at a point. It can be seen that there are nine components of stress acting at a point.

The nine stresses are arranged in a stress tensor as

$$\sigma = \begin{bmatrix} \sigma_{11} & \sigma_{12} & \sigma_{13} \\ \sigma_{21} & \sigma_{22} & \sigma_{23} \\ \sigma_{31} & \sigma_{32} & \sigma_{33} \end{bmatrix} \tag{5.3}$$

It should be noted that a stress component has two indices: the first index indicates the direction of the normal to the plane on which it acts and the second refers to the direction of the stress component. A stress component is positive if it acts on a positive face in the positive direction or on a negative face in the negative direction.

The stress tensor is symmetric, since by taking moments about the axis passing through the point (or the center of the "cube" as shown), it can be shown that

$$\sigma_{12} = \sigma_{21} \qquad \sigma_{13} = \sigma_{31} \qquad \sigma_{23} = \sigma_{32} \qquad (5.4)$$

In general,

$$\sigma_{ij} = \sigma_{ji} \qquad (5.5)$$

This shows that the stress tensor contains only six independent components.

5.2.2 STRESS TENSOR–STRESS VECTOR RELATIONSHIPS

In order to study the transformation of stress, let us isolate an infinitesimal tetrahedron, as shown in Figure 5.3. The plane ABC is perpendicular to an arbitrary-oriented normal $\vec{n}$ written in vector matrix notation as

$$\{n\} = \{n_1, n_2, n_3\}^T \qquad (5.6)$$

The components $\{n_i\}$ are the direction cosines of the normal $\vec{n}$. If ΔS is the area of the surface ABC, then the areas of the other surfaces can be expressed as

$$
\begin{aligned}
&\text{for} \quad \text{COB} \quad \Delta S_1 = n_1 \Delta S \\
&\text{for} \quad \text{AOC} \quad \Delta S_2 = n_2 \Delta S \\
&\text{for} \quad \text{BOA} \quad \Delta S_3 = n_3 \Delta S
\end{aligned}
\qquad (5.7)
$$

These expressions denote that the faces are the projections of the oblique face onto the coordinates planes. Figure 5.3 also shows the stress vectors $\vec{T}_1^*, \vec{T}_2^*, \vec{T}_3^*$, which are the components of the $\vec{T}^*$ on

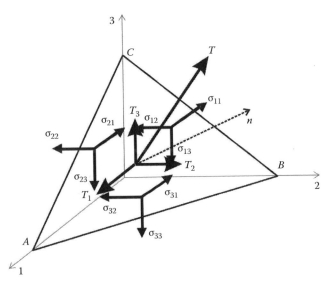

FIGURE 5.3 Stress components on a tetrahedron.

the three mutually orthogonal planes. These stress vectors are resolved along the coordinate axes 1, 2, and 3 as

$$\sigma_{11}^*, \sigma_{12}^*, \sigma_{13}^* \quad \text{for} \quad \vec{T}_1^*$$

$$\sigma_{21}^*, \sigma_{22}^*, \sigma_{23}^* \quad \text{for} \quad \vec{T}_2^* \qquad (5.8)$$

$$\sigma_{31}^*, \sigma_{32}^*, \sigma_{33}^* \quad \text{for} \quad \vec{T}_3^*$$

The asterisk (*) indicates that we are dealing with average values. Remember that a stress vector is a point quantity.

The body force vector $\vec{q}^*$ that acts throughout the body is also shown. Resolution of $\vec{q}^*$ and $\vec{T}^*$ into the directions of the coordinate axis yields

$$\{q^*\} = \{q_1^*, q_2^*, q_3^*\}^T \qquad (5.9)$$

$$\{T^*\} = \{T_1^*, T_2^*, T_3^*\}^T \qquad (5.10)$$

Equilibrium requires the vector sum of all forces acting on the tetrahedron to be zero. To obtain the forces acting on the tetrahedron, the stress components must be multiplied by the respective areas on which they act. Requiring equilibrium in the x-direction yields

$$T_1^* \Delta S + q_1^* \Delta V - \sigma_{11}^* n_1 \Delta S - \sigma_{21}^* n_2 \Delta S - \sigma_{31}^* n_3 \Delta S = 0 \qquad (5.11)$$

The volume of the tetrahedron can be written as

$$\Delta V = \frac{1}{3} h \Delta S \qquad (5.12)$$

where h is the perpendicular distance from point O to the base ABC. Substituting for V and dividing by S in the equilibrium equation yields

$$T_1^* + q_1^* \frac{1}{3} h - \sigma_{11}^* n_1 - \sigma_{21}^* n_2 - \sigma_{31}^* n_3 = 0 \qquad (5.13)$$

Now let the tetrahedron shrink to a point by taking the limit as $h \to 0$, and noting in this process that the starred (average) quantities take on the actual values of those quantities at a point, results in

$$T_1 = \sigma_{11} n_1 + \sigma_{21} n_2 + \sigma_{31} n_3 \qquad (5.14)$$

Similarly, we obtain for the y and z directions

$$T_2 = \sigma_{12} n_1 + \sigma_{22} n_2 + \sigma_{32} n_3$$

$$T_3 = \sigma_{13} n_1 + \sigma_{23} n_2 + \sigma_{33} n_3 \qquad (5.15)$$

These expressions can be grouped in a matrix form as follows:

$$\begin{Bmatrix} T_1 \\ T_2 \\ T_3 \end{Bmatrix} = \begin{bmatrix} \sigma_{11} & \sigma_{21} & \sigma_{31} \\ \sigma_{12} & \sigma_{22} & \sigma_{32} \\ \sigma_{13} & \sigma_{23} & \sigma_{33} \end{bmatrix} \begin{Bmatrix} n_1 \\ n_2 \\ n_3 \end{Bmatrix} \qquad (5.16)$$

or simply as

$$\{T\} = [\sigma]^T \{n\} \qquad (5.17)$$

or in index notation as

$$T_i = \sigma_{ji} n_j \qquad (5.18)$$

5.2.3 Transformation of the Stress Tensor

If the components of the stress tensor σ in the basis $(\vec{e}_1, \vec{e}_2, \vec{e}_3)$ are known, let us find the components of the same tensor in another basis $(\vec{e}'_1, \vec{e}'_2, \vec{e}'_3)$ obtained from a rotation of axes around the origin. Since the stress tensor is a second-order tensor, it obeys the same transformation laws for second-order tensor as detailed in Appendix C. Therefore, the components of the stress tensor in the new basis are obtained respectively in index and matrix notations as

$$\sigma'_{km} = l_{ki}l_{mj}\sigma_{ij} \tag{5.19}$$

$$[\sigma'] = [Q][\sigma][Q]^T \tag{5.20}$$

The components l_{ij} or Q_{ij} are the cosines of the angles formed by the unit vectors $(\vec{e}'_i, \vec{e}_j)$. The inverse transformations are obtained as

$$\sigma_{ij} = l_{ki}l_{mj}\sigma'_{km}$$
$$[\sigma] = [Q]^T[\sigma'][Q] \tag{5.21}$$

5.2.4 Equilibrium Equations

Equilibrium of a small cube of material that is removed from a larger body subject to external forces requires that the resultant force and moment acting on the cube must be equal to zero.

In Figure 5.4, the components of stress acting on the positive faces of the element are shown. The components acting on the negative faces are omitted for the sake of clarity of the figure. The omitted components are σ_{11}, σ_{12}, σ_{13} on face 1; σ_{21}, σ_{22}, σ_{23} on face 2; and σ_{31}, σ_{32}, σ_{33} on face 3.

The stresses vary throughout the body, and it is assumed that their components and derivatives are continuous functions of the coordinates. To express this variation, the well-known rules of differential calculus can be used:

$$\sigma_{11}(x + dx) = \sigma_{11}(x) + \frac{\partial \sigma_{11}}{\partial x} \tag{5.22}$$

In addition to the stress components acting on the body, body forces such as the ones due to gravity are also present and have intensities b_x, b_y, and b_z or simply b_i. When the stress components are

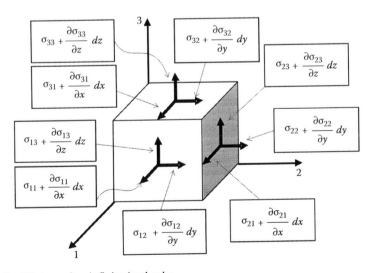

FIGURE 5.4 Equilibrium of an infinitesimal cube.

multiplied by the area on which they act, force components are obtained. Requiring equilibrium in the x direction leads to

$$-\sigma_{11}dydz + \left(\sigma_{11} + \frac{\partial \sigma_{11}}{\partial x}dx\right)dydz - \sigma_{21}dxdz + \left(\sigma_{21} + \frac{\partial \sigma_{21}}{\partial x}dx\right)dxdz$$

$$- \sigma_{31}dxdy + \left(\sigma_{31} + \frac{\partial \sigma_{31}}{\partial x}dx\right)dxdy + b_xdxdydz = 0 \qquad (5.23)$$

After rearranging, Equation (5.23) becomes

$$\frac{\partial \sigma_{11}}{\partial x} + \frac{\partial \sigma_{21}}{\partial y} + \frac{\partial \sigma_{31}}{\partial z} + b_x = 0 \qquad (5.24)$$

Requiring equilibrium in y and z directions as well results in

$$\frac{\partial \sigma_{12}}{\partial x} + \frac{\partial \sigma_{22}}{\partial y} + \frac{\partial \sigma_{32}}{\partial z} + b_y = 0$$

$$\frac{\partial \sigma_{13}}{\partial x} + \frac{\partial \sigma_{23}}{\partial y} + \frac{\partial \sigma_{33}}{\partial z} + b_z = 0 \qquad (5.25)$$

Noticing that x, y, and z are actually the first, second, and third directions, Equations (5.24) and (5.25) can be simply written as

$$\sigma_{ij,i} + b_j = 0 \qquad (5.26)$$

or because of the symmetry of the stress tensor as

$$\sigma_{ij,j} + b_i = 0 \qquad (5.27)$$

The comma "," in expressions (5.26) and (5.27) indicates derivative with respect to a direction designated by the index following the comma ",".

5.2.5 PRINCIPAL STRESSES

Since the stress tensor is a second-order tensor, the calculation of the principal stress values and their associated principal directions is exactly the same as for a general second-order tensor detailed in Appendix C.

In other words, in the basis $(\vec{e}_1, \vec{e}_2, \vec{e}_3)$, the stress vector $\vec{T} = \sigma\vec{n}$ on the cutting plane $P(n)$ is not parallel to the normal $\vec{n}$, the problem is to find the cutting plane $P(n')$ whose normal $\vec{n}'$ is parallel to $\vec{T}$ such that $\vec{T} = \sigma\vec{n}' = \lambda\vec{n}'$ where λ is a scalar. This plane, together with two other planes, which are all mutually perpendicular, forms a basis called the principal basis of the tensor (Figure 5.5). This new basis is made of the principal directions of the tensor. In this basis, the tensor reduces to its diagonal form

$$\sigma = \begin{bmatrix} \sigma_1 & 0 & 0 \\ 0 & \sigma_2 & 0 \\ 0 & 0 & \sigma_3 \end{bmatrix} \qquad (5.28)$$

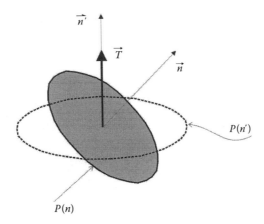

FIGURE 5.5 Principal directions of a stress tensor.

where σ_1, σ_2, and σ_3 are the principal stresses and roots of the characteristic equation of the tensor

$$\sigma^3 - I_1\sigma^2 + I_2\sigma - I_3 = 0 \qquad (5.29)$$

where I_1, I_2, and I_3 are the stress invariants, which are independent of the coordinates system. They are obtained as

$$
\begin{aligned}
I_1 &= \sigma_{ii} \\
I_2 &= \frac{1}{2}(\sigma_{ii}\sigma_{jj} - \sigma_{ij}\sigma_{ij}) \\
I_3 &= |\sigma_{ij}| = det([\sigma])
\end{aligned}
\qquad (5.30)
$$

These invariants can also be expressed in terms of σ_1, σ_2, and σ_3, which are invariants themselves:

$$
\begin{aligned}
I_1 &= \sigma_1 + \sigma_2 + \sigma_3 \\
I_2 &= \sigma_1\sigma_2 + \sigma_2\sigma_3 + \sigma_3\sigma_1 \\
I_3 &= \sigma_1\sigma_2\sigma_3
\end{aligned}
\qquad (5.31)
$$

5.2.6 VON MISES STRESS

What is referred to as von Mises stress is another form of invariant of the stress tensor. As the reader will find out in subsequent chapters, Abaqus by default plots a contour of the von Mises stress. This quantity is very useful when plastic yielding of a material is present. Indeed, it is possible for a material to yield under a given combination of the principal stresses even though none of them exceeds the yield stress of the material. The von Mises stress is a formula combining the principal stresses into an equivalent stress that can be compared to the yield stress of the material, and it is given as

$$(\sigma_1 - \sigma_2)^2 + (\sigma_2 - \sigma_3)^2 + (\sigma_3 - \sigma_1)^2 = 2\sigma_e^2 \qquad (5.32)$$

5.2.7 NORMAL AND TANGENTIAL COMPONENTS OF THE STRESS VECTOR

In a basis formed by the principal directions of the stress tensor, the stress vector may be resolved into a normal and tangential component as shown in Figure 5.6.

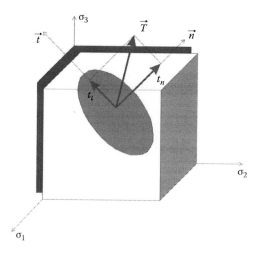

FIGURE 5.6   Tangential and normal components of the stress vector.

By definition, the stress vector is expressed respectively in vector, index, and matrix notations as

$$\vec{T} = \sigma\vec{n}$$
$$T_i = \sigma_{ij}n_j \qquad (5.33)$$
$$\{T\} = [\sigma]\{n\}$$

The normal is the scalar or dot product of $\vec{T}$ with $\vec{n}$ written in vector, index, and matrix notations as

$$\sigma_n = \vec{T} \cdot \vec{n}$$
$$\sigma_n = n_i n_j \sigma_{ij} \qquad (5.34)$$
$$\sigma_n = \{n\}^T[\sigma]\{n\}$$

In the principal basis, the components of the stress vector can also be expressed as

$$T_1 = \sigma_1 n_1$$
$$T_2 = \sigma_2 n_2 \qquad (5.35)$$
$$T_3 = \sigma_3 n_3$$

Substituting Equations (5.35) in any equation of (5.34) yields

$$\sigma_n = \sigma_1 n_1^2 + \sigma_2 n_2^2 + \sigma_3 n_3^2 \qquad (5.36)$$

Using Pythagoras theorem gives the tangential or shear component as

$$\sigma_s^2 = T_i T_i - \sigma_n^2 \qquad (5.37)$$

Notice that the term $T_i T_i$ represents the modulus of the stress vector $\vec{T}$; it is actually the scalar product of $\vec{T}$ by itself. Substituting (5.35) and (5.36) in (5.37) yields

$$\sigma_s^2 = \sigma_1^2 n_1^2 + \sigma_2^2 n_2^2 + \sigma_3^2 n_3^2 - (\sigma_1 n_1^2 + \sigma_2 n_2^2 + \sigma_3 n_3^2) \qquad (5.38)$$

When the principal stresses are ordered according to $\sigma_1 \geq \sigma_2 \geq \sigma_3$, the maximum shear stress is given as

$$\sigma_s = \frac{1}{2}(\sigma_1 - \sigma_3) \tag{5.39}$$

Combining Equations (5.38) and (5.37) with the identity $n_1^2 + n_2^2 + n_3^2 = 1$ and solving for the direction cosines n_i, we obtain

$$n_1^2 = \frac{(\sigma_n - \sigma_2)(\sigma_n - \sigma_3) + \sigma_s^2}{(\sigma_1 - \sigma_2)(\sigma_1 - \sigma_3)}$$

$$n_2^2 = \frac{(\sigma_n - \sigma_1)(\sigma_n - \sigma_3) + \sigma_s^2}{(\sigma_2 - \sigma_1)(\sigma_2 - \sigma_3)} \tag{5.40}$$

$$n_3^2 = \frac{(\sigma_n - \sigma_1)(\sigma_n - \sigma_2) + \sigma_s^2}{(\sigma_3 - \sigma_1)(\sigma_3 - \sigma_2)}$$

These equations serve as the basis for Mohr's circle of stress.

5.2.8 MOHR'S CIRCLES FOR STRESS

Mohr's circles provide a convenient graphical two-dimensional representation of the three-dimensional state of stress. Mohr's circles are drawn in the (σ_n, σ_s) stress space. Given the ordering $\sigma_1 \geq \sigma_2 \geq \sigma_3$, it can be seen that the numerator of the right-hand-side of Equation (5.40) is positive; that is,

$$(\sigma_n - \sigma_2)(\sigma_n - \sigma_3) + \sigma_s^2 \geq 0 \tag{5.41}$$

This equation represents stress points in the (σ_n, σ_s) stress space that are on or outside the circle C_1, shown in Figure 5.7, which has for equation:

$$(\sigma_n - (\sigma_2 + \sigma_3)/2)^2 + \sigma_s^2 = ((\sigma_2 - \sigma_3)/2)^2 \tag{5.42}$$

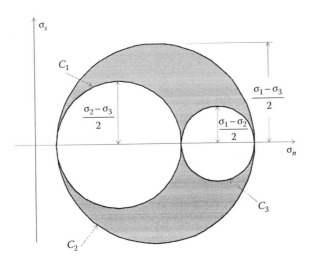

FIGURE 5.7 Mohr's circles.

The same approach can be used to draw two other circles C_2 and C_3 represented by the following equations:

$$(\sigma_n - (\sigma_3 + \sigma_1)/2)^2 + \sigma_s^2 = ((\sigma_3 - \sigma_1)/2)^2 \tag{5.43}$$

$$(\sigma_n - (\sigma_1 + \sigma_2)/2)^2 + \sigma_s^2 = ((\sigma_1 - \sigma_2)/2)^2 \tag{5.44}$$

5.2.9 ENGINEERING REPRESENTATION OF STRESS

Previously, it was shown that the stress tensor is symmetric and therefore possesses only six independent components. For this reason, engineers more often write the stress tensor as a vector with six components:

$$\begin{Bmatrix} \sigma_1 \\ \sigma_2 \\ \sigma_3 \\ \sigma_4 \\ \sigma_5 \\ \sigma_6 \end{Bmatrix} \equiv \begin{Bmatrix} \sigma_{11} \\ \sigma_{22} \\ \sigma_{33} \\ \sigma_{12} \\ \sigma_{23} \\ \sigma_{13} \end{Bmatrix} \equiv \begin{Bmatrix} \sigma_{xx} \\ \sigma_{yy} \\ \sigma_{zz} \\ \sigma_{xy} \\ \sigma_{yz} \\ \sigma_{xz} \end{Bmatrix} \tag{5.45}$$

With this notation, the transformation law for stress in the case of a rotation around the axis 3, or axis z, by an angle ψ is written as

$$\begin{Bmatrix} \sigma'_{11} \\ \sigma'_{22} \\ \sigma'_{33} \\ \sigma'_{12} \\ \sigma'_{23} \\ \sigma'_{13} \end{Bmatrix} = \begin{bmatrix} \cos^2\psi & \sin\psi & 0 & 2\sin\psi\cos\psi & 0 & 0 \\ \sin^2\psi & \cos^2\psi & 0 & -2\sin\psi\cos\psi & 0 & 0 \\ 0 & 0 & 1 & 0 & 0 & 0 \\ -\sin\psi\cos\psi & \sin\psi\cos\psi & \cos^2\psi - \sin^2\psi & 0 & 0 & 0 \\ 0 & 0 & 0 & 0 & \cos\psi & -\sin\psi \\ 0 & 0 & 0 & 0 & \sin\psi & \cos\psi \end{bmatrix}$$

$$\times \begin{Bmatrix} \sigma_{11} \\ \sigma_{22} \\ \sigma_{33} \\ \sigma_{12} \\ \sigma_{23} \\ \sigma_{13} \end{Bmatrix} \tag{5.46}$$

5.3 DEFORMATION AND STRAIN

5.3.1 DEFINITION

The term deformation refers to a change in shape of the body between some initial undeformed configuration and some final deformed configuration, as represented in Figure 5.8. After deformation, point M moves to M^* and point N moves to N^*. The segment MN not only undergoes a change in length but also a change in its direction. Most often, deformation is not just a function of the spatial

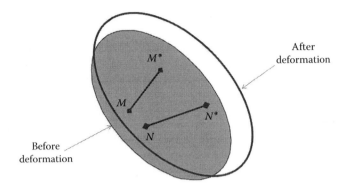

FIGURE 5.8 Schematic representation of the deformation of a solid body.

coordinates, but it is also a function of time. Some deformation processes such as creep in concrete or relaxation in prestressing tendons occur over very long periods of time. Often the configuration at time $t = 0$ is chosen as the reference configuration, and the current configuration refers to the configuration which the body occupies at current time t.

5.3.2 LAGRANGIAN AND EULERIAN DESCRIPTIONS

During deformation, the particles of a body move along various paths. Relative to a Cartesian coordinate system, a particle that originally occupied a position (X, Y, Z) in the undeformed configuration occupies the position (x, y, z) in the deformed configuration. This motion may be expressed by the equations

$$x = x(X, Y, Z, t)$$
$$y = y(X, Y, Z, t) \tag{5.47}$$
$$z = z(X, Y, Z, t)$$

or more compactly in index notation as

$$x_i = x_i(X_1, X_2, X_3, t) \tag{5.48}$$

Equations (5.47) and (5.48) can be thought of as a mapping of the initial configuration into the current configuration. This description of motion is known as the Lagrangian description. For instance, when a body undergoes deformation, a quantity associated with a particle such as temperature changes with time. Such changes in temperature can be expressed according to Equation (5.48) as

$$\theta = \theta(X_1, X_2, X_3, t) \tag{5.49}$$

The Lagrangian description is also known as the material description or reference description.

On the other hand, the motion may be given in the form

$$X_i = X_i(x_1, x_2, x_3, t) \tag{5.50}$$

Given the current position of a particle, this description can be thought of as one that provides a tracing to the original position of the particle. This description is known as the Eulerian description.

The triples (X, Y, Z) and (x, y, z) are also known respectively as material and spatial coordinates.

The Lagrangian description seems the most suitable in solid mechanics, since in these problems there is usually an easy way to identify a reference configuration for which all information is known.

However, it is of little use in fluid mechanics, because in nonsteady flow the reference position at time $t = 0$ of a particle is generally not known. In this book, since we are primarily dealing with solid mechanics, we will use the Lagrangian description. The coordinates of a particle in the initial configuration are labeled (X, Y, Z).

5.3.3 DISPLACEMENT

Relative to a Cartesian coordinate system, let $(\vec{e}_1, \vec{e}_2, \vec{e}_3)$ be the unit vectors in the directions of the superposed coordinates (X_1, X_2, X_3) and (x_1, x_2, x_3). The position of the particle M at time $t = 0$ can be described by the vector $\overrightarrow{OM}$, as shown in Figure 5.9:

$$\overrightarrow{OM} = X_1\vec{e}_1 + X_2\vec{e}_2 + X_3\vec{e}_3 \tag{5.51}$$

The particle originally at M moves to M^* in the current configuration at time t. Its new position is described by the vector $\overrightarrow{OM^*}$:

$$\overrightarrow{OM^*} = x_1\vec{e}_1 + x_2\vec{e}_2 + x_3\vec{e}_3 \tag{5.52}$$

The equation $x_i = x_i(X_1, X_2, X_3, t)$ describes the path of the particle, which at time $t = 0$ is located at M. The vector $\overrightarrow{MM^*}$ is the displacement vector from the reference to the current configuration obtained as

$$\overrightarrow{MM^*} = \overrightarrow{OM^*} - \overrightarrow{OM} \tag{5.53}$$

which, after substitution of Equations (5.51) and (5.52), becomes

$$\overrightarrow{MM^*} = (x_1 - X_1)\vec{e}_1 + (x_2 - X_2)\vec{e}_2 + (x_3 - X_3)\vec{e}_3 \tag{5.54}$$

This equation is normally written as

$$\overrightarrow{MM^*} = \vec{u} = u_1\vec{e}_1 + u_2\vec{e}_2 + u_3\vec{e}_3 \tag{5.55}$$

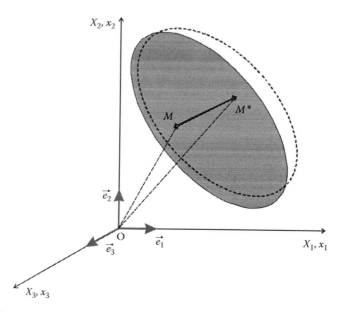

FIGURE 5.9 Reference and current configurations.

The terms u_i are the components of the displacement vector, and they are assumed to be continuous functions of the coordinates X_i; $u_i = u_i(X_1, X_2, X_3, t)$.

5.3.4 DISPLACEMENT AND DEFORMATION GRADIENTS

Once more, let us consider the deformed and undeformed configuration in a Cartesian coordinate system where the unit vectors $(\vec{e}_1, \vec{e}_2, \vec{e}_3)$ are the directions of the superposed coordinates (X_1, X_2, X_3) and (x_1, x_2, x_3).

Figure 5.10 represents the deformation process of an infinitesimal element originally at MN in the undeformed configuration, which moves to the position M^*N^* in the deformed configuration. During deformation, point M moves to M^*, and its new position is given by

$$\overrightarrow{OM^*} = \overrightarrow{OM} + \overrightarrow{MM^*} \tag{5.56}$$

The vector $\overrightarrow{MM^*}$ represents the displacement of point M and is noted $\vec{u}(M)$.

Point N also moves to N^*, and its new position is given by the vector position:

$$\overrightarrow{ON^*} = \overrightarrow{ON} + \overrightarrow{NN^*} \tag{5.57}$$

Again, the vector $\overrightarrow{NN^*}$ represents the displacement of point N noted $\vec{u}(N)$.

The relative position between points N and M after deformation is expressed as

$$\overrightarrow{M^*N^*} = \overrightarrow{ON^*} - \overrightarrow{OM^*} = \overrightarrow{ON} - \overrightarrow{OM} + \vec{u}(N) - \vec{u}(M) \tag{5.58}$$

Since points M and N are very close to each other, MN is an infinitesimal element with length dS. It follows therefore that

$$\overrightarrow{MN} = \overrightarrow{ON} - \overrightarrow{OM} = d(\overrightarrow{OM}) = d\vec{S} \tag{5.59}$$

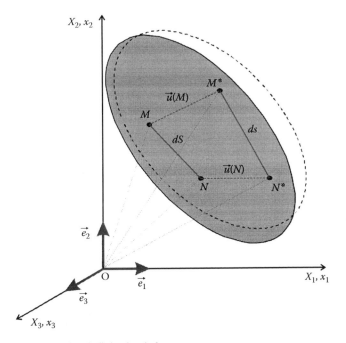

FIGURE 5.10 Deformations of an infinitesimal element.

The same can be said about points M^* and N^*:

$$\overrightarrow{M^*N^*} = \overrightarrow{ON^*} - \overrightarrow{OM^*} = d(\overrightarrow{OM^*}) = d\vec{s} \tag{5.60}$$

Substituting (5.59) and (5.60), expression (5.58) is rewritten as

$$d\vec{s} = d\vec{S} + d\vec{u}(M) \tag{5.61}$$

Introducing the Cartesian components of the vectors,

$$d\vec{S} = dX_1\vec{e}_1 + dX_2\vec{e}_2 + dX_3\vec{e}_3 \tag{5.62}$$

$$d\vec{s} = dx_1\vec{e}_1 + dx_2\vec{e}_2 + dx_3\vec{e}_3 \tag{5.63}$$

$$d\vec{u} = du_1\vec{e}_1 + du_2\vec{e}_2 + du_3\vec{e}_3 \tag{5.64}$$

Equation (5.62) becomes

$$dx_i = dX_i + \frac{\partial u_i}{\partial X_j}dX_j = (\delta_{ij} + u_{i,j})dX_j = F_{ij}dX_j \tag{5.65}$$

or in matrix notation as

$$\begin{Bmatrix} dx_1 \\ dx_2 \\ dx_3 \end{Bmatrix} = \begin{Bmatrix} dX_1 \\ dX_2 \\ dX_3 \end{Bmatrix} + \begin{bmatrix} \dfrac{\partial u_1}{\partial X_1} & \dfrac{\partial u_1}{\partial X_2} & \dfrac{\partial u_1}{\partial X_3} \\[2mm] \dfrac{\partial u_2}{\partial X_1} & \dfrac{\partial u_2}{\partial X_2} & \dfrac{\partial u_2}{\partial X_3} \\[2mm] \dfrac{\partial u_3}{\partial X_1} & \dfrac{\partial u_3}{\partial X_2} & \dfrac{\partial u_3}{\partial X_3} \end{bmatrix} \begin{Bmatrix} dX_1 \\ dX_2 \\ dX_3 \end{Bmatrix} \tag{5.66}$$

and, in a more compact form, as

$$\{dx\} = ([I] + [\nabla u]\{dX\} = [F]\{dX\} \tag{5.67}$$

The matrices $[F]$ and $[\nabla u]$ are respectively called the deformation gradient matrix and the displacement gradient matrix.

5.3.5 GREEN LAGRANGE STRAIN MATRIX

The variation of the square of the distances between points M and N in the undeformed configuration and points M^* and N^* in the deformed configuration is given as

$$\|\overrightarrow{M^*N^*}\|^2 - \|\overrightarrow{MN}\|^2 = \|\overrightarrow{dOM^*}\|^2 - \|\overrightarrow{dOM}\|^2 = \|d\vec{s}\|^2 - \|d\vec{S}\|^2 \tag{5.68}$$

Substituting for $d\vec{s}$ from Equation (5.61), Equation (5.68) becomes

$$\|d\vec{s}\|^2 - \|d\vec{S}\|^2 = d\vec{S}d\vec{u} + d\vec{u}d\vec{S} + \|d\vec{u}\|^2 \tag{5.69}$$

Introducing the Cartesian components of the vectors $d\vec{S}$ and $d\vec{u}$, Equation (5.69) becomes

$$\|d\vec{s}\|^2 - \|d\vec{S}\|^2 = \frac{\partial u_i}{\partial X_j}dX_jdX_i + \frac{\partial u_j}{\partial X_i}dX_idX_j + \frac{\partial u_k}{\partial X_j}\frac{\partial u_k}{\partial X_i}dX_idX_j$$

$$= \left(\frac{\partial u_i}{\partial X_j} + \frac{\partial u_j}{\partial X_i} + \frac{\partial u_k}{\partial X_j}\frac{\partial u_k}{\partial X_i} \right)dX_idX_j$$

$$= 2E_{ij}dX_idX_j \tag{5.70}$$

where

$$E_{ij} = \frac{1}{2}\left(\frac{\partial u_i}{\partial X_j} + \frac{\partial u_j}{\partial X_i} + \frac{\partial u_k}{\partial X_j}\frac{\partial u_k}{\partial X_i}\right) \tag{5.71}$$

Equations (5.70) and (5.71) can be written in matrix notation as

$$\|d\vec{s}\|^2 - \|d\vec{S}\|^2 = \{dX\}^T[\nabla u]\{dX\} + \{dX\}^T[\nabla u]^T\{dX\} + \{dX\}^T[\nabla u]^T[\nabla u]\{dX\}$$
$$= \{dX\}^T([\nabla u] + [\nabla u]^T + [\nabla u]^T[\nabla u])\{dX\}$$
$$= 2\{dX\}^T[E]\{dX\} \tag{5.72}$$

with

$$[E] = \frac{1}{2}([\nabla u] + [\nabla u]^T + [\nabla u]^T[\nabla u]) \tag{5.73}$$

The tensor E_{ij} or $[E]$ is called the Green Lagrange strain tensor or matrix.

Using Equation (5.67), $[F] = [I] + [\nabla u]$, the Green Lagrange strain matrix can be expressed as

$$[E] = \frac{1}{2}([F]^T[F] - [I]) \tag{5.74}$$

The Green Lagrange strain is symmetric, and this can be easily verified from Equation (5.74).

The nine components of the tensor when expanded using Equations (5.71) or (5.73) become

$$E_{11} = \frac{\partial u_1}{\partial X_1} + \frac{1}{2}\left(\left(\frac{\partial u_1}{\partial X_1}\right)^2 + \left(\frac{\partial u_2}{\partial X_1}\right)^2 + \left(\frac{\partial u_3}{\partial X_1}\right)^2\right)$$

$$E_{22} = \frac{\partial u_2}{\partial X_2} + \frac{1}{2}\left(\left(\frac{\partial u_1}{\partial X_2}\right)^2 + \left(\frac{\partial u_2}{\partial X_2}\right)^2 + \left(\frac{\partial u_3}{\partial X_2}\right)^2\right) \tag{5.75}$$

$$E_{33} = \frac{\partial u_3}{\partial X_3} + \frac{1}{2}\left(\left(\frac{\partial u_1}{\partial X_3}\right)^2 + \left(\frac{\partial u_2}{\partial X_3}\right)^2 + \left(\frac{\partial u_3}{\partial X_3}\right)^2\right)$$

$$E_{12} = E_{21} = \frac{1}{2}\left[\frac{\partial u_1}{\partial X_2} + \frac{\partial u_2}{\partial X_1} + \left(\frac{\partial u_1}{\partial X_1}\frac{\partial u_1}{\partial X_2} + \frac{\partial u_2}{\partial X_1}\frac{\partial u_2}{\partial X_2} + \frac{\partial u_3}{\partial X_1}\frac{\partial u_3}{\partial X_2}\right)\right]$$

$$E_{23} = E_{32} = \frac{1}{2}\left[\frac{\partial u_2}{\partial X_3} + \frac{\partial u_3}{\partial X_2} + \left(\frac{\partial u_1}{\partial X_2}\frac{\partial u_1}{\partial X_3} + \frac{\partial u_2}{\partial X_2}\frac{\partial u_2}{\partial X_3} + \frac{\partial u_3}{\partial X_2}\frac{\partial u_3}{\partial X_3}\right)\right]$$

$$E_{31} = E_{13} = \frac{1}{2}\left[\frac{\partial u_1}{\partial X_3} + \frac{\partial u_3}{\partial X_1} + \left(\frac{\partial u_1}{\partial X_1}\frac{\partial u_1}{\partial X_3} + \frac{\partial u_2}{\partial X_1}\frac{\partial u_2}{\partial X_3} + \frac{\partial u_3}{\partial X_1}\frac{\partial u_3}{\partial X_3}\right)\right]$$

5.3.6 SMALL DEFORMATION THEORY

5.3.6.1 Infinitesimal Strain

In small deformation theory, it is assumed that the first derivatives of displacements are so small that the squares and products of these derivatives are negligible compared to the linear terms. It follows therefore that the terms $(\partial u_k/\partial x_i)(\partial u_k/\partial x_j)$ in Equation (5.71) and $[\nabla u]^T[\nabla u]$ in Equation (5.73) are negligible and equal to zero. As a result, the Green Lagrange strain tensor reduces to the infinitesimal strain tensor, which is written in both index and matrix notations as

$$\epsilon_{ij} = \frac{1}{2}\left(\frac{\partial u_i}{\partial X_j} + \frac{\partial u_j}{\partial X_i}\right) \tag{5.76}$$

$$[\epsilon] = \frac{1}{2}([\nabla u] + [\nabla u]^T) \tag{5.77}$$

Within the context of small deformation theory, Equation (5.70) is rewritten as

$$\|d\vec{s}\|^2 - \|d\vec{S}\|^2 = 2\epsilon_{ij}dX_idX_j \tag{5.78}$$

Further, assuming that $dS \approx ds$ for small deformations, this equation may be put in the form

$$\frac{ds - dS}{dS} = \epsilon_{ij}\frac{dX_i}{dS}\frac{dX_j}{dS} \tag{5.79}$$

or in matrix form as

$$\frac{ds - dS}{dS} = \frac{1}{dS^2}\begin{Bmatrix} dX_1 \\ dX_2 \\ dX_3 \end{Bmatrix}^T [\epsilon] \begin{Bmatrix} dX_1 \\ dX_2 \\ dX_3 \end{Bmatrix} \tag{5.80}$$

The left-hand-side of Equations (5.79) or (5.80) is recognized as the change in length of the differential element and is called the normal strain for the element originally having direction cosines dX_i/dS. Introducing the direction cosines $\alpha_i = dX_i/dS$, Equations (5.79) and (5.80) become

$$\frac{ds - dS}{dS} = \epsilon_{ij}\alpha_i\alpha_j \tag{5.81}$$

$$= \frac{1}{dS^2}\begin{Bmatrix} \alpha_1 \\ \alpha_2 \\ \alpha_3 \end{Bmatrix}^T [\epsilon] \begin{Bmatrix} \alpha_1 \\ \alpha_2 \\ \alpha_3 \end{Bmatrix} \tag{5.82}$$

In addition, in small deformation theory, there is very little difference between the material (X_1, X_2, X_3) and spatial (x_1, x_2, x_3) coordinates. Hence, it is immaterial whether the infinitesimal strain tensor is written as

$$\frac{1}{2}\left(\frac{\partial u_i}{\partial X_j} + \frac{\partial u_j}{\partial X_i}\right) \quad \text{or} \quad \frac{1}{2}\left(\frac{\partial u_i}{\partial x_j} + \frac{\partial u_j}{\partial x_i}\right)$$

5.3.6.2 Geometrical Interpretation of the Terms of the Strain Tensor

In the context of small deformation theory, let us consider the deformation behavior of two orthogonal infinitesimal elements MN and ML respectively parallel to the axis x_1 and x_2, as shown in Figure 5.11:

$$\overrightarrow{MN} = dS_1\vec{e}_1$$

$$\overrightarrow{ML} = dS_2\vec{e}_2$$

After deformation, points M, N, and L move respectively to M', N', and L':

$$\overrightarrow{M'N'} = \overrightarrow{MN} + \overrightarrow{NN'} - \overrightarrow{MM'} \tag{5.83}$$

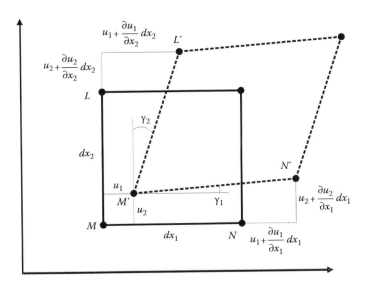

FIGURE 5.11   Geometrical representation of the components of strain at a point.

which can be written using Cartesian components as

$$\overrightarrow{M'N'} = dx_1\vec{e}_1 + \left(u_1 + \frac{\partial u_1}{\partial x_1}dx_1\right)\vec{e}_1 + \left(u_2 + \frac{\partial u_2}{\partial x_1}dx_1\right)\vec{e}_2 - u_1\vec{e}_1 - u_2\vec{u}_2$$

$$= \left(dx_1 + \frac{\partial u_1}{\partial x_1}dx_1\right)\vec{e}_1 + \left(\frac{\partial u_2}{\partial x_1}dx_1\right)\vec{e}_2 \tag{5.84}$$

The term $(\partial u_1/dx_1)dx_1 = \epsilon_{11}dx_1$ represents the change of length of the infinitesimal element MN in the direction x_1. It follows therefore that ϵ_{11} represents the straining at point M in the direction x_1, and the term $(\partial u_2/dx_1)dx_1 = \epsilon_{21}dx_1$ represents its distortion in the direction x_2.

The same reasoning can be carried out for the infinitesimal element ML:

$$\overrightarrow{M'L'} = dx_2\vec{e}_2 + \left(u_2 + \frac{\partial u_2}{\partial x_2}dx_2\right)\vec{e}_2 + \left(u_1 + \frac{\partial u_1}{\partial x_2}dx_2\right)\vec{e}_1 - u_1\vec{e}_1 - u_2\vec{u}_2$$

$$= \left(\frac{\partial u_1}{\partial x_2}dx_2\right)\vec{e}_1 + \left(dx_2 + \frac{\partial u_2}{\partial x_2}dx_2\right)\vec{e}_2 \tag{5.85}$$

Again, the term $(\partial u_2/dx_2)dx_2 = \epsilon_{22}dx_2$ represents the change in length of the infinitesimal element ML in the direction x_2. It follows therefore that ϵ_{22} represents the straining at point M in the direction x_2, and the term $(\partial u_1/\partial x_2)dx_2 = \epsilon_{12}dx_2$ represents its distortion in the direction x_1.

The angle γ_1 between the directions $\overrightarrow{MN}$ and $\overrightarrow{M'N'}$ before and after deformation is such that

$$\tan\gamma_1 = \frac{\dfrac{\partial u_2}{\partial x_1}dx_1}{dx_1 + \dfrac{\partial u_1}{\partial x_1}dx_1} = \frac{\dfrac{\partial u_2}{\partial x_1}}{1 + \dfrac{\partial u_1}{\partial x_1}} \approx \frac{\partial u_2}{\partial x_1} \tag{5.86}$$

since

$$\frac{\partial u_1}{\partial x_1} \ll 1$$

The same can be said for the angle γ_2 between the directions $\overrightarrow{ML}$ and $\overrightarrow{M'L'}$:

$$\tan \gamma_2 = \frac{\partial u_1}{\partial x_2} \tag{5.87}$$

It follows therefore that the angle between $\overrightarrow{MN}$ and $\overrightarrow{ML}$ that was equal to $\pi/2$ before deformation reduces by $\gamma_{12} = \gamma_1 + \gamma_2$ after deformation has taken place.

The angle γ_{12} is called the engineering shear strain at point M written as

$$\gamma_{12} = 2\epsilon_{12} \tag{5.88}$$

5.3.6.3 Compatibility Conditions

The strain tensor contains six independent components. Integration of these six components should lead to the three displacement components (u_1, u_2, u_3). However, the solution is not unique unless the six components of strain verify the following compatibility equations:

$$\frac{\partial^2 \epsilon_{ii}}{\partial x_j^2} + \frac{\partial^2 \epsilon_{jj}}{\partial x_i^2} = 2\frac{\partial^2 \epsilon_{ij}}{\partial x_i \partial x_j} \quad \text{for } i \neq j \tag{5.89}$$

$$\frac{\partial^2 \epsilon_{ii}}{\partial x_j \partial x_k} = \left(-\frac{\partial \epsilon_{jk}}{\partial x_i} + \frac{\partial \epsilon_{ik}}{\partial x_j} + \frac{\partial \epsilon_{ij}}{\partial x_k} \right) \quad \text{for } i \neq j \neq k \tag{5.90}$$

5.3.7 PRINCIPAL STRAINS

In terms of components, the strain tensors E and ϵ bear some resemblance to the stress tensor. Therefore, the entire development for principal strains, principal strain directions and strain invariants, may be carried out exactly as was done for the stress tensor.

In particular, in the basis made of the principal directions, the strain tensor reduces to its diagonal form

$$\epsilon = \begin{bmatrix} \epsilon_1 & 0 & 0 \\ 0 & \epsilon_2 & 0 \\ 0 & 0 & \epsilon_3 \end{bmatrix} \tag{5.91}$$

where ϵ_1, ϵ_2, and ϵ_3 are the principal stresses and roots of the characteristic equation of the tensor:

$$\epsilon^3 - I_1\epsilon^2 + I_2\epsilon - I_3 = 0 \tag{5.92}$$

where

$$I_1 = \epsilon_{ii}$$
$$I_2 = \frac{1}{2}(\epsilon_{ii}\epsilon_{jj} - \epsilon_{ij}\epsilon_{ij}) \tag{5.93}$$
$$I_3 = |\epsilon_{ij}| = det([\epsilon])$$

These invariants can also be expressed in terms of ϵ_1, ϵ_2, and ϵ_3, which are invariants themselves, as

$$I_1 = \epsilon_1 + \epsilon_2 + \epsilon_3$$
$$I_2 = \epsilon_1\epsilon_2 + \epsilon_2\epsilon_3 + \epsilon_3\epsilon_1 \tag{5.94}$$
$$I_3 = \epsilon_1\epsilon_2\epsilon_3$$

5.3.8 Transformation of the Strain Tensor

Like the stress tensor, the strain tensor transforms according to the transformation law of second-order tensors. If the components of the strain tensor ϵ are known in the basis $(\vec{e}_1, \vec{e}_2, \vec{e}_3)$, then its components in the basis $(\vec{e}'_1, \vec{e}'_2, \vec{e}'_3)$ are obtained in both index and matrix notations as

$$\epsilon'_{km} = l_{ki} l_{mj} \epsilon_{ij} \tag{5.95}$$

$$[\epsilon'] = [Q][\epsilon][Q]^T \tag{5.96}$$

The components l_{ij} or Q_{ij} are the cosines of the angles formed by the unit vectors $(\vec{e}'_i, \vec{e}_j)$. The inverse transformations are obtained as

$$\epsilon_{ij} = l_{ki} l_{mj} \epsilon'_{km}$$
$$[\epsilon] = [Q]^T [\epsilon'][Q] \tag{5.97}$$

5.3.9 Engineering Representation of Strain

Like the stress tensor, the strain tensor is symmetric and therefore possesses only six independent components. Engineers also prefer to substitute for the shear strains the engineering shear strains as

$$
\begin{Bmatrix} \epsilon_1 \\ \epsilon_2 \\ \epsilon_3 \\ \epsilon_4 \\ \epsilon_5 \\ \epsilon_6 \end{Bmatrix} \equiv
\begin{Bmatrix} \epsilon_{11} \\ \epsilon_{22} \\ \epsilon_{33} \\ \gamma_{12} \\ \gamma_{23} \\ \gamma_{13} \end{Bmatrix} \equiv
\begin{Bmatrix} \epsilon_{xx} \\ \epsilon_{yy} \\ \epsilon_{zz} \\ \gamma_{xy} \\ \gamma_{yz} \\ \gamma_{xz} \end{Bmatrix} \equiv
\begin{Bmatrix} \epsilon_{xx} \\ \epsilon_{yy} \\ \epsilon_{zz} \\ 2\epsilon_{xy} \\ 2\epsilon_{yz} \\ 2\epsilon_{xz} \end{Bmatrix} \tag{5.98}
$$

With this notation, the transformation law for strain in the case of a rotation around the axis 3, or axis z, is written as

$$
\begin{Bmatrix} \epsilon'_{11} \\ \epsilon'_{22} \\ \epsilon'_{33} \\ \gamma'_{12} \\ \gamma'_{23} \\ \gamma'_{13} \end{Bmatrix} =
\begin{bmatrix}
\cos^2\psi & \sin\psi & 0 & \sin\psi\cos\psi & 0 & 0 \\
\sin\psi & \cos^2\psi & 0 & -\sin\psi\cos\psi & 0 & 0 \\
0 & 0 & 1 & 0 & 0 & 0 \\
-2\sin\psi\cos\psi & 2\sin\psi\cos\psi & \cos^2\psi - \sin^2\psi & 0 & 0 & 0 \\
0 & 0 & 0 & 0 & \cos\psi & -\sin\psi \\
0 & 0 & 0 & 0 & \sin\psi & \cos\psi
\end{bmatrix}
$$

$$
\times
\begin{Bmatrix} \epsilon_{11} \\ \epsilon_{22} \\ \epsilon_{33} \\ \gamma_{12} \\ \gamma_{23} \\ \gamma_{13} \end{Bmatrix} \tag{5.99}
$$

5.4 STRESS–STRAIN CONSTITUTIVE RELATIONS

5.4.1 GENERALIZED HOOKE'S LAW

The stress tensor is related to the strain tensor through the generalized Hooke's law, which is given in index notation as

$$\sigma_{ij} = D_{ijkl}\epsilon_{kl} \tag{5.100}$$

where D_{ijkl} is the stiffness tensor. This is a fourth-order tensor with 81 components. Equation (5.100) represents actually nine equations of which the first one is given as

$$\sigma_{11} = D_{1111}\epsilon_{11} + D_{1112}\epsilon_{12} + D_{1113}\epsilon_{13} + D_{1121}\epsilon_{21}$$
$$+ D_{1122}\epsilon_{22} + D_{1123}\epsilon_{23} + D_{1131}\epsilon_{31} + D_{1132}\epsilon_{32} + D_{1133}\epsilon_{33} \tag{5.101}$$

Luckily, in practice the equations are much simpler and not all the 81 components are independent. The symmetry of both the stress and strain tensors introduces some simplifications into the constitutive equations:

$$D_{ijkl} = D_{ijlk} = D_{jikl} = D_{jilk} \tag{5.102}$$

In addition, the assumption of linear elastic material behavior implies the existence of a strain energy density function. Omitting the proof, this energy density function is given as

$$dU = \sigma_{ij}\epsilon_{ij} = \sigma_{11}\epsilon_{11} + \sigma_{22}\epsilon_{22} + \sigma_{33}\epsilon_{33} + \sigma_{12}\epsilon_{12} + \sigma_{23}\epsilon_{23} + \sigma_{13}\epsilon_{13} \tag{5.103}$$

According to Equations (5.101) and (5.103), it follows that

$$\frac{\partial U}{\partial \epsilon_{11}} = \sigma_{11} = D_{1111}\epsilon_{11} + D_{1112}\epsilon_{12} + D_{1113}\epsilon_{13} + D_{1121}\epsilon_{21} + D_{1122}\epsilon_{22} + D_{1123}\epsilon_{23}$$
$$+ D_{1131}\epsilon_{31} + D_{1132}\epsilon_{32} + D_{1133}\epsilon_{33} \tag{5.104}$$

and

$$\frac{\partial U}{\partial \epsilon_{22}} = \sigma_{22} = D_{2211}\epsilon_{11} + D_{2212}\epsilon_{12} + D_{2213}\epsilon_{13} + D_{2221}\epsilon_{21} + D_{2222}\epsilon_{22} + D_{2223}\epsilon_{23}$$
$$+ D_{2231}\epsilon_{31} + D_{2232}\epsilon_{32} + D_{2233}\epsilon_{33} \tag{5.105}$$

Hence,

$$\frac{\partial^2 U}{\partial \epsilon_{11}\partial \epsilon_{22}} = D_{1122} = D_{2211} \tag{5.106}$$

and, in general,

$$\frac{\partial^2 U}{\partial \epsilon_{kl}\partial \epsilon_{mn}} = D_{klmn} = D_{mnkl} \tag{5.107}$$

Equation (5.107) shows that the fourth-order tensor D_{ijkl} is symmetric. In other words, the number of independent elastic coefficients is reduced from 36 to 21.

The generalized Hooke's law for an anisotropic material can now be written using engineering matrix notation as

$$
\begin{Bmatrix} \sigma_{11} \\ \sigma_{22} \\ \sigma_{33} \\ \sigma_{12} \\ \sigma_{23} \\ \sigma_{13} \end{Bmatrix} = \begin{bmatrix} D_{1111} & D_{1122} & D_{1133} & D_{1112} & D_{1123} & D_{1113} \\ D_{2211} & D_{2222} & D_{2233} & D_{2212} & D_{2223} & D_{2213} \\ D_{3311} & D_{3322} & D_{3333} & D_{3312} & D_{3323} & D_{3313} \\ D_{1211} & D_{1222} & D_{1233} & D_{1212} & D_{1223} & D_{1213} \\ D_{2311} & D_{2322} & D_{2333} & D_{2312} & D_{2323} & D_{2313} \\ D_{1311} & D_{1322} & D_{1333} & D_{1312} & D_{1323} & D_{1313} \end{bmatrix} \begin{Bmatrix} \epsilon_{11} \\ \epsilon_{22} \\ \epsilon_{33} \\ \gamma_{12} = 2\epsilon_{12} \\ \gamma_{23} = 2\epsilon_{23} \\ \gamma_{13} = 2\epsilon_{13} \end{Bmatrix}
$$

(5.108)

with $D_{klmn} = D_{mnkl}$.

In practice, it is sometimes more useful to express observed strains in terms of applied stresses, using the compliance tensor obtained by inverting (5.108)

$$
\begin{Bmatrix} \epsilon_{11} \\ \epsilon_{22} \\ \epsilon_{33} \\ \gamma_{12} \\ \gamma_{23} \\ \gamma_{13} \end{Bmatrix} = \begin{bmatrix} C_{1111} & C_{1122} & C_{1133} & C_{1112} & C_{1123} & C_{1113} \\ C_{2211} & C_{2222} & C_{2233} & C_{2212} & C_{2223} & C_{2213} \\ C_{3311} & C_{3322} & C_{3333} & C_{3312} & C_{3323} & C_{3313} \\ C_{1211} & C_{1222} & C_{1233} & C_{1212} & C_{1223} & C_{1213} \\ C_{2311} & C_{2322} & C_{2333} & C_{2312} & C_{2323} & C_{2313} \\ C_{1311} & C_{1322} & C_{1333} & C_{1312} & C_{1323} & C_{1313} \end{bmatrix} \begin{Bmatrix} \sigma_{11} \\ \sigma_{22} \\ \sigma_{33} \\ \sigma_{12} \\ \sigma_{23} \\ \sigma_{13} \end{Bmatrix}
$$

(5.109)

Further simplifications in the number of constants can be achieved if certain symmetries exist in the material. But, before investigating these material symmetries, it is important to know how a fourth-order tensor is transformed. Since the components of stress and strain are functions of the system of reference axes, the elastic coefficients in Equation (5.108) are also functions of this orientation.

If the components of the stiffness tensor D_{ijkl} in the basis $(\vec{e}_1, \vec{e}_2, \vec{e}_3)$ are known, its components in the basis $(\vec{e}_1', \vec{e}_2', \vec{e}_3')$ are obtained according to the following transformation rule:

$$
D'_{prst} = l_{pi} l_{rj} l_{sk} l_{tl} D_{ijkl}
$$

(5.110)

5.4.2 Material Symmetries

5.4.2.1 Symmetry with respect to a Plane

A material that exhibits symmetry of its elastic properties to one plane is called a monoclinic material. This symmetry is expressed by the requirement that the material constants do not change under a change from the basis $(\vec{e}_1, \vec{e}_2, \vec{e}_3)$ to $(\vec{e}_1', \vec{e}_2', \vec{e}_3')$ such as the one represented in Figure 5.12.

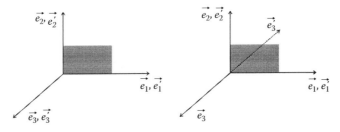

FIGURE 5.12 Monoclinic material.

The direction cosines of the primed axis with respect to the unprimed axis are given as

$$l_{ij} = \cos(\vec{e}'_i, \vec{e}_j) = \begin{pmatrix} 1 & 0 & 0 \\ 0 & 1 & 0 \\ 0 & 0 & -1 \end{pmatrix} \tag{5.111}$$

It follows therefore that

$$D'_{1111} = l_{1k}l_{1l}l_{1m}l_{1n}D_{klmn} = l_{11}l_{11}l_{11}l_{11}D_{1111} = D_{1111} \tag{5.112}$$

In a similar way, for this type of symmetry, it follows that when $(\vec{e}_3)$ and $(\vec{e}'_3)$ are in the same direction, we obtain

$$D'_{1113} = l_{1k}l_{1l}l_{1m}l_{3n}D_{klmn} = l_{11}l_{11}l_{11}l_{33}D_{1113}D_{1113} = D_{1113} \tag{5.113}$$

and when they are in the opposite direction, we obtain

$$D'_{1113} = l_{1k}l_{1l}l_{1m}l_{3n}D_{klmn} = l_{11}l_{11}l_{11}l_{33}D_{1113}D_{1113} = -D_{1113} \tag{5.114}$$

which is impossible. It follows therefore that $D_{1113} = 0$.

In a similar fashion, it can be shown that the number of elements is reduced from 21 to 13; that is, the elastic matrix is written as follows:

$$[D] = \begin{bmatrix} D_{1111} & D_{1122} & D_{1133} & D_{1112} & 0 & 0 \\ D_{2211} & D_{2222} & D_{2233} & D_{2212} & 0 & 0 \\ D_{3311} & D_{3322} & D_{3333} & D_{3312} & 0 & 0 \\ D_{1211} & D_{1222} & D_{1233} & D_{1212} & 0 & 0 \\ 0 & 0 & 0 & 0 & D_{2323} & D_{2313} \\ 0 & 0 & 0 & 0 & D_{1323} & D_{1313} \end{bmatrix} \tag{5.115}$$

Similarly, the compliance matrix becomes

$$[C] = \begin{bmatrix} C_{1111} & C_{1122} & C_{1133} & C_{1112} & 0 & 0 \\ C_{2211} & C_{2222} & C_{2233} & C_{2212} & 0 & 0 \\ C_{3311} & C_{3322} & C_{3333} & C_{3312} & 0 & 0 \\ C_{1211} & C_{1222} & C_{1233} & C_{1212} & 0 & 0 \\ 0 & 0 & 0 & 0 & C_{2323} & C_{2313} \\ 0 & 0 & 0 & 0 & C_{1323} & C_{1313} \end{bmatrix} \tag{5.116}$$

5.4.2.2 Symmetry with respect to Three Orthogonal Planes

A material that exhibits symmetry of its elastic planes with respect to three orthogonal planes is called an orthotropic material. Following the same reasoning as for the symmetry with respect to a single plane, and equating terms to zero where contradictions arise, the elastic matrix reduces from 13 terms to 9:

$$[D] = \begin{bmatrix} D_{1111} & D_{1122} & D_{1133} & 0 & 0 & 0 \\ D_{2211} & D_{2222} & D_{2233} & 0 & 0 & 0 \\ D_{3311} & D_{3322} & D_{3333} & 0 & 0 & 0 \\ 0 & 0 & 0 & D_{1212} & 0 & 0 \\ 0 & 0 & 0 & 0 & D_{2323} & D_{2313} \\ 0 & 0 & 0 & 0 & D_{1323} & D_{1313} \end{bmatrix} \tag{5.117}$$

5.4.2.3 Symmetry of Rotation with respect to One Axis

A material that posseses an axis of symmetry, in the sense that all rays at right angle to this axis have the same elastic properties, is called a transversely isotropic material. If this axis is for example $\vec{e}_3$, as shown in Figure 5.13, then a change of basis obtained by rotation around $\vec{e}_3$ will leave the elastic properties unaltered. Making use of this property leads to

$$D_{1111} = D_{2222} \quad D_{2323} = D_{1313}$$
$$D_{1133} = D_{2233} \quad D_{1212} = \frac{1}{2}(D_{1111} - D_{1122}) \tag{5.118}$$

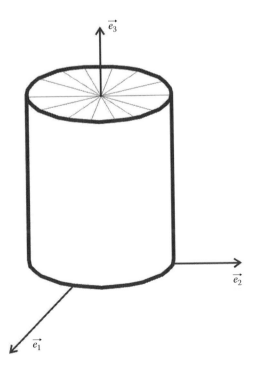

FIGURE 5.13 Symmetry of rotation.

The number of independent coefficient in the elastic matrix is now reduced to 5:

$$[D] = \begin{bmatrix} D_{1111} & D_{1122} & D_{1133} & 0 & 0 & 0 \\ D_{1122} & D_{1111} & D_{1133} & 0 & 0 & 0 \\ D_{1133} & D_{1133} & D_{3333} & 0 & 0 & 0 \\ 0 & 0 & 0 & \frac{1}{2}(D_{1111} - D_{1122}) & 0 & 0 \\ 0 & 0 & 0 & 0 & D_{1313} & 0 \\ 0 & 0 & 0 & 0 & 0 & D_{1313} \end{bmatrix} \tag{5.119}$$

The compliance matrix is obtained as

$$[C] = \begin{bmatrix} C_{1111} & C_{1122} & C_{1133} & 0 & 0 & 0 \\ C_{1122} & C_{1111} & C_{1133} & 0 & 0 & 0 \\ C_{1133} & C_{1133} & C_{3333} & 0 & 0 & 0 \\ 0 & 0 & 0 & 2(C_{1111} - C_{1122}) & 0 & 0 \\ 0 & 0 & 0 & 0 & C_{1313} & 0 \\ 0 & 0 & 0 & 0 & 0 & C_{1313} \end{bmatrix} \tag{5.120}$$

5.4.3 Isotropic Material

A material is isotropic if its elastic properties are the same in any direction and therefore do not depend on the choice of the coordinates system. The elastic and compliance matrices remain unaltered by any change of orthonormal basis. The use of these properties leads to

$$D_{1313} = \frac{1}{2}(D_{1111} - D_{1122}) \tag{5.121}$$

$$D_{3333} = D_{1111}$$

$$D_{1133} = D_{1122} \tag{5.122}$$

The elastic matrix is written as

$$[D] = \begin{bmatrix} D_{1111} & D_{1122} & D_{1122} & 0 & 0 & 0 \\ D_{1122} & D_{1111} & D_{1122} & 0 & 0 & 0 \\ D_{1122} & D_{1122} & D_{1111} & 0 & 0 & 0 \\ 0 & 0 & 0 & \frac{1}{2}(D_{1111} - D_{1122}) & 0 & 0 \\ 0 & 0 & 0 & 0 & \frac{1}{2}(D_{1111} - D_{1122}) & 0 \\ 0 & 0 & 0 & 0 & 0 & \frac{1}{2}(D_{1111} - D_{1122}) \end{bmatrix} \tag{5.123}$$

and the compliance matrix as

$$[C] = \begin{bmatrix} C_{1111} & C_{1122} & C_{1122} & 0 & 0 & 0 \\ C_{1122} & C_{1111} & C_{1122} & 0 & 0 & 0 \\ C_{1122} & C_{1122} & C_{1111} & 0 & 0 & 0 \\ 0 & 0 & 0 & 2(C_{1111} - C_{1122}) & 0 & 0 \\ 0 & 0 & 0 & 0 & 2(C_{1111} - C_{1122}) & 0 \\ 0 & 0 & 0 & 0 & 0 & 2(C_{1111} - C_{1122}) \end{bmatrix} \qquad (5.124)$$

In fact, the elastic matrix possesses only two independent components.

Introducing the elastic properties λ and μ known as the Lamé's constants, the stress–strain relations for an isotropic material become

$$\begin{Bmatrix} \sigma_{11} \\ \sigma_{22} \\ \sigma_{33} \\ \sigma_{12} \\ \sigma_{23} \\ \sigma_{13} \end{Bmatrix} = \begin{bmatrix} \lambda + 2\mu & \lambda & \lambda & 0 & 0 & 0 \\ \lambda & \lambda + 2\mu & \lambda & 0 & 0 & 0 \\ \lambda & \lambda & \lambda + 2\mu & 0 & 0 & 0 \\ 0 & 0 & 0 & \mu & 0 & 0 \\ 0 & 0 & 0 & 0 & \mu & 0 \\ 0 & 0 & 0 & 0 & 0 & \mu \end{bmatrix} \begin{Bmatrix} \epsilon_{11} \\ \epsilon_{22} \\ \epsilon_{33} \\ \gamma_{12} \\ \gamma_{23} \\ \gamma_{13} \end{Bmatrix} \qquad (5.125)$$

In index notation, the previous relationship is written as

$$\sigma_{ij} = \lambda \delta_{ij} \epsilon_{kk} + 2\mu \epsilon_{ij} \qquad (5.126)$$

The compliance matrix is given as

$$\begin{Bmatrix} \epsilon_{11} \\ \epsilon_{22} \\ \epsilon_{33} \\ \gamma_{12} \\ \gamma_{23} \\ \gamma_{13} \end{Bmatrix} = \begin{bmatrix} \dfrac{\lambda + 2\mu}{\mu(3\lambda + 2\mu)} & \dfrac{-\lambda}{2\mu(3\lambda + 2\mu)} & \dfrac{-\lambda}{2\mu(3\lambda + 2\mu)} & 0 & 0 & 0 \\ \dfrac{-\lambda}{2\mu(3\lambda + 2\mu)} & \dfrac{\lambda + 2\mu}{\mu(3\lambda + 2\mu)} & \dfrac{-\lambda}{2\mu(3\lambda + 2\mu)} & 0 & 0 & 0 \\ \dfrac{-\lambda}{2\mu(3\lambda + 2\mu)} & \dfrac{-\lambda}{2\mu(3\lambda + 2\mu)} & \dfrac{\lambda + 2\mu}{\mu(3\lambda + 2\mu)} & 0 & 0 & 0 \\ 0 & 0 & 0 & \dfrac{1}{\mu} & 0 & 0 \\ 0 & 0 & 0 & 0 & \dfrac{1}{\mu} & 0 \\ 0 & 0 & 0 & 0 & 0 & \dfrac{1}{\mu} \end{bmatrix} \begin{Bmatrix} \sigma_{11} \\ \sigma_{22} \\ \sigma_{33} \\ \sigma_{12} \\ \sigma_{23} \\ \sigma_{13} \end{Bmatrix} \qquad (5.127)$$

which can also be written in index notation as

$$\epsilon_{ij} = \frac{-\lambda \delta_{ij}}{2\mu(3\lambda + 2\mu)} \sigma_{nn} + \frac{1}{2\mu} \sigma_{ij} \qquad (5.128)$$

Notice that in index notation the engineering shear strain γ_{ij} is not used.

5.4.3.1 Modulus of Elasticity

Let us consider a uniaxial tension or compression test. In this case, the only stress that is different from zero is σ_{11}. From Equation (5.127), it can be seen that all the shear strains γ_{ij} are equal to zero. The strain in the direction of the test is given as

$$\epsilon_{11} = \frac{-\lambda\delta_{ij}}{2\mu(3\lambda + 2\mu)}\sigma_{11} + \frac{1}{2\mu}\sigma_{11} \tag{5.129}$$

This relation can be rearranged to give

$$\sigma_{11} = \frac{\mu(3\lambda + 2\mu)}{(\lambda + \mu)}\epsilon_{11} = E\epsilon_{11} \tag{5.130}$$

which is the well-known Hooke's law. Equation (5.130) shows the relationship between the elastic modulus E and the Lamé constants λ and μ.

5.4.3.2 Poisson's Ratio

From Equation (5.127), when only σ_{11} is different from zero, it can also be seen that the strains in the directions 2 and 3 are given as

$$\epsilon_{22} = \epsilon_{33} = \frac{\lambda}{2\mu(3\lambda + 2\mu)}\sigma_{11} = \frac{-\lambda}{2(\lambda + \mu)}\epsilon_{11} = -\nu\epsilon_{11} = -\frac{E}{\nu}\sigma_{11} \tag{5.131}$$

The coefficient $\nu = -\lambda/(2(\lambda+\mu))$ is called Poisson's ratio. Equation (5.131) gives the relationships between Poisson's ratio and the Lamé constants.

5.4.3.3 Shear Modulus

Let us consider a pure shear test in the plane $(\vec{e}_1, \vec{e}_2)$ made by the directions 1 and 2. The only stress that is different from zero is $\sigma_{12} = \tau$. The stress–strain relations can be written as

$$\sigma_{12} = 2\mu\epsilon_{12} = \mu\gamma_{12} \tag{5.132}$$

The coefficient μ is called the shear modulus. It is much better known as G.

5.4.3.4 Bulk Modulus

Another test to consider is the application of hydrostatic compression or tension $\sigma_{12} = \sigma_{23} = \sigma_{13} = 0$. In this test,

$$\sigma_{11} = \sigma_{22} = \sigma_{33} = \frac{1}{3}\sigma_{ii} = p \tag{5.133}$$

where p stands for hydrostatic pressure.

As a result of this test, the strains are also spherical

$$\epsilon_{11} = \epsilon_{22} = \epsilon_{33} = \frac{1}{3}\epsilon_{ii} = \epsilon_v \tag{5.134}$$

where ϵ_v stands for volumetric strain.

TABLE 5.1
Relationships between the Coefficients of Elasticity

	λ, μ	E, ν	E, G
λ	λ	$\dfrac{E\nu}{(1+\nu)(1-2\nu)}$	$\dfrac{G(E-2G)}{3G-E}$
μ	μ	$\dfrac{E}{2(1+\nu)}$	G
E	$\dfrac{\mu(3\lambda+2\mu)}{\lambda+\mu}$	E	E
ν	$\dfrac{\lambda}{2(\lambda+\mu)}$	ν	$\dfrac{E-2G}{2G}$
K	$\lambda+\dfrac{2}{3}\mu$	$\dfrac{E}{3(1-2\nu)}$	$\dfrac{GE}{3(3G-E)}$

It follows that

$$p = \left(\lambda + \frac{2}{3}\mu\right)\epsilon_v = K\epsilon_v \tag{5.135}$$

The coefficient K is called the bulk modulus or the compressibility modulus.

Table 5.1 gives the relationships between the coefficients of elasticity.

Finally, the stress–strain relationships for an isotropic material can be written in terms of E and ν as

$$\begin{Bmatrix} \sigma_{11} \\ \sigma_{22} \\ \sigma_{33} \\ \sigma_{12} \\ \sigma_{23} \\ \sigma_{13} \end{Bmatrix} = \frac{E}{(1+\nu)(1-2\nu)} \begin{bmatrix} 1-\nu & \nu & \nu & 0 & 0 & 0 \\ \nu & 1-\nu & \nu & 0 & 0 & 0 \\ \nu & \nu & 1-\nu & 0 & 0 & 0 \\ 0 & 0 & 0 & \dfrac{1-2\nu}{2} & 0 & 0 \\ 0 & 0 & 0 & 0 & \dfrac{1-2\nu}{2} & 0 \\ 0 & 0 & 0 & 0 & 0 & \dfrac{1-2\nu}{2} \end{bmatrix} \begin{Bmatrix} \epsilon_{11} \\ \epsilon_{22} \\ \epsilon_{33} \\ \gamma_{12} \\ \gamma_{23} \\ \gamma_{13} \end{Bmatrix} \tag{5.136}$$

for the elastic matrix and as

$$\begin{Bmatrix} \epsilon_{11} \\ \epsilon_{22} \\ \epsilon_{33} \\ \gamma_{12} \\ \gamma_{23} \\ \gamma_{13} \end{Bmatrix} = \frac{1}{E} \begin{bmatrix} 1 & -\nu & -\nu & 0 & 0 & 0 \\ -\nu & 1 & -\nu & 0 & 0 & 0 \\ -\nu & -\nu & 1 & 0 & 0 & 0 \\ 0 & 0 & 0 & 2(1+\nu) & 0 & 0 \\ 0 & 0 & 0 & 0 & 2(1+\nu) & 0 \\ 0 & 0 & 0 & 0 & 0 & 2(1+\nu) \end{bmatrix} \begin{Bmatrix} \sigma_{11} \\ \sigma_{22} \\ \sigma_{33} \\ \sigma_{12} \\ \sigma_{23} \\ \sigma_{13} \end{Bmatrix} \tag{5.137}$$

for the compliance matrix.

5.4.4 PLANE STRESS AND PLANE STRAIN

In reality, all solids are three dimensional. Fortunately, for many problems that are of practical interest, some simplifying assumptions can be made regarding the stress or strain distributions, and solutions can be carried out in a relatively simpler manner. A solid with one dimension relatively small compared to the two others and loaded in its plane can be analyzed using the plane stress approach. The surfaces of the beam, shown in Figure 5.14 ($z = \pm t/2$), are free of forces and therefore the stress components σ_{33}, σ_{13}, and σ_{23} are equal to zero. If the beam is thin, it can be reasonably assumed that these components are zero throughout the thickness of the beam, and the other stress components σ_{11}, σ_{22}, and σ_{12} remain practically constant.

The nonzero stresses are σ_{11}, σ_{22}, and σ_{12}. Therefore, Equation (5.137) becomes

$$\begin{Bmatrix} \epsilon_{11} \\ \epsilon_{22} \\ \gamma_{12} \end{Bmatrix} = \frac{1}{E} \begin{bmatrix} 1 & -\nu & 0 \\ -\nu & 1 & 0 \\ 0 & 0 & 2(1+\nu) \end{bmatrix} \begin{Bmatrix} \sigma_{11} \\ \sigma_{22} \\ \sigma_{12} \end{Bmatrix} \tag{5.138}$$

Inverting expression (5.138) yields

$$\begin{Bmatrix} \sigma_{11} \\ \sigma_{22} \\ \sigma_{12} \end{Bmatrix} = \frac{E}{1-\nu^2} \begin{bmatrix} 1 & \nu & 0 \\ \nu & 1 & 0 \\ 0 & 0 & \dfrac{(1-\nu)}{2} \end{bmatrix} \begin{Bmatrix} \epsilon_{11} \\ \epsilon_{22} \\ \gamma_{12} \end{Bmatrix} \tag{5.139}$$

It should be pointed out that in plane stress ϵ_{33} is not equal to zero and is given as

$$\epsilon_{33} = \frac{-\nu}{E}(\sigma_{11} + \sigma_{22}) \tag{5.140}$$

Plane strain, on the other hand, occurs in a three-dimensional solid subject to a uniform loading acting constantly along its length. A typical example is a very long strip footing subject to a uniformly distributed load, as shown in Figure 5.15. In these conditions, change of thickness is

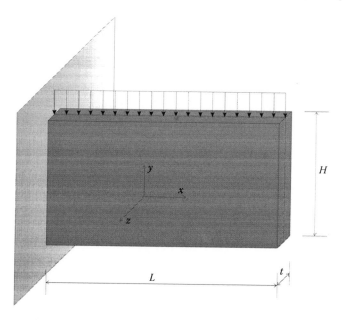

FIGURE 5.14   A state of plane stress.

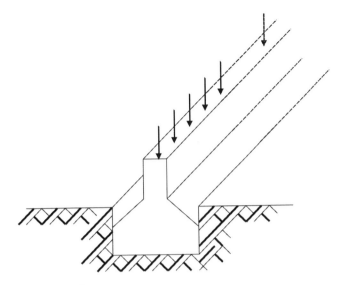

FIGURE 5.15   State of plane strain.

prevented. Therefore, the ends of the footing are prevented from moving in the z-direction; that is, the displacement w of each face in the z-direction is equal to zero. By symmetry, at the mid-section of the footing w must be also equal to zero. In such a case, the components of strain ϵ_{33}, γ_{13}, and γ_{23} are equal to zero.

The nonzero stresses are ϵ_{11}, ϵ_{22}, and γ_{12}. Therefore, Equation (5.136) becomes

$$\begin{Bmatrix} \sigma_{11} \\ \sigma_{22} \\ \sigma_{12} \end{Bmatrix} = \frac{E}{(1+\nu)(1-2\nu)} \begin{bmatrix} 1-\nu & \nu & 0 \\ \nu & 1-\nu & 0 \\ 0 & 0 & \dfrac{1-2\nu}{2} \end{bmatrix} \begin{Bmatrix} \epsilon_{11} \\ \epsilon_{22} \\ \gamma_{12} \end{Bmatrix} \tag{5.141}$$

Inverting expression (5.141) yields

$$\begin{Bmatrix} \epsilon_{11} \\ \epsilon_{22} \\ \gamma_{12} \end{Bmatrix} = \frac{1+\nu}{E} \begin{bmatrix} 1-\nu & -\nu & 0 \\ -\nu & 1-\nu & 0 \\ 0 & 0 & 2 \end{bmatrix} \begin{Bmatrix} \sigma_{11} \\ \sigma_{22} \\ \sigma_{12} \end{Bmatrix} \tag{5.142}$$

Note also that in a state of plane strain σ_{33} is not equal to zero but it is given as

$$\sigma_{33} = \nu(\epsilon_{11} + \epsilon_{22}) \tag{5.143}$$

5.5   SOLVED PROBLEMS

5.5.1   PROBLEM 5.1

The stress tensor at a point P is given as

$$\sigma = \begin{pmatrix} 2 & 4 & 3 \\ 4 & 0 & 0 \\ 3 & 0 & -1 \end{pmatrix}$$

Find the stress vector on a plane that passes through P and is parallel to the plane $x + 2y + 2z - 6 = 0$.

Solution

The function defining the surface of the plane can be written as

$$f(xyz) = x + 2y + 2z - 6 = 0.$$

The vector normal to the plane $\vec{V}$ is obtained as

$$\vec{V} = \frac{\partial f}{\partial x}\vec{e}_1 + \frac{\partial f}{\partial y}\vec{e}_2 + \frac{\partial f}{\partial z}\vec{e}_3 = 1\vec{e}_1 + 2\vec{e}_2 + 2\vec{e}_3$$

The normal unit vector $\vec{n}$ to the plane is therefore obtained as

$$\vec{n} = \frac{1\vec{e}_1 + 2\vec{e}_2 + 2\vec{e}_3}{\|\vec{V}\|} = \frac{1}{3}\vec{e}_1 + \frac{2}{3}\vec{e}_2 + \frac{2}{3}\vec{e}_3$$

Hence,

$$T_1 = \sigma_{11}n_1 + \sigma_{21}n_2 + \sigma_{31}n_3 = 2 \times \frac{1}{3} + 4 \times \frac{2}{3} + 3 \times \frac{2}{3} = \frac{16}{3}$$

$$T_2 = \sigma_{12}n_1 + \sigma_{22}n_2 + \sigma_{23}n_3 = 4 \times \frac{1}{3} + 0 \times \frac{2}{3} + 0 \times \frac{2}{3} = \frac{4}{3}$$

$$T_3 = \sigma_{13}n_1 + \sigma_{23}n_2 + \sigma_{33}n_3 = 3 \times \frac{1}{3} + 0 \times \frac{2}{3} - 1 \times \frac{2}{3} = \frac{1}{3}$$

5.5.2 PROBLEM 5.2

The state of stress at point is given with respect to the Cartesian axes (o, x, y, z) by the stress matrix

$$\sigma = \begin{pmatrix} 2 & -2 & 0 \\ -2 & \sqrt{2} & 0 \\ 0 & 0 & -\sqrt{2} \end{pmatrix}$$

- Determine the stress tensor σ' in the Cartesian axes (o, x', y', z') obtained by rotating the axes (o, x, y, z) around z by 45° anticlockwise.
- Check the result using the engineering notation of stress.

Solution

Index and matrix notations
The basis $(\vec{e}_1', \vec{e}_2', \vec{e}_3')$ is obtained from the basis $(\vec{e}_1, \vec{e}_2, \vec{e}_3)$ by a rotation of 45° around $\vec{e}_3$, as shown in Figure 5.16. The transformation tensor l_{ij} (or matrix $[Q]$) are respectively given as

$$l_{ij} = \cos(\vec{e}_i', \vec{e}_j) = \begin{pmatrix} \sqrt{2}/2 & \sqrt{2}/2 & 0 \\ -\sqrt{2}/2 & \sqrt{2}/2 & 0 \\ 0 & 0 & 1 \end{pmatrix}$$

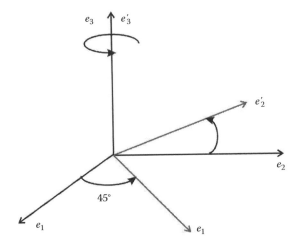

FIGURE 5.16 Change of basis.

and

$$[Q] = [\cos(\vec{e}_i', \vec{e}_j)] = \begin{bmatrix} \sqrt{2}/2 & \sqrt{2}/2 & 0 \\ -\sqrt{2}/2 & \sqrt{2}/2 & 0 \\ 0 & 0 & 1 \end{bmatrix}$$

The stress tensor σ' in the basis $(\vec{e}_1', \vec{e}_2', \vec{e}_3')$ is obtained as

$$\sigma'_{km} = l_{ki} l_{mj} \sigma_{ij}$$

Let us consider the first component σ'_{11}. It is obtained as

$$\begin{aligned} \sigma'_{11} &= l_{1i} l_{1j} \sigma_{ij} \\ &= l_{11} l_{11} \sigma_{11} + l_{11} l_{12} \sigma_{12} + l_{11} l_{13} \sigma_{13} \\ &\quad + l_{12} l_{11} \sigma_{21} + l_{12} l_{12} \sigma_{22} + l_{12} l_{13} \sigma_{23} \\ &\quad + l_{13} l_{11} \sigma_{31} + l_{13} l_{12} \sigma_{32} + l_{13} l_{13} \sigma_{33} \end{aligned}$$

$$\begin{aligned} \sigma'_{11} &= \frac{\sqrt{2}}{2} \frac{\sqrt{2}}{2} 2 + \frac{\sqrt{2}}{2} \frac{\sqrt{2}}{2} (-2) + \frac{\sqrt{2}}{2} \times 0 \times 0 \\ &\quad + \frac{\sqrt{2}}{2} \frac{\sqrt{2}}{2} (-2) + \frac{\sqrt{2}}{2} \frac{\sqrt{2}}{2} \sqrt{2} + \frac{\sqrt{2}}{2} \times 0 \times 0 \\ &\quad + 0 \frac{\sqrt{2}}{2} \times 0 + 0 \frac{\sqrt{2}}{2} \times 0 + 0 \times 0(-\sqrt{2}) = \frac{\sqrt{2}}{2} - 1 \end{aligned}$$

Repeating the same process for all the other terms we obtain

$$\sigma' = \begin{pmatrix} \dfrac{\sqrt{2}}{2} - 1 & \dfrac{\sqrt{2}}{2} - 1 & 0 \\ \dfrac{\sqrt{2}}{2} - 1 & \dfrac{\sqrt{2}}{2} + 3 & 0 \\ 0 & 0 & -\sqrt{2} \end{pmatrix}$$

In matrix notation, the transformation is carried out as

$$[\sigma'] = [Q][\sigma][Q]^T$$

$$= \begin{bmatrix} \sqrt{2}/2 & \sqrt{2}/2 & 0 \\ -\sqrt{2}/2 & \sqrt{2}/2 & 0 \\ 0 & 0 & 1 \end{bmatrix} \begin{bmatrix} 2 & -2 & 0 \\ -2 & \sqrt{2} & 0 \\ 0 & 0 & -\sqrt{2} \end{bmatrix} \begin{bmatrix} \sqrt{2}/2 & -\sqrt{2}/2 & 0 \\ \sqrt{2}/2 & \sqrt{2}/2 & 0 \\ 0 & 0 & 1 \end{bmatrix}$$

$$= \begin{bmatrix} \dfrac{\sqrt{2}}{2} - 1 & \dfrac{\sqrt{2}}{2} - 1 & 0 \\ \dfrac{\sqrt{2}}{2} - 1 & \dfrac{\sqrt{2}}{2} + 3 & 0 \\ 0 & 0 & -\sqrt{2} \end{bmatrix}$$

Engineering notation
According to the engineering notation, the stress transformation law is given as

$$\begin{Bmatrix} \sigma'_{11} \\ \sigma'_{22} \\ \sigma'_{33} \\ \sigma'_{12} \\ \sigma'_{23} \\ \sigma'_{13} \end{Bmatrix} = \begin{bmatrix} \cos^2\psi & \sin\psi & 0 & 2\sin\psi\cos\psi & 0 & 0 \\ \sin\psi & \cos^2\psi & 0 & -2\sin\psi\cos\psi & 0 & 0 \\ 0 & 0 & 1 & 0 & 0 & 0 \\ -\sin\psi\cos\psi & \sin\psi\cos\psi & \cos^2\psi - \sin^2\psi & 0 & 0 & 0 \\ 0 & 0 & 0 & 0 & \cos\psi & -\sin\psi \\ 0 & 0 & 0 & 0 & \sin\psi & \cos\psi \end{bmatrix}$$

$$\times \begin{Bmatrix} \sigma_{11} \\ \sigma_{22} \\ \sigma_{33} \\ \sigma_{12} \\ \sigma_{23} \\ \sigma_{13} \end{Bmatrix}$$

Introducing the numerical values we obtain

$$\begin{Bmatrix} \sigma'_{11} \\ \sigma'_{22} \\ \sigma'_{33} \\ \sigma'_{12} \\ \sigma'_{23} \\ \sigma'_{13} \end{Bmatrix} = \begin{bmatrix} 0.5 & 0.5 & 0 & 1 & 0 & 0 \\ 0.5 & 0.5 & 0 & -1 & 0 & 0 \\ 0 & 0 & 1 & 0 & 0 & 0 \\ -0.5 & 0.5 & 0 & 0 & 0 & 0 \\ 0 & 0 & 0 & 0 & \dfrac{\sqrt{2}}{2} & -\dfrac{\sqrt{2}}{2} \\ 0 & 0 & 0 & 0 & \dfrac{\sqrt{2}}{2} & \dfrac{\sqrt{2}}{2} \end{bmatrix} \begin{Bmatrix} 2 \\ \sqrt{2} \\ -\sqrt{2} \\ -2 \\ 0 \\ 0 \end{Bmatrix} = \begin{Bmatrix} \dfrac{\sqrt{2}}{2} - 1 \\ \dfrac{\sqrt{2}}{2} + 3 \\ -\sqrt{2} \\ \dfrac{\sqrt{2}}{2} - 1 \\ 0 \\ 0 \end{Bmatrix}$$

The results compare very well.

5.5.3 PROBLEM 5.3

The Lagrangian description of the deformation of a body is given by

$$x_1 = X_1$$
$$x_2 = X_2 + 0.2X_3$$
$$x_3 = X_3 + 0.2X_2$$

- Determine the deformation gradients $[F]$ and the Green Lagrange strain matrix $[E]$.
- Calculate the change in squared length of the lines OA, AC, and the diagonal OC for the small undeformed rectangle shown in Figure 5.17.

Solution

The deformation gradient is given by

$$F_{ij} = \frac{\partial x_i}{\partial X_j} \equiv [F] = \begin{bmatrix} \dfrac{\partial x_1}{\partial X_1} & \dfrac{\partial x_1}{\partial X_2} & \dfrac{\partial x_1}{\partial X_3} \\ \dfrac{\partial x_2}{\partial X_1} & \dfrac{\partial x_2}{\partial X_2} & \dfrac{\partial x_2}{\partial X_3} \\ \dfrac{\partial x_3}{\partial X_1} & \dfrac{\partial x_3}{\partial X_2} & \dfrac{\partial x_3}{\partial X_3} \end{bmatrix} = \begin{bmatrix} 1 & 0 & 0 \\ 0 & 1 & 0.2 \\ 0 & 0.2 & 1 \end{bmatrix}$$

The Green Lagrange strain matrix is given as

$$[E] = \frac{1}{2}\left([F]^T[F] - [I]\right)$$

$$= \frac{1}{2}\left(\begin{bmatrix} 1 & 0 & 0 \\ 0 & 1 & 0.2 \\ 0 & 0.2 & 1 \end{bmatrix} \begin{bmatrix} 1 & 0 & 0 \\ 0 & 1 & 0.2 \\ 0 & 0.2 & 1 \end{bmatrix} - \begin{bmatrix} 1 & 0 & 0 \\ 0 & 1 & 0 \\ 0 & 0 & 1 \end{bmatrix} \right)$$

$$= \begin{bmatrix} 0 & 0 & 0 \\ 0 & 0.02 & 0.2 \\ 0 & 0.2 & 0.02 \end{bmatrix}$$

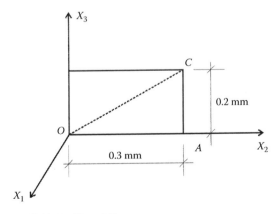

FIGURE 5.17 Displacement field (Problem 5.3).

The change in squared length of an infinitesimal element is given in index notation as

$$\|d\vec{s}\|^2 - \|d\vec{S}\|^2 = 2E_{ij}dX_i dX_j$$

or in matrix notation as

$$\|d\vec{s}\|^2 - \|d\vec{S}\|^2 = 2\{dX\}^T[E]\{dX\}$$

The change in the squared length of the segment OA is obtained as

$$2\begin{bmatrix} 0 & 0.3 & 0 \end{bmatrix}\begin{bmatrix} 0 & 0 & 0 \\ 0 & 0.02 & 0.2 \\ 0 & 0.2 & 0.02 \end{bmatrix}\begin{bmatrix} 0 \\ 0.3 \\ 0 \end{bmatrix} = 0.036 \text{ mm}^2$$

The change in the squared length of the segment AC is obtained as

$$2\begin{bmatrix} 0 & 0 & 0.2 \end{bmatrix}\begin{bmatrix} 0 & 0 & 0 \\ 0 & 0.02 & 0.2 \\ 0 & 0.2 & 0.02 \end{bmatrix}\begin{bmatrix} 0 \\ 0 \\ 0.2 \end{bmatrix} = 0.0004 \text{ mm}^2$$

The change in the squared length of the segment OC is obtained as

$$2\begin{bmatrix} 0 & 0.3 & 0.2 \end{bmatrix}\begin{bmatrix} 0 & 0 & 0 \\ 0 & 0.02 & 0.2 \\ 0 & 0.2 & 0.02 \end{bmatrix}\begin{bmatrix} 0 \\ 0.3 \\ 0.2 \end{bmatrix} = 0.0532 \text{ mm}^2$$

5.5.4 Problem 5.4

Assuming small strain theory, determine the linear strain tensor $[\epsilon]$ for the displacement field given by

$$\vec{u} = (x_1 - x_3)^2\vec{e}_1 + (x_2 + x_3)\vec{e}_2 - x_1 x_2\vec{e}_3$$

At point $P(0, 2, -1)$, determine

- The engineering normal strain in the direction $8\vec{e}_1 - 1\vec{e}_2 + 4\vec{e}_3$
- The change in right angle between $\vec{v}_1 = 8\vec{e}_1 - 1\vec{e}_2 + 4\vec{e}_3$ and $\vec{v}_2 = 4\vec{e}_1 + 4\vec{e}_2 - 7\vec{e}_3$

Solution

The linear strain tensor is given as

$$\epsilon_{ij} = \frac{1}{2}\left(\frac{\partial u_i}{\partial x_j} + \frac{\partial u_j}{\partial x_i}\right)$$

or in matrix form as

$$[\epsilon] = \frac{1}{2}\left([\nabla u] + [\nabla u]^T\right)$$

The displacement gradient is given as

$$\frac{\partial u_i}{\partial x_j} \equiv [\nabla u] = \begin{bmatrix} \dfrac{\partial u_1}{\partial x_1} & \dfrac{\partial u_1}{\partial x_2} & \dfrac{\partial u_1}{\partial x_3} \\[2mm] \dfrac{\partial u_2}{\partial x_1} & \dfrac{\partial u_2}{\partial x_2} & \dfrac{\partial u_2}{\partial x_3} \\[2mm] \dfrac{\partial u_3}{\partial x_1} & \dfrac{\partial u_3}{\partial x_2} & \dfrac{\partial u_3}{\partial x_3} \end{bmatrix} = \begin{bmatrix} 2(x_1 - x_3) & 0 & -2(x_1 - x_3) \\ 0 & 2(x_2 + x_3) & 2(x_2 + x_3) \\ -x_2 & -x_1 & 0 \end{bmatrix}$$

The linear strain tensor is therefore obtained as

$$[\epsilon] = \frac{1}{2}\left(\begin{bmatrix} 2(x_1 - x_3) & 0 & -2(x_1 - x_3) \\ 0 & 2(x_2 + x_3) & 2(x_2 + x_3) \\ -x_2 & -x_1 & 0 \end{bmatrix} + \begin{bmatrix} 2(x_1 - x_3) & 0 & -x_2 \\ 0 & 2(x_2 + x_3) & -x_1 \\ -2(x_1 - x_3) & 2(x_2 + x_3) & 0 \end{bmatrix} \right)$$

$$= \begin{bmatrix} 2(x_1 - x_3) & 0 & -(x_1 - x_3) - \dfrac{x_2}{2} \\[2mm] 0 & 2(x_2 + x_3) & (x_2 + x_3) - \dfrac{x_1}{2} \\[2mm] -(x_1 - x_3) - \dfrac{x_2}{2} & (x_2 + x_3) - \dfrac{x_1}{2} & 0 \end{bmatrix}$$

At $P(0, 2, -1)$, the strain tensor is given as

$$[\epsilon] = \begin{bmatrix} 2 & 0 & -2 \\ 0 & 2 & 1 \\ -2 & 1 & 0 \end{bmatrix}$$

The unit vector in the direction $\vec{v}_1 = 8\vec{e}_1 - 1\vec{e}_2 + 4\vec{e}_3$ is given by

$$\frac{\vec{v}_1}{\|\vec{v}_1\|} = \frac{8}{9}\vec{e}_1 - \frac{1}{9}\vec{e}_2 + \frac{4}{9}\vec{e}_3$$

The engineering normal strain in this direction is given as

$$e = \begin{bmatrix} \dfrac{8}{9} & -\dfrac{1}{9} & \dfrac{4}{9} \end{bmatrix} \begin{bmatrix} 2 & 0 & -2 \\ 0 & 2 & 1 \\ -2 & 1 & 0 \end{bmatrix} \begin{bmatrix} \dfrac{8}{9} \\[2mm] -\dfrac{1}{9} \\[2mm] \dfrac{4}{9} \end{bmatrix} = \frac{-6}{81}$$

The unit vector in the direction $\vec{v}_2 = 4\vec{e}_1 + 4\vec{e}_2 - 7\vec{e}_3$ is given by

$$\frac{\vec{v}_2}{\|\vec{v}_2\|} = \frac{4}{9}\vec{e}_1 + \frac{4}{9}\vec{e}_2 - \frac{7}{9}\vec{e}_3$$

The change of right angle between $\vec{v}_1$ and $\vec{v}_2$ is given as

$$\gamma_{12} = 2 \begin{bmatrix} \dfrac{8}{9} & -\dfrac{1}{9} & \dfrac{4}{9} \end{bmatrix} \begin{bmatrix} 2 & 0 & -2 \\ 0 & 2 & 1 \\ -2 & 1 & 0 \end{bmatrix} \begin{bmatrix} \dfrac{4}{9} \\ \dfrac{4}{9} \\ \dfrac{7}{9} \\ -\dfrac{7}{9} \end{bmatrix} = \dfrac{318}{81}$$

5.5.5 PROBLEM 5.5

A two-dimensional solid is deformed as shown in Figure 5.18. Under the restriction of small deformation theory, determine the linear strain tensor. The solid lines represent the undeformed state. Deduce the engineering form of the strain tensor. The dimensions are given in mm.

Solution

Comparing with Figure 5.11, it can be clearly seen that

$$\epsilon_{11} = \frac{\partial u_1}{\partial x_1} \equiv \frac{0.02}{2} = 0.01$$

$$\epsilon_{22} = \frac{\partial u_2}{\partial x_2} \equiv \frac{0.036}{3} = 0.012$$

$$\gamma_1 = \frac{\partial u_2}{\partial x_1} \equiv \frac{0.010}{2} = 0.005$$

$$\gamma_2 = \frac{\partial u_1}{\partial x_2} \equiv \frac{0.012}{3} = 0.004$$

$$\gamma_{12} = \gamma_1 + \gamma_2 = 0.009$$

$$\epsilon_{12} = \frac{\gamma_{12}}{2} = 0.0045$$

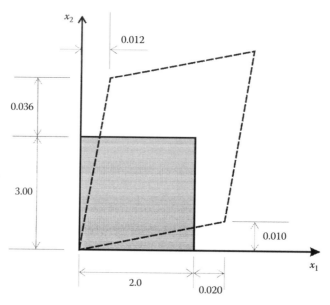

FIGURE 5.18   Displacement field (Problem 5.5).

The strain tensor is given as

$$[\epsilon] = \begin{bmatrix} 0.01 & 0.0045 \\ 0.0045 & 0.012 \end{bmatrix}$$

Using engineering notation the strain tensor is given in a vector form as

$$\{\epsilon\} = \begin{Bmatrix} \epsilon_{11} \\ \epsilon_{22} \\ \gamma_{12} \end{Bmatrix} = \begin{Bmatrix} \epsilon_{xx} \\ \epsilon_{yy} \\ \gamma_{xy} \end{Bmatrix} = \begin{Bmatrix} 0.01 \\ 0.012 \\ 0.009 \end{Bmatrix}$$

5.5.6 PROBLEM 5.6

A 45° strain rosette measures longitudinal strain along the axes shown in Figure 5.19. The following readings are obtained at point P: $\epsilon_{11} = 0.005$, $\epsilon'_{11} = 0.004$, and $\epsilon_{22} = 0.007$ mm/mm. Determine the shear strain γ_{12} at the point.

Solution

The unit vector in the direction X'_1 is given as

$$\vec{n} = \frac{\sqrt{2}}{2}\vec{e}_1 + \frac{\sqrt{2}}{2}\vec{e}_2 + 0\vec{e}_3$$

The stretch or engineering normal strain in the direction X'_1 is given as

$$e = \begin{bmatrix} \dfrac{\sqrt{2}}{2} & \dfrac{\sqrt{2}}{2} & 0 \end{bmatrix} \begin{bmatrix} 0.005 & \epsilon_{12} & 0 \\ \epsilon_{12} & 0.007 & 0 \\ 0 & 0 & 0 \end{bmatrix} \begin{bmatrix} \dfrac{\sqrt{2}}{2} \\ \dfrac{\sqrt{2}}{2} \\ 0 \end{bmatrix} = 0.004$$

It follows therefore that

$$\frac{2}{4}(2\epsilon_{12} + 0.012) = 0.004 \Rightarrow \gamma_{12} = -0.004$$

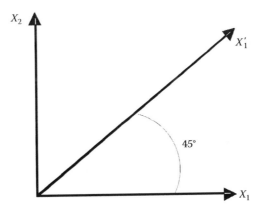

FIGURE 5.19 Strain rosette.

5.5.7 PROBLEM 5.7

Consider a cube of an isotropic linear elastic body whose edges are 10 mm long sitting in a rigid mold with a gap of 0.02 mm between the faces of the mold and that of the cube (Figure 5.20).

Determine the pressure on the lateral faces and the maximum shearing stress in the cube when the uniform pressure applied in the z-direction reaches 1200 MPa. Take $E = 60,000$ MPa, and $\nu = 0.3$.

Solution

First check that the pressure is causing enough lateral strains for the cube to reach the wall.

If we assume that the walls are inexistent, then only nonzero stress is σ_{33}. It follows from the strain–stress relations that

$$
\begin{Bmatrix} \epsilon_{11} \\ \epsilon_{22} \\ \epsilon_{33} \\ \gamma_{12} \\ \gamma_{23} \\ \gamma_{13} \end{Bmatrix} = \frac{1}{E} \begin{bmatrix} 1 & -\nu & -\nu & 0 & 0 & 0 \\ -\nu & 1 & -\nu & 0 & 0 & 0 \\ -\nu & -\nu & 1 & 0 & 0 & 0 \\ 0 & 0 & 0 & 2(1+\nu) & 0 & 0 \\ 0 & 0 & 0 & 0 & 2(1+\nu) & 0 \\ 0 & 0 & 0 & 0 & 0 & 2(1+\nu) \end{bmatrix} \begin{Bmatrix} 0 \\ 0 \\ \sigma_{33} \\ 0 \\ 0 \\ 0 \end{Bmatrix}
$$

which yields

$$
\epsilon_{11} = \epsilon_{22} = -\frac{\mu}{E}\sigma_{33} = -\frac{0.3}{60000}(-1{,}200) = 0.006
$$

The displacement of the lateral faces is $\Delta L = 0.006 \times 10 = 0.06$ mm, which is the total displacement. Therefore, each face is displaced by 0.03 mm, which is greater than the 0.02 mm gap. As a result, lateral forces will develop preventing the lateral faces from expanding more than 0.02 mm (Figure 5.21).

At contact, the lateral strains in the cube will be equal to

$$
\epsilon_{11} = \epsilon_{22} = -\frac{0.04}{10} = 0.004
$$

Since the loading is in the principal directions, the only nonzero strains and stresses are ϵ_{11}, ϵ_{22}, ϵ_{33}, σ_{11}, σ_{22}, and σ_{33}, of which ϵ_{11}, ϵ_{22}, σ_{33} are known, and σ_{11}, σ_{22}, ϵ_{33} are the unknowns.

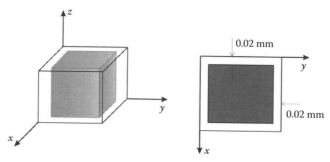

FIGURE 5.20   Problem 5.7.

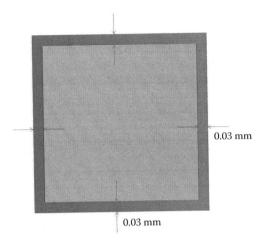

0.03 mm

0.03 mm

FIGURE 5.21 Displacements without the rigid walls.

The stress–strain relations can therefore be written as

$$
\begin{Bmatrix} 0.004 \\ 0.004 \\ \epsilon_{33} \\ 0 \\ 0 \\ 0 \end{Bmatrix} = \frac{1}{E} \begin{bmatrix} 1 & -\nu & -\nu & 0 & 0 & 0 \\ -\nu & 1 & -\nu & 0 & 0 & 0 \\ -\nu & -\nu & 1 & 0 & 0 & 0 \\ 0 & 0 & 0 & 2(1+\nu) & 0 & 0 \\ 0 & 0 & 0 & 0 & 2(1+\nu) & 0 \\ 0 & 0 & 0 & 0 & 0 & 2(1+\nu) \end{bmatrix} \begin{Bmatrix} \sigma_{11} \\ \sigma_{22} \\ -1200 \\ 0 \\ 0 \\ 0 \end{Bmatrix}
$$

which yield three equations

$$0.004 = \frac{1}{60000}(\sigma_{11} - 0.3\sigma_{22} + 0.3 \times 1{,}200)$$

$$0.004 = \frac{1}{60000}(-0.3\sigma_{11} + \sigma_{22} + 0.3 \times 1{,}200)$$

$$\epsilon_{33} = \frac{1}{60000}(-0.3\sigma_{11} - 0.3\sigma_{22} - 1{,}200)$$

Solving the system of equations yields

$$\sigma_{11} = \sigma_{22} = -171.43\,\text{MPa}$$

and

$$\epsilon_{33} = -0.018$$

The maximum shear stress is given as

$$\sigma_s = \frac{1}{2}(\sigma_{11} - \sigma_{33}) = \frac{1}{2}(-171.43 + 1200) = 514.3\,\text{MPa}$$

5.5.8 Problem 5.8

The displacement field of a circular bar that is being twisted by equal and opposite end moments is given by

$$u_1 = 0$$

$$u_2 = -2 \times 10^{-5} x_1 x_3$$

$$u_3 = 2 \times 10^{-5} x_1 x_3$$

The length of the bar is 2000 mm and the diameter is 400 mm. If the bar is made of an isotropic linear elastic material with $E = 2.1 \times 10^5$ MPa, and $\nu = 0.3$, using small deformation theory, determine the state of stress in the points (2000, 100, 100) and (1000, 100, 100). What can be concluded about the variation of the stress along the length of the beam?

Solution

The displacement gradient is given as

$$[\nabla u] = \begin{bmatrix} 0 & 0 & 0 \\ -2 \times 10^{-5} x_3 & 0 & -2 \times 10^{-5} x_1 \\ 2 \times 10^{-5} x_2 & 2 \times 10^{-5} x_1 & 0 \end{bmatrix}$$

The strain tensor (small deformations) is given as

$$[\epsilon] = \frac{1}{2}\left([\nabla u] + [\nabla u]^T\right) = \begin{bmatrix} 0 & -1 \times 10^{-5} x_3 & 1 \times 10^{-5} x_2 \\ -1 \times 10^{-5} x_3 & 0 & 0 \\ 1 \times 10^{-5} x_2 & 0 & 0 \end{bmatrix}$$

It can be seen that the strain tensor is not a function of the x_1 coordinate:

$$[\epsilon(2000, 100, 1000)] = [\epsilon(1000, 100, 1000)] = \begin{bmatrix} 0 & -1 \times 10^{-3} & 1 \times 10^{-3} \\ -1 \times 10^{-3} & 0 & 0 \\ 1 \times 10^{-3} & 0 & 0 \end{bmatrix}$$

The stresses do not vary along the length of the beam:

$$[\sigma] = \begin{bmatrix} 0 & -161.54 & 161.54 \\ -161.54 & 0 & 0 \\ 161.54 & 0 & 0 \end{bmatrix} \text{MPa}$$

6 Weighted Residual Methods

6.1 INTRODUCTION

In Chapters 2 and 3, we used well-known methods of structural analysis to develop the stiffness matrices of the bar and beam elements. The reason being that these elements are one-dimensional, and the exact solutions of the differential equations governing their behaviors are well known. For other structural problems in two and three dimensions, such direct approaches are inexistent for the obvious reason that it is not possible to find analytical solutions to the differential equations governing their behavior, except in the case of very simple geometries. The alternative is to replace the differential equations by approximate algebraic equations. This is achieved by using weighted residual methods.

6.2 GENERAL FORMULATION

Given a physical problem (be it structural or not) whose behavior is governed by a set of differential equations:

$$\mathbb{B}(\{u\}) = 0 \ \text{ on } \ \Omega \tag{6.1}$$

where
  $\mathbb{B}(\ )$ represents a linear differential operator
  $\{u\}$ is the unknown function
  Ω is the geometrical domain

Since the variable $\{u\}$ is unknown, we may try to substitute for it a trial or approximate function of our choosing, say $\{\bar{u}\}$ given as a polynomial function:

$$\{\bar{u}\} = \sum_{i=1}^{n} \alpha_i P_i(\{x\}) \tag{6.2}$$

where
  the coefficients α_i are general parameters
  $P_i(\{x\})$ is a polynomial base

Substituting $\{\bar{u}\}$ for $\{u\}$ will not in general satisfy the differential equation (6.1) and will result in a residual over the domain Ω; that is,

$$\mathbb{B}(\{\bar{u}\}) \neq 0 \ \text{ on } \ \Omega \tag{6.3}$$

The essence of the weighted residual methods is to force the residual to zero in some average over the whole domain Ω. To do so, we multiply the residual by a weighting function ψ and force the integral of the weighted residual to zero over the whole domain; that is,

$$\{W\} = \int_{\Omega} \psi \mathbb{B}(\{\bar{u}\}) \, d\Omega = 0 \tag{6.4}$$

There is a variety of residual methods such as collocation method, subdomain method, least-squares method, method of moments, and Galerkin method. They all differ in the choice of the weighting function ψ. The most popular however is the Galerkin method, and it is the only one described in this chapter.

6.3 GALERKIN METHOD

In the Galerkin method, the weighting function is simply the variation of the trial function itself; that is,

$$\psi = \delta\{\bar{u}\} = \sum_{i=1}^{n} \delta\alpha_i P_i(\{x\}) \tag{6.5}$$

Substituting for ψ and $\{\bar{u}\}$, Equation (6.4) becomes

$$\{W\} = \int_{\Omega} \sum_{i=1}^{n} \delta\alpha_i P_i(\{x\}) \mathbb{B}\left(\sum_{i=1}^{n} \alpha_i P_i(\{x\})\right) d\Omega = 0$$

$$= \{\delta\alpha_i\}^T \int_{\Omega} P_i(\{x\}) \left(\mathbb{B}\left(\sum_{i=1}^{n} \alpha_i P_i(\{x\})\right)\right) d\Omega = 0 \tag{6.6}$$

Since the preceding relation must equal zero for any arbitrary $\delta\alpha_i$, it can be written as

$$W_1 = \int_{\Omega} P_1(\{x\}) \left(\mathbb{B}\left(\sum_{i=1}^{n} \alpha_i P_i(\{x\})\right)\right) d\Omega = 0$$

$$W_2 = \int_{\Omega} P_2(\{x\}) \left(\mathbb{B}\left(\sum_{i=1}^{n} \alpha_i P_i(\{x\})\right)\right) d\Omega = 0$$

$$\vdots = \vdots \tag{6.7}$$

$$W_n = \int_{\Omega} P_n(\{x\}) \left(\mathbb{B}\left(\sum_{i=1}^{n} \alpha_i P_i(\{x\})\right)\right) d\Omega = 0$$

The system of Equations (6.7) can be solved for the unknown coefficients α_i.

Example

Let us consider the following differential equation:

$$\mathbb{B}(u(x)) = \frac{d^2 u(x)}{dx^2} + u(x) \text{ on } \Omega = [0, 1] \tag{6.8}$$

with boundary conditions

$$u(x = 0) = 1$$

$$u(x = 1) = 0 \tag{6.9}$$

This differential equation has an exact solution given by

$$u(x) = 1 - \frac{\sin(x)}{\sin(1)} \tag{6.10}$$

Let us solve the differential equation using the method of Galerkin. We choose the approximating function $\bar{u}(x)$ in the form of a polynomial:

$$\bar{u}(x) = \alpha_0 + \alpha_1 x + \alpha_2 x^2 \tag{6.11}$$

To ensure that the trial function $\bar{u}(x)$ approximate the exact function $u(x)$ as best as possible, we need to make sure that it is derivable as many times as required by the differential operator and satisfies the boundary conditions; that is,

$$\bar{u}(x = 0) = 1 \Rightarrow \alpha_0 = 1$$
$$\bar{u}(x = 1) = 0 \Rightarrow 1 + \alpha_1 + \alpha_2 = 0 \tag{6.12}$$
$$\Rightarrow \alpha_1 = -(1 + \alpha_2)$$

The trial function therefore becomes

$$\bar{u}(x) = \alpha_2 (x^2 - x) - x + 1 \tag{6.13}$$

It is twice derivable and satisfies the boundary conditions. Substituting $\bar{u}(x)$ in Equation (6.8), the residual is written as

$$\mathbb{R}(\bar{u}(x)) = \frac{d^2 \bar{u}(x)}{dx^2} + \bar{u}(x)$$
$$= \alpha_2 (x^2 - x + 2) - x \tag{6.14}$$

The corresponding weighting function is obtained as

$$\psi = \delta \bar{u}(x) = \delta \alpha_2 (x^2 - x) \tag{6.15}$$

Integrating the product of the weighted residual over the domain yields

$$W = \int_0^{+1} \delta \alpha_2 (x^2 - x) \times (\alpha_2 (x^2 - x + 2) - x) \, dx = 0 \tag{6.16}$$

Since $\delta \alpha_2 \neq 0$, it follows

$$W = \int_0^{+1} (x^2 - x) \times (\alpha_2 (x^2 - x + 2) - x) \, dx = 0 \tag{6.17}$$

Evaluating the integral leads to an algebraic equation of the form

$$\frac{1}{12} - \frac{3}{10} \alpha_2 = 0 \Rightarrow \alpha_2 = \frac{5}{18} \tag{6.18}$$

The final approximation is then written as

$$\bar{u}(x) = \frac{5}{18} (x^2 - x) - x + 1 \tag{6.19}$$

Figure 6.1 shows a graphical comparison between the exact solution, Equation (6.10), and the approximate solution, Equation (6.19). With only one parameter α_2, the approximate solution is very acceptable.

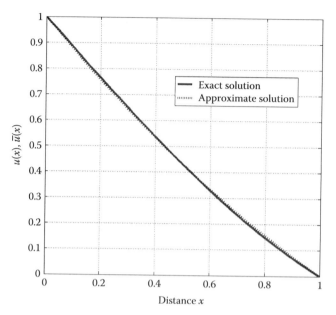

FIGURE 6.1 Graphical comparison of exact and approximate solution.

6.4 WEAK FORM

Given the following differential equation

$$\mathbb{B}(u(x)) = \frac{d^2u(x)}{dx^2} + u(x) + x = 0 \ \text{ on } \ \Omega = [0, 1] \tag{6.20}$$

with boundary conditions

$$u(x = 0) = g \ \text{ essential}$$
$$\frac{du}{dx}(x = 1) = p \ \text{ natural} \tag{6.21}$$

with g and p being real constants. The first boundary conditions imposed on $u(x)$ is termed essential, while the second boundary condition imposed on its derivative is termed natural. If we apply the weighted residual Equation (6.4) to the differential Equation (6.20), we obtain

$$\int_0^1 \psi \left(\frac{d^2\bar{u}(x)}{dx^2} + \bar{u}(x) + x \right) dx = 0 \tag{6.22}$$

We could also do the same thing to the natural boundary condition given in the form of a differential equation; that is,

$$\left[\left(\frac{d\bar{u}}{dx} - p \right) \psi \right]_{(x=1)} = 0 \tag{6.23}$$

Since both expressions (6.22) and (6.23) are equal to zero, we can write

$$\int_0^1 \psi \left(\frac{d^2\bar{u}(x)}{dx^2} + \bar{u}(x) + x \right) dx = \left[\left(\frac{d\bar{u}}{dx} - p \right) \psi \right]_{(x=1)} \tag{6.24}$$

Equation (6.24) is an integral form of the differential Equation (6.20) and its natural boundary condition.

In Equation (6.24), the trial function $\bar{u}(x)$ must not only satisfy the essential boundary condition but it should also be derivable twice as required by the differential operator in order to approach the exact function $u(x)$. On the other hand, the function ψ does not need to be continuous at all.

Now, let us integrate Equation (6.24) by part once:

$$\int_0^1 \left((\bar{u}(x) + x)\psi - \frac{d\bar{u}(x)}{dx}\frac{d\psi}{dx} \right)dx + \left[p\psi \right]_0^1 = 0 \tag{6.25}$$

Notice that both the functions $\bar{u}(x)$ and ψ must be only derivable once. In other words, we have alleviated the condition of continuity imposed on $\bar{u}(x)$ by one and increased that imposed on ψ by one as well.

If we continue to integrate by part, we obtain

$$\int_0^1 \left((\bar{u}(x) + x)\psi + \bar{u}(x)\frac{d^2\psi}{dx^2} \right)dx + \left[p\psi - \bar{u}(x)\frac{d\psi}{dx} \right]_0^1 = 0 \tag{6.26}$$

We end up with an identical problem to Equation (6.24); this time the function ψ needs to be derivable twice, while the function $\bar{u}(x)$ does not have to be continuous at all. It follows therefore that Equation (6.25) is the most appropriate. It is called the weak form. In addition, when the Galerkin method is used, the functions $\bar{u}(x)$ and ψ have the same degree of continuity since $\psi = \delta\bar{u}(x)$.

6.5 INTEGRATING BY PART OVER TWO AND THREE DIMENSIONS (GREEN THEOREM)

In the previous section, it was shown that the order of the derivative was lowered by integrating by part the residual. Integration by part is relatively easy to carry out over one dimension. However, many engineering problems of practical importance are defined over two or three dimensions. Integrating by parts over such domains is more challenging. Fortunately, it can be done by means of the Green theorem.

Let us evaluate by part the following integral

$$\iint_\Omega \Phi \frac{\partial \Psi}{\partial x} dxdy \tag{6.27}$$

over the domain Ω represented in Figure 6.2.

First let us integrate by part with respect to the variable x using the well-known formula

$$\int_{X_L}^{X_R} U dV = (UV_{x=X_R} - UV_{x=X_L}) - \int_{X_L}^{X_R} V dU \tag{6.28}$$

It follows therefore that

$$\iint_\Omega \Phi \frac{\partial \Psi}{\partial x} dxdy = -\iint_\Omega \frac{\partial \Phi}{\partial x}\Psi dxdy + \int_{Y_B}^{Y_T} [(\phi\psi)_{x=X_R} - (\phi\psi)_{x=X_L}] dy \tag{6.29}$$

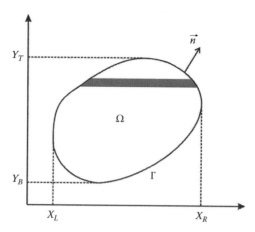

FIGURE 6.2 Integration by parts in two and three dimensions.

If we consider an infinitesimal element of the boundary, $d\Gamma$, on the right side, we can write

$$dy = n_x d\Gamma \qquad (6.30)$$

where n_x is the director cosine of the angle formed by the normal $\vec{n}$ with the axis x, as shown in Figure 6.3. On the left side of the boundary, we have

$$dy = -n_x d\Gamma \qquad (6.31)$$

It follows therefore that the last term of Equation (6.29) can be written in the form of a curvilinear integral as

$$\oint_\Gamma \Phi \Psi n_x d\Gamma \qquad (6.32)$$

Finally, the integral in (6.27) is rewritten as

$$\iint_\Omega \Phi \frac{\partial \Psi}{\partial x} dx dy = -\iint_\Omega \frac{\partial \Phi}{\partial x} \Psi dx dy + \oint_\Gamma \Phi \Psi n_x d\Gamma \qquad (6.33)$$

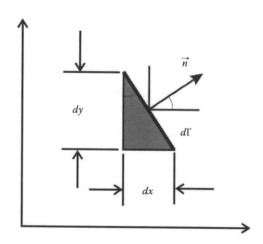

FIGURE 6.3 Infinitesimal element of the boundary.

In the same manner, if we integrate along the direction y, we obtain

$$\iint_{\Omega} \Phi \frac{\partial \Psi}{\partial y} dxdy = -\iint_{\Omega} \frac{\partial \Phi}{\partial y} \Psi dxdy + \oint_{\Gamma} \Phi \Psi n_y d\Gamma \tag{6.34}$$

Remark: Γ represents a surface when Ω is a three-dimensional space.

Example: Weak form of equilibrium equations

The equilibrium equations of a deformable body in three-dimensions are given as

$$\frac{\partial \sigma_{xx}}{\partial x} + \frac{\partial \tau_{xy}}{\partial y} + \frac{\partial \tau_{xz}}{\partial z} + b_x = 0$$

$$\frac{\partial \tau_{xy}}{\partial x} + \frac{\partial \sigma_{yy}}{\partial y} + \frac{\partial \tau_{yz}}{\partial z} + b_y = 0 \tag{6.35}$$

$$\frac{\partial \tau_{xz}}{\partial x} + \frac{\partial \tau_{yz}}{\partial y} + \frac{\partial \sigma_{zz}}{\partial z} + b_z = 0$$

or in a more compact form as

$$\mathbb{L}([\sigma]) = \mathbb{F}([\sigma]) + \{b\} = 0 \tag{6.36}$$

with

$$\mathbb{F}([\sigma]) = [\nabla([\sigma])]^T$$

$$\nabla = \left[\frac{\partial}{\partial x}, \frac{\partial}{\partial y}, \frac{\partial}{\partial z} \right]$$

and

$$[\sigma] = \begin{bmatrix} \sigma_{xx} & \tau_{xy} & \tau_{xz} \\ \tau_{xy} & \sigma_{yy} & \tau_{yz} \\ \tau_{xz} & \tau_{yz} & \sigma_{zz} \end{bmatrix}$$

Let us apply the equation of the weighted residuals to (6.36), and as a weighting function, we choose virtual displacements such that $\psi = \delta\{U\} = \delta[u, v, w]^T$; that is,

$$\int_V \delta\{U\}^T \{\mathbb{F}([\sigma]) + \{b\}\} dV = 0 \tag{6.37}$$

where V designates the volume of the solid. Equation (6.37) can be developed in the following form:

$$\int_V \left[\delta u \left(\frac{\partial \sigma_{xx}}{\partial x} + \frac{\partial \tau_{xy}}{\partial y} + \frac{\partial \tau_{xz}}{\partial z} + b_x \right) + \delta v \left(\frac{\partial \tau_{xy}}{\partial x} + \frac{\partial \sigma_{yy}}{\partial y} + \frac{\partial \tau_{yz}}{\partial z} + b_y \right) \right.$$

$$\left. + \delta w \left(\frac{\partial \tau_{xz}}{\partial x} + \frac{\partial \tau_{yz}}{\partial y} + \frac{\partial \sigma_{zz}}{\partial z} + b_z \right) \right] dV \tag{6.38}$$

To obtain the weak form of expression (6.38), we will integrate it by part using the theorem of Green. First let us consider only the first term:

$$\int_V \delta u \frac{\partial \sigma_{xx}}{\partial x} dV = \int_A \delta u \sigma_{xx} l_x dA - \int_V \sigma_{xx} \frac{\partial(\delta u)}{\partial x} dV \tag{6.39}$$

where A is the surface representing the boundary of the domain V.

After repeating the integration by part for all the terms of Equation (6.38), we obtain

$$
\int_V \left(\sigma_{xx}\frac{\partial(\delta u)}{\partial x} + \tau_{xy}\left(\frac{\partial(\delta u)}{\partial y} + \frac{\partial(\delta v)}{\partial x}\right) + \tau_{xz}\left(\frac{\partial(\delta u)}{\partial z} + \frac{\partial(\delta w)}{\partial x}\right) + \cdots + \sigma_{zz}\frac{\partial(\delta w)}{\partial z} \right.
$$

$$
\left. - \delta u b_x - \delta v b_y - \delta w b_z \right) dV + \int_A \left(\delta u(\sigma_{xx}l_x + \tau_{xy}l_y + \tau_{xz}l_z) \right.
$$

$$
\left. + \delta v(\tau_{xy}l_x + \sigma_{yy}l_y + \tau_{yz}l_z) + \delta w(\tau_{xz}l_x + \tau_{yz}l_y + \sigma_{zz}l_z) \right) dA = 0 \qquad (6.40)
$$

The operator $\delta(\ )$ is linear and has the following properties:

$$
\frac{\partial(\delta u)}{\partial x} = \frac{\delta(\partial u)}{\partial x} = \delta\frac{(\partial u)}{\partial x} = \delta\epsilon_{xx} \qquad (6.41)
$$

It follows therefore that the variations of the partial derivatives of the displacements in Equation (6.40) can be grouped as

$$
\delta\{\epsilon\}^T = \left[\delta\frac{(\partial u)}{\partial x}, \delta\frac{(\partial v)}{\partial y}, \ldots, \left(\delta\frac{(\partial w)}{\partial x} + \delta\frac{(\partial u)}{\partial z}\right) \right]^T \qquad (6.42)
$$

The first nine terms of the integral of the volume of Equation (6.40) can be grouped as

$$
\int_V \delta\{\epsilon\}^T\{\sigma\}\, dV \qquad (6.43)
$$

and the remaining terms as

$$
-\int_V \delta\{U\}^T\{b\}\, dV \qquad (6.44)
$$

As to the terms resulting from the integral over the area, they can be grouped as

$$
-\int_A \delta\{U\}^T\{t\}\, dA \qquad (6.45)
$$

where $\{t\}$ is the stress vector given as

$$
\{t\} = \begin{Bmatrix} t_x \\ t_y \\ t_z \end{Bmatrix} = \begin{Bmatrix} \sigma_{xx}l_x + \tau_{xy}l_y + \tau_{xz}l_z \\ \tau_{xy}l_x + \sigma_{yy}l_y + \tau_{yz}l_z \\ \tau_{xz}l_x + \tau_{yz}l_y + \sigma_{zz}l_z \end{Bmatrix}
$$

Equation (6.40) can then be rewritten as

$$
\int_V \delta\{\epsilon\}^T\{\sigma\}\, dV = \int_V \delta\{U\}^T\{b\}\, dV + \int_A \delta\{U\}^T\{t\}\, dA \qquad (6.46)
$$

Expression (6.46) is nothing but the expression of the theorem of virtual work, which states: *If a deformable body in equilibrium is subjected to an arbitrary virtual displacement field associated with a compatible deformation of the body, the virtual work of external forces on the body is equal to the virtual strain energy of the internal stresses.* It can be therefore concluded that the theorem of virtual work is the weak form of the equilibrium equations.

6.6 RAYLEIGH RITZ METHOD

6.6.1 DEFINITION

A functional Π is a function of a set of functions and their derivatives:

$$\Pi = \Pi\left(u, \frac{\partial u}{\partial x}, \frac{\partial^2 u}{\partial x^2}, \cdots\right) \tag{6.47}$$

The first variation of Π is defined as

$$\delta\Pi = \frac{\partial\Pi}{\partial u}\delta u + \frac{\partial\Pi}{\partial\left(\frac{\partial u}{\partial x}\right)}\delta\left(\frac{\partial u}{\partial x}\right) + \cdots \tag{6.48}$$

where δu and $\delta\left(\frac{\partial u}{\partial x}\right)$ are arbitrary variations of u and $\frac{\partial u}{\partial x}$.

6.6.2 FUNCTIONAL ASSOCIATED WITH AN INTEGRAL FORM

Consider Equation (6.25). If we adopt the method of Galerkin and substituting $\delta\bar{u}(x)$ for ψ, we obtain

$$\int_0^1 \left((\bar{u}(x) + x)\delta\bar{u}(x) - \frac{d\bar{u}(x)}{dx}\frac{d\delta\bar{u}(x)}{dx}\right)dx + \left[p\delta\bar{u}(x)\right]_0^1 = 0 \tag{6.49}$$

which can be rewritten as

$$\delta\left[\frac{1}{2}\int_0^1 \bar{u}(x)^2 dx - \frac{1}{2}\int_0^1 \left(\frac{d\bar{u}(x)}{dx}\right)^2 dx + \int_0^1 x\bar{u}(x)dx + \left[p\bar{u}(x)\right]_{x=1}\right] = 0 \tag{6.50}$$

or simply as

$$\delta\Pi = 0 \tag{6.51}$$

where Π is a functional given by

$$\Pi = \frac{1}{2}\int_0^1 \bar{u}(x)^2 dx - \frac{1}{2}\int_0^1 \left(\frac{d\bar{u}(x)}{dx}\right)^2 dx + \int_0^1 x\bar{u}(x)dx + \left[p\bar{u}(x)\right]_{x=1} \tag{6.52}$$

It can be clearly seen that Π is a function of $\bar{u}(x)$ and its derivatives.

6.6.3 RAYLEIGH RITZ METHOD

If the functional is known, then the Rayleigh Ritz method can be used to discretize it; that is, to replace it with algebraic equations. The method consists in finding trial functions such as the one given by Equation (6.2) that satisfy the essential boundary conditions and minimize the functional:

$$\delta\Pi = 0 \tag{6.53}$$

If we substitute for $\bar{u}(x)$ using Equation (6.2), the variation of the functional becomes

$$\delta\Pi = \frac{\partial\Pi}{\partial\alpha_1}\delta\alpha_1 + \frac{\partial\Pi}{\partial\alpha_2}\delta\alpha_2 + \cdots + \frac{\partial\Pi}{\partial\alpha_n}\delta\alpha_n = 0 \tag{6.54}$$

Since $\delta\Pi$ must be equal to zero for any arbitrary $\delta\alpha_i$, it follows

$$\frac{\partial\Pi}{\partial\alpha_1} = 0$$

$$\frac{\partial\Pi}{\partial\alpha_2} = 0$$

$$\vdots \quad \vdots \tag{6.55}$$

$$\frac{\partial\Pi}{\partial\alpha_n} = 0$$

which constitutes a system of n equations that could be solved for the parameters α_i.

Example

Consider the following functional

$$\Pi = \frac{1}{2}\int_0^1 \bar{u}(x)^2\, dx - \frac{1}{2}\int_0^1 \left(\frac{d\bar{u}(x)}{dx}\right)^2 dx + \int_0^1 x\bar{u}(x)\, dx \tag{6.56}$$

which is associated to the following differential equation

$$\mathbb{B}(u(x)) = \frac{d^2 u(x)}{dx^2} + u(x) + x = 0 \text{ on } \Omega = [0, 1] \tag{6.57}$$

with essential conditions

$$u(x = 0) = 0$$
$$u(x = 1) = 0 \tag{6.58}$$

The analytical solution for the aforementioned differential equation is given by

$$u(x) = \frac{\sin(x)}{\sin(1)} - x \tag{6.59}$$

Applying the method of Rayleigh Ritz to expression (6.56) consists first in finding trial functions that satisfy the essential boundary conditions; that is,

$$u_1 = x(x - 1)\alpha_1 \quad \text{One parameter}$$
$$u_2 = x(x - 1)(\alpha_1 + \alpha_2 x) \quad \text{Two parameters} \tag{6.60}$$

Substituting the first trial function in expression (6.56) leads to

$$\Pi_1 = \frac{1}{2}\int_0^1 \left(\alpha_1^2(x^2 - x)^2 - \alpha_1^2(2x - 1)^2 + 2\alpha_1 x^2(x - 1)\right)dx$$

$$= \frac{1}{2}\int_0^1 \left(\alpha_1^2 x^4 + (\alpha_1 - \alpha_1^2)2x^3 + (3\alpha_1^2 + 2\alpha_1)x^2 + 4\alpha_1^2 x - \alpha_1^2\right)dx \tag{6.61}$$

Evaluating the integral yields

$$\Pi_1 = -\frac{3}{20}\alpha_1^2 - \frac{1}{12}\alpha_1 \tag{6.62}$$

Taking the first variation of Π_1, we obtain

$$\delta\Pi_1 = \frac{\partial\Pi_1}{\partial\alpha_1}\delta\alpha_1 = 0 = -\frac{6}{20}\alpha_1 - \frac{1}{12} \tag{6.63}$$

Solving for α_1, we obtain

$$\alpha_1 = \frac{-5}{18} \tag{6.64}$$

The trial function can therefore be written as

$$u_1(x) = \frac{-5}{18}x(x - 1) \tag{6.65}$$

Substituting the second trial function with two parameters in expression (6.56) and integrating leads to

$$\Pi_2 = -\frac{3}{20}\alpha_1^2 - \frac{13}{210}\alpha_2 - \frac{3}{20}\alpha_1\alpha_2 - \frac{1}{12}\alpha_1 - \frac{1}{20}\alpha_2 \tag{6.66}$$

Taking the first variation of Π_2 with respect to α_1 and α_2, we obtain

$$\delta\Pi_2 = \frac{\partial\Pi_2}{\partial\alpha_1}\delta\alpha_1 = -\frac{3}{10}\alpha_1 - \frac{3}{20}\alpha_2 - \frac{1}{12} = 0$$
$$\delta\Pi_2 = \frac{\partial\Pi_2}{\partial\alpha_2}\delta\alpha_2 = -\frac{3}{20}\alpha_1 - \frac{13}{105}\alpha_2 - \frac{1}{20} = 0 \tag{6.67}$$

Solving for α_1 and α_2, we obtain

$$\alpha_1 = \frac{-71}{369}$$
$$\alpha_2 = \frac{-7}{41} \tag{6.68}$$

The trial function can therefore be written as

$$u_2(x) = x(x - 1)\left(-\frac{71}{369} - \frac{7x}{41}\right) \tag{6.69}$$

Figure 6.4 shows a graphical comparison between the exact solution, Equation (6.59), and the approximate solutions, Equation (6.65) with one parameter and Equation (6.69) with two parameters. The approximate solution with two parameters is more precise.

6.6.4 Example of a Natural Functional

The total potential energy of a structure or solid in equilibrium is defined as the sum of the internal energy (strain energy) and the external energy (the potential energy of the externally applied forces); that is,

$$\Pi = U_i + U_e \tag{6.70}$$

For conservative systems (no dissipation of energy), the loss in external potential energy must be equal to the work, W, done by the external forces on the system:

$$-U_e = W \tag{6.71}$$

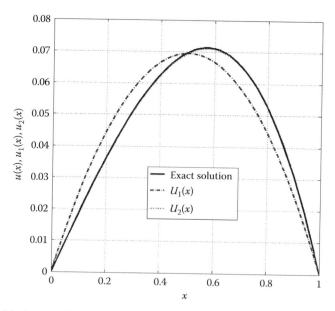

FIGURE 6.4   Graphical comparison of the exact and approximate solutions.

Therefore, the total energy can be written as

$$\Pi = U_i - W \tag{6.72}$$

Furthermore, Π is a functional since it is a function of functions (displacements) and their derivatives (strains). The minimum of potential energy requires that Π becomes minimal for a stable equilibrium configuration of the structure to exist:

$$\delta\Pi = \delta U_i - \delta W = 0 \tag{6.73}$$

The term δU_i represents the variation in strain energy and is given as

$$\delta U_i = \int_V \delta\{\epsilon\}^T\{\sigma\}\, dV \tag{6.74}$$

The term δW represents the work done by the external forces on the system, which comprises the work done by the body forces $\{b\}$, the surface tractions $\{t\}$ and any concentrated forces $\{P_i\}, i = 1, n$:

$$\delta W = \int_V \delta\{U\}^T\{b\}\, dV + \int_A \delta\{U\}^T\{t\}\, dA + \Sigma_i \delta\{U\}^T_{(\{x\}=\{x\}_i)}\{P_i\} \tag{6.75}$$

Finally, the variation of the total energy can be written as

$$\delta\Pi = \int_V \delta\{\epsilon\}^T\{\sigma\}\, dV - \int_V \delta\{U\}^T\{b\}\, dV - \int_A \delta\{U\}^T\{t\}\, dA - \Sigma_i \delta\{U\}^T_{(\{x\}=\{x\}_i)}\{P_i\} = 0 \tag{6.76}$$

Example

Use the Rayleigh Ritz method to derive the stiffness matrix of the beam element shown in Figure 3.3.

The total energy of the beam element shown in Figure 3.3 is given as

$$\Pi = \int_0^L \frac{EI}{2}\left(\frac{d^2 w}{dx^2}\right)^2 dx - \int_0^L w(x)q(x)\,dx - F_1 w_{(x=0)} - F_2 w_{(x=L)}$$

$$- M_1\left(\frac{dw}{dx}\right)_{(x=0)} - M_2\left(\frac{dw}{dx}\right)_{(x=L)} \tag{6.77}$$

Expression (6.77) is also the functional associated with the weak form of the fourth-order differential equation of beam flexure given by Equation (3.5).

It is interesting to note that in Equation (6.77) the highest order derivative is only of order 2. It is recommended therefore to use a trial function that is at least twice derivable and satisfies the essential boundary conditions imposed on $w(x)$ and its first derivative dw/dx. The function $w(x)$ takes on the values of w_1 at $x = 0$, and the value of w_2 at $x = L$, while the derivative (the slope) dw/dx takes the values of θ_1 at $x = 0$, and the value of θ_2 at $x = L$. A suitable trial function for the problem at hand would be

$$\overline{w}(x) = \alpha_1 + \alpha_2 \times x + \alpha_3 \times x^2 + \alpha_4 \times x^3 \tag{6.78}$$

Expression (6.78) can be rewritten in a matrix form as

$$\overline{w}(x) = \begin{bmatrix} 1 & x & x^2 & x^3 \end{bmatrix} \begin{Bmatrix} \alpha_1 \\ \alpha_2 \\ \alpha_3 \\ \alpha_4 \end{Bmatrix} \tag{6.79}$$

Note that $\overline{\theta}(x)$ is obtained by formally deriving $\overline{w}(x)$ with respect to x:

$$\overline{\theta}(x) = \alpha_2 + 2\alpha_3 \times x + 3\alpha_4 \times x^2 \tag{6.80}$$

There are four parameters α_1, α_2, α_3, and α_4, which can be identified using the four nodal values $\{w_1, \theta_1, w_2, \theta_2\}^T$. Evaluating $\overline{w}(x)$ and $\overline{\theta}(x)$ at nodes 1 and 2, where x is respectively equal to 0 and L, results in

$$\overline{w}(x = 0) = \alpha_1 = w_1$$

$$\overline{\theta}(x = 0) = \alpha_2 = \theta_1$$

$$\overline{w}(x = L) = \alpha_1 + \alpha_2 \times L + \alpha_3 \times L^2 + \alpha_4 \times L^3 = w_2$$

$$\overline{\theta}(x = L) = \alpha_2 + 2\alpha_3 \times L + 3\alpha_4 \times L^2 = \theta_2$$

Solving for the parameters α_i and rearranging the results in a matrix form yields

$$\begin{Bmatrix} \alpha_1 \\ \alpha_2 \\ \alpha_3 \\ \alpha_4 \end{Bmatrix} = \begin{bmatrix} 1 & 0 & 0 & 0 \\ 0 & 1 & 0 & 0 \\ -\dfrac{3}{L^2} & -\dfrac{2}{L} & \dfrac{3}{L^2} & -\dfrac{1}{L} \\ \dfrac{2}{L^2} & \dfrac{1}{L} & -\dfrac{2}{L^3} & \dfrac{1}{L^2} \end{bmatrix} \begin{Bmatrix} w_1 \\ \theta_1 \\ w_2 \\ \theta_2 \end{Bmatrix} \tag{6.81}$$

Substituting for α_i in (6.78) results in

$$\overline{w}(x) = \begin{bmatrix} 1 & x & x^2 & x^3 \end{bmatrix} \begin{bmatrix} 1 & 0 & 0 & 0 \\ 0 & 1 & 0 & 0 \\ -\dfrac{3}{L^2} & -\dfrac{2}{L} & \dfrac{3}{L^2} & -\dfrac{1}{L} \\ \dfrac{2}{L^2} & \dfrac{1}{L} & -\dfrac{2}{L^3} & \dfrac{1}{L^2} \end{bmatrix} \begin{Bmatrix} w_1 \\ \theta_1 \\ w_2 \\ \theta_2 \end{Bmatrix} \tag{6.82}$$

Carrying out the matrix multiplication yields

$$\overline{w}(x) = \begin{bmatrix} N_1(x) & N_2(x) & N_3(x) & N_4(x) \end{bmatrix} \begin{Bmatrix} w_1 \\ \theta_1 \\ w_2 \\ \theta_2 \end{Bmatrix} \tag{6.83}$$

with

$$N_1(x) = \left(1 - 3x^2/L^2 + 2x^3/L^3\right) \tag{6.84}$$

$$N_2(x) = \left(x - 2x^2/L + x^3/L^2\right) \tag{6.85}$$

$$N_3(x) = \left(3x^2/L^2 - 2x^3/L^3\right) \tag{6.86}$$

$$N_4(x) = \left(-x^2/L + x^3/L^2\right) \tag{6.87}$$

In a more compact form, Equation (6.83) may be rewritten as

$$\overline{w}(x) = [N]\{d_e\} \tag{6.88}$$

As opposed to expression (6.78), which is a general approximation with general parameters α_i, expression (6.83) is a nodal approximation. Nodal approximations will be treated in more detail in Chapter 7.

From engineering beam theory, the bending moment $M(x)$ is the resultant of the stresses acting above and below the neutral axis and is related to the curvature $\chi(x)$ through Equation (3.3). Substituting for $\overline{w}(x)$ using Equation (6.83), the curvature can be approximated as

$$\overline{\chi} = \frac{d^2\overline{w}(x)}{dx^2} = \begin{bmatrix} \dfrac{d^2 N_1(x)}{dx^2} & \dfrac{d^2 N_2(x)}{dx^2} & \dfrac{d^2 N_3(x)}{dx^2} & \dfrac{d^2 N_4(x)}{dx^2} \end{bmatrix} \begin{Bmatrix} w_1 \\ \theta_1 \\ w_2 \\ \theta_2 \end{Bmatrix} \tag{6.89}$$

which is usually written as

$$\overline{\chi} = \frac{d^2\overline{w}(x)}{dx^2} = [B]\{d_e\} \tag{6.90}$$

The matrix $[B]$ contains the second derivatives of the functions $N_i(x)$:

$$[B] = \left[\left(-\frac{6}{L^2} + \frac{12x}{L^3}\right) \quad \left(-\frac{4}{L} + \frac{6x}{L^2}\right) \quad \left(\frac{6}{L^2} - \frac{12x}{L^3}\right) \quad \left(-\frac{2}{L} + \frac{6x}{L^2}\right) \right] \tag{6.91}$$

Taking the first variation of expression (6.77) and equating it to zero yields

$$\delta\Pi = \int_0^L (\delta\overline{\chi})EI(\overline{\chi}) \, dx - \int_0^L \delta\overline{w}(x)q(x) \, dx - \delta\overline{w}_{(x=0)}F_1 - \delta\overline{w}_{(x=L)}F_2$$

$$- \delta\left(\frac{d\overline{w}_{(x=0)}}{dx}\right)M_1 - \delta\left(\frac{d\overline{w}_{(x=L)}}{dx}\right)M_2 = 0 \tag{6.92}$$

Substituting in (6.92) for $\overline{w}(x)$ and $\overline{\chi}$ using respectively Equations (6.83) and (6.90) yields

$$\int_0^L \delta\{d_e\}^T[B]^T EI[B]\{d_e\} \, dx = \int_0^L \delta\{d_e\}^T[N]^T q(x) \, dx + \delta\{d_e\}^T[N_{(x=0)}]^T F_1 + \delta\{d_e\}^T[N_{(x=L)}]^T F_2$$

$$+ \delta\{d_e\}^T \left(\frac{d[N_{(x=0)}]^T}{dx}\right)M_1 + \delta\{d_e\}^T \left(\frac{d[N_{(x=L)}]^T}{dx}\right)M_2 \tag{6.93}$$

After evaluating the derivatives and taking into account that $\delta\{d_e\}$ is independent of the coordinates x, Equation (6.93) can be rewritten as

$$\left[\int_0^L [B]^T EI[B]\,dx\right]\{d_e\} = \int_0^L \begin{Bmatrix} N_1(x) \\ N_2(x) \\ N_3(x) \\ N_4(x) \end{Bmatrix} q(x)\,dx + \begin{Bmatrix} 1 \\ 0 \\ 0 \\ 0 \end{Bmatrix} F_1 + \begin{Bmatrix} 0 \\ 0 \\ 1 \\ 0 \end{Bmatrix} F_2 + \begin{Bmatrix} 0 \\ 1 \\ 0 \\ 0 \end{Bmatrix} M_1 + \begin{Bmatrix} 0 \\ 0 \\ 0 \\ 1 \end{Bmatrix} M_2 \qquad (6.94)$$

Substituting for $[B]$ using Equation (6.91) and evaluating the integral in the left-hand side of Equation (6.94) yields the stiffness matrix of the beam element as

$$[K_e] = \begin{bmatrix} 12EI/L^3 & 6EI/L^2 & -12EI/L^3 & 6EI/L^2 \\ 6EI/L^2 & 4EI/L & -6EI/L^2 & 2EI/L \\ -12EI/L^3 & -6EI/L^2 & 12EI/L^3 & -6EI/L^2 \\ 6EI/L^2 & 2EI/L & -6EI/L^2 & 4EI/L \end{bmatrix} \qquad (6.95)$$

Note that the matrix $[K_e]$ is exactly the same as the stiffness matrix given in expression (3.30).

Substituting for $N_i(x)$ using Equation (6.84), evaluating the integral on the right-hand side of Equation (6.94), and assuming $q(x) = q$ constant yields

$$\int_0^L \begin{Bmatrix} N_1(x) \\ N_2(x) \\ N_3(x) \\ N_4(x) \end{Bmatrix} q\,dx = \begin{Bmatrix} qL/2 \\ qL^2/12 \\ qL/2 \\ -qL^2/12 \end{Bmatrix} \qquad (6.96)$$

As can be noticed, Equation (6.96) transforms a uniformly distributed load into statically equivalent nodal loads.

7 Finite Element Approximation

7.1 INTRODUCTION

In Chapters 2 through 4, we dealt with skeletal structures whose discretization into an assembly of elements was relatively easy. Whether it is a truss, beam, or portal frame, intuitively the structure can be represented as an assembly of one-dimensional members, for which the exact solutions to the differential equations for each member are well known. However, with a solid continuum such as a reinforced concrete shell or a gravity concrete dam, such an intuitive approach does not exist. For example, a planar surface can be discretized with any element belonging to the triangular or quadrilateral families of elements, a three-dimensional solid can be discretized with any element belonging to the tetrahedron, rectangular prism, or brick families of elements. The choice of the element type is a matter for the analyst. However, the exact solutions to the differential equations governing the behavior of such elements are not known. To establish the matrix relationship between the forces and the nodal displacements at the nodes, the weighted residual methods, in particular the theorem of virtual work or the principle of minimum potential energy, introduced in Chapter 6, will be used. The nodal displacements at nodes are obtained through a nodal interpolation of the field variable (displacement field) over the element. Such an interpolation has already been used to derive the matrix relationship between forces and nodal displacements for the beam element; see the example in Section 6.6.4, where the concepts of general and nodal approximations were introduced briefly. In this chapter, they will be treated in more detail for a variety of finite elements.

7.2 GENERAL AND NODAL APPROXIMATIONS

Given a thick wall surrounding a furnace such as the one represented in Figure 7.1. Five thermocouples are embedded in the wall to measure the temperature variation across. Now, suppose that we want to estimate the temperature at any point in the wall. The easiest approach is to fit the data points to a fourth-order polynomial such as

$$\overline{T(x)} = \alpha_1 + \alpha_2 \times x + \alpha_3 \times x^2 + \alpha_4 \times x^3 + \alpha_5 \times x^4 \tag{7.1}$$

Having five data points, it is relatively easy to identify the five parameters α_i of the polynomial. Equation (7.1) can be rewritten as

$$\overline{T(x)} = \begin{bmatrix} 1 & x & x^2 & x^3 & x^4 \end{bmatrix} \times \begin{Bmatrix} \alpha_1 \\ \alpha_2 \\ \alpha_3 \\ \alpha_4 \\ \alpha_5 \end{Bmatrix} \tag{7.2}$$

The coefficients α_i are called the general parameters of the approximation, and they do not have any physical meaning. However, they could be given one if we make the polynomial approximation

191

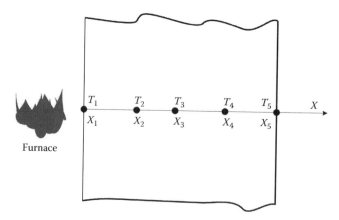

FIGURE 7.1   Thick wall with embedded thermocouples.

$\overline{T(x)}$ coincide with the exact solution at the five data points x_i called nodes. It follows

$$\overline{T(x_1)} = \alpha_1 + \alpha_2 \times x_1 + \alpha_3 \times x_1{}^2 + \alpha_4 \times x_1{}^3 + \alpha_5 \times x_1{}^4 = T_1$$

$$\overline{T(x_2)} = \alpha_1 + \alpha_2 \times x_2 + \alpha_3 \times x_2{}^2 + \alpha_4 \times x_2{}^3 + \alpha_5 \times x_2{}^4 = T_2$$

$$\overline{T(x_3)} = \alpha_1 + \alpha_2 \times x_3 + \alpha_3 \times x_3{}^2 + \alpha_4 \times x_3{}^3 + \alpha_5 \times x_3{}^4 = T_3 \qquad (7.3)$$

$$\overline{T(x_4)} = \alpha_1 + \alpha_2 \times x_4 + \alpha_3 \times x_4{}^2 + \alpha_4 \times x_4{}^3 + \alpha_5 \times x_4{}^4 = T_4$$

$$\overline{T(x_5)} = \alpha_1 + \alpha_2 \times x_5 + \alpha_3 \times x_5{}^2 + \alpha_4 \times x_5{}^3 + \alpha_5 \times x_5{}^4 = T_5$$

which can be rewritten in a matrix form as

$$
\begin{bmatrix}
1 & x_1 & x_1{}^2 & x_1{}^3 & x_1{}^4 \\
1 & x_2 & x_2{}^2 & x_2{}^3 & x_2{}^4 \\
1 & x_3 & x_3{}^2 & x_3{}^3 & x_3{}^4 \\
1 & x_4 & x_4{}^2 & x_1{}^3 & x_4{}^4 \\
1 & x_5 & x_5{}^2 & x_5{}^3 & x_5{}^4
\end{bmatrix}
\times
\begin{Bmatrix}
\alpha_1 \\ \alpha_2 \\ \alpha_3 \\ \alpha_4 \\ \alpha_5
\end{Bmatrix}
=
\begin{Bmatrix}
T_1 \\ T_2 \\ T_3 \\ T_4 \\ T_5
\end{Bmatrix}
\qquad (7.4)
$$

or simply as

$$[A]\{\alpha\} = \{T\} \qquad (7.5)$$

If the matrix $[A]$ is regular, that is, all the points x_i are distinct, then it is possible to write

$$\{\alpha\} = [A]^{-1} \times \{T\} \qquad (7.6)$$

Substituting for $\{\alpha\}$ using Equation (7.6), Equation (7.2) becomes

$$\overline{T(x)} = \begin{bmatrix} 1 & x & x^2 & x^3 & x^4 \end{bmatrix} \times [A]^{-1} \times \{T\} \qquad (7.7)$$

which, after rearranging, becomes

$$\overline{T(x)} = \begin{bmatrix} N_1(x) & N_2(x) & N_3(x) & N_4(x) & N_5(x) \end{bmatrix} \times \begin{Bmatrix} T_1 \\ T_2 \\ T_3 \\ T_4 \\ T_5 \end{Bmatrix} \qquad (7.8)$$

or simply as

$$\overline{T(x)} = [N] \times \{T\} \tag{7.9}$$

Since from Equation (7.3) $\overline{T(x_i)} = T_i$, it follows from (7.9) that

$$N_i(x_j) = \begin{cases} 1 \\ 0 \end{cases} \quad \text{if} \quad \begin{matrix} i = j \\ i \neq j \end{matrix} \tag{7.10}$$

Contrarily to approximation (7.1), which is a general approximation, approximation (7.8) is called a nodal approximation, since the general parameters $\{\alpha_i\}$, as the unknowns, are replaced by the values $\{T\}$ of the function at the nodes. The functions $N_i(x)$ are called the shape functions and they satisfy relation (7.10). In this particular case, they are also polynomial functions of order 4.

7.3 FINITE ELEMENT APPROXIMATION

Now, suppose that we have a large number of data points (say 100), and we would like to construct a nodal approximation over the whole domain for a given function $V(x)$. Such a trial function will have 100 shape functions

$$\overline{V(x)} = \begin{bmatrix} N_1(x) & N_2(x) & \cdots & \cdots & N_{100}(x) \end{bmatrix} \times \begin{Bmatrix} V_1 \\ V_2 \\ \vdots \\ \vdots \\ V_{100} \end{Bmatrix} \tag{7.11}$$

and each one of them will be a polynomial of order 99. Polynomials of high order are known to be very unstable as their derivatives change sign frequently. Not only are they cumbersome but also very difficult to handle particularly from a computational point of view.

To avoid dealing with high-order polynomials, the alternative is to subdivide the domain into subdomains called elements and construct the trial function over each element. This process is called finite element approximation.

Given a function $V(x)$ defined over a domain $\Omega : [x_1, x_n]$ as represented in Figure 7.2. Let us construct an approximation for $V(x)$ based on the principle of finite element approximation. It involves dividing the domain into elements connected by nodes. The details of the discretization are given as follows:

- *Nodes*: $1, 2, \ldots, n-1, n$
- *Nodal coordinates*: $x_1, x_2, \ldots, x_{n-1}, x_n$

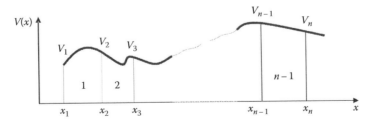

FIGURE 7.2 Finite element discretization.

- *Nodal values*: $V_1, V_2, \ldots, V_{n-1}, V_n$
- *Elements*: $\Omega_1 : [x_1, x_2]$, $\Omega_2 : [x_2, x_3], \ldots, \Omega_{n-1} : [x_{n-1}, x_n]$

First, let us start with the construction of a general approximation $\overline{V(x)}^{(1)}$ for element 1. Since we have only two points x_1 and x_2, we choose an approximation with two parameters α_1 and α_2:

$$\overline{V(x)}^{(1)} = \alpha_1 + \alpha_2 \times x \tag{7.12}$$

Making the trial function coincide with $V(x)$ at x_1 and x_2 yields

$$\overline{V(x_1)}^{(1)} = \alpha_1 + \alpha_2 \times x_1 = V_1$$
$$\overline{V(x_2)}^{(1)} = \alpha_1 + \alpha_2 \times x_2 = V_2 \tag{7.13}$$

Solving for α_1 and α_2 yields

$$\begin{Bmatrix} \alpha_1 \\ \alpha_2 \end{Bmatrix} = \frac{1}{x_2 - x_1} \times \begin{bmatrix} x_2 & x_1 \\ -1 & 1 \end{bmatrix} \times \begin{Bmatrix} V_1 \\ V_2 \end{Bmatrix} \tag{7.14}$$

Substituting for $\{\alpha\}$ in Equation (7.12), the trial function becomes

$$\overline{V(x)}^{(1)} = \frac{1}{x_2 - x_1} \times \begin{bmatrix} 1 & x \end{bmatrix} \times \begin{bmatrix} x_2 & x_1 \\ -1 & 1 \end{bmatrix} \times \begin{Bmatrix} V_1 \\ V_2 \end{Bmatrix} \tag{7.15}$$

Multiplying and rearranging yields

$$\overline{V(x)}^{(1)} = \begin{bmatrix} \dfrac{x_2 - x}{x_2 - x_1} & \dfrac{-x_1 + x}{x_2 - x_1} \end{bmatrix} \times \begin{Bmatrix} V_1 \\ V_2 \end{Bmatrix} = N_1(x)^{(1)} \times V_1 + N_2(x)^{(1)} \times V_2 \tag{7.16}$$

with

$$N_1(x)^{(1)} = \frac{x_2 - x}{x_2 - x_1}$$
$$N_2(x)^{(1)} = \frac{-x_1 + x}{x_2 - x_1} \tag{7.17}$$

In this case, the shape functions $N_1(x)^{(1)}$ and $N_2(x)^{(1)}$ are first-order polynomials in x because only two points were used.

Now, if we are to construct trial functions $\overline{V(x)}^{(2)}$ and $\overline{V(x)}^{(n-1)}$ for elements 2 to $n-1$, the process will be exactly the same; that is,

- *Element 2*: $\overline{V(x)}^{(2)} = N_1(x)^{(2)} \times V_2 + N_2(x)^{(2)} \times V_3$
- $\ldots$
- *Element $n-1$*: $\overline{V(x)}^{(n-1)} = N_1(x)^{(n-1)} \times V_{n-1} + N_2(x)^{(n-1)} \times V_n$

The shape functions $N_1(x)^{(e)}$ and $N_2(x)^{(e)}$ have the same form over each element. The only thing that differentiates them from element to element are the coordinates of the nodes associated to the element. For example, for element 2, $N_1(x)^{(2)}$ is obtained as

$$N_1(x)^{(2)} = \frac{x_3 - x}{x_3 - x_2} \tag{7.18}$$

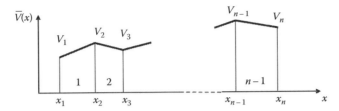

FIGURE 7.3 Finite element approximation.

whereas for element $n - 1$ it is given as

$$N_1(x)^{(n-1)} = \frac{x_n - x}{x_n - x_{n-1}} \tag{7.19}$$

This property is very interesting since it is repetitive, therefore making the programming easy on a digital computer.

Finally, the approximation over the global domain is obtained by adding the approximations over the elements $\overline{V}(x) = \sum_{e=1}^{n-1} \overline{V}^e(x)$, as shown in Figure 7.3. Notice that the approximation is linear over each element. It is also continuous at the nodes, that connect the elements.

The finite element nodal approximation can be extended to functions with many variables. However, the geometrical definition of the elements and the construction of the shape functions become more problematic as we will see in the following sections.

7.4 BASIC PRINCIPLES FOR THE CONSTRUCTION OF TRIAL FUNCTIONS

7.4.1 COMPATIBILITY PRINCIPLE

The construction of the trial solution over a finite element must essentially satisfy the requirements of the problem to solve and the geometry of the element. To illustrate this statement, consider the bar and the beam problems shown respectively in Figure 7.4a and b. Under the effect of the applied force P, every cross section A of the bar is subject to a constant stress $\sigma = P/A$. As a result, the bar is under a constant strain $\epsilon = \sigma/E$, where E represents the elastic modulus of the material. In a one-dimensional context, the normal strain ϵ is actually given as a direct derivative of the displacement $u(x)$; that is, $\epsilon = du(x)/dx$. Since the strain is constant all over the bar, it follows that

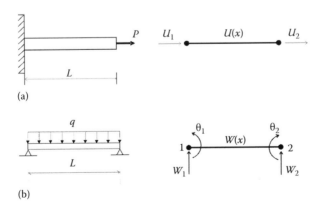

FIGURE 7.4 Geometrical illustration of the compatibility principle: (a) bar element, (b) beam element.

the displacement $u(x)$ is a linear function of x. As a result, it is possible to construct a trial function $\bar{u}(x)$ for the displacement using a linear polynomial

$$\bar{u}(x) = \alpha_1 + \alpha_2 x \tag{7.20}$$

The parameters α_1 and α_2 are identified using the two end nodal values U_1 and U_2. The bar problem is classified as a C^0 problem. The trial solution must be continuous and its derivative must exist.

Now let us consider the beam problem. Under the applied uniformly distributed loading, every cross section of the beam is subject to a vertical displacement $w(x)$ and a rotation $\theta(x)$. From the engineering beam theory, the rotation $\theta(x)$ is obtained as the first derivative of the deflection $w(x)$; that is, $\theta(x) = dw(x)/dx$. The slope $\theta(x)$ must be continuous, otherwise the beam would develop "kinks" in its deflected shape. Therefore, if we are about to construct a trial function $\bar{w}(x)$ for the deflection, then both the trial function and its first derivative must be continuous. The second derivative, which represents the curvature of the beam, must exist. A suitable trial function that satisfies these requirements would be

$$\bar{w}(x) = \alpha_1 + \alpha_2 \times x + \alpha_3 \times x^2 + \alpha_4 \times x^3 \tag{7.21}$$

The four parameters α_1, α_2, α_3, and α_4 can be identified using the two end nodal values for the deflection, w_1, w_2, and the two end values for the slope, θ_1 and θ_2. The beam problem is classified as a C^1 problem. The trial solution and its first derivative must be continuous, the second derivative must exist.

In general, the compatibility principle can be formulated as follows:

- For a class C^0 problem (continuity C^0), the trial solution must be continuous across the boundary of the elements but not necessarily its derivatives.
- For a class C^1 problem (continuity C^1), both the trial solution and its first-order derivatives must be continuous across the boundary of the elements but not necessarily its second-order derivatives.
- For a class C^n problem (continuity C^n), the trial solution and its $(n-1)$th order derivatives must be continuous across the boundary of the elements but not necessarily its nth order derivatives.

7.4.2 COMPLETENESS PRINCIPLE

Again, consider the bar problem in Figure 7.4a. If the applied force P is different from zero, then the displacement $u(x)$ has a finite value different from zero at any point x belonging to the bar except at $x = 0$, where a displacement equal to zero is imposed (boundary condition). If we choose to discretize the bar with a linear two-nodded element, then the adopted trial function given in Equation (7.20) will make a suitable choice since if the size of the elements shrinks to zero, that is, $\lim_{x \to 0} u(x) = \alpha_1$, which is a constant representing the actual value of the displacement at that point. However, if the trial function did not contain a constant term, $\lim_{x \to 0} u(x)$ will be equal to zero, which actually does not represent the real case. Furthermore, the constant term is necessary for the trial function to be able to represent a rigid body motion. In this case, all points must have the same displacement $u(x) = \alpha$. In addition, we have $du(x)/dx = \alpha_2$, which represents the real case of the bar with a constant deformation. This leads to the definition of the completeness principle, which can be stated as follows. When the size of the element shrinks to zero, the trial function must be able to represent:

- For a class C^0 problem (continuity C^0), a constant value of the exact function as well as constant values of its first-order derivatives.

- For a class C^1 problem (continuity C^1), a constant value of the exact function as well as constant values of its first- and second-order derivatives.
- For a class C^n problem (continuity C^n), a constant value of the exact function as well as constant values of its derivatives up to the nth order.

These conditions, as stated by the principles of compatibility and completeness, are sufficient to ensure that the finite element solution converges to the exact solution. Luckily, nowadays we do not need to observe these principles every time we solve a problem with the finite element method. All the common elements that are in use in practice have been developed and checked according to these principles, and more... Both their geometrical and analytical formulations are supplied in element libraries in most finite element analysis software. However, it is never enough to reiterate that solutions obtained with the finite element method are only approximations to the exact solution. Therefore, it is worthwhile to understand these principles in order to assess the accuracy or make a diagnosis of a finite element model.

7.5 TWO-DIMENSIONAL FINITE ELEMENT APPROXIMATION

7.5.1 PLANE LINEAR TRIANGULAR ELEMENT FOR C^0 PROBLEMS

7.5.1.1 Shape Functions

Given a class C^0 problem defined over a two-dimensional domain Ω. The unknown function for which we propose to construct an approximation will be referred to as $F(x, y)$. The function must be continuous all over the domain but not necessarily its derivatives. However they should exist. Given the complexity of the domain, such as the one represented in Figure 7.5, it is not possible to construct the approximation over the whole domain. We will therefore proceed by constructing the approximation over an element of simple geometry such as a triangle. In virtue of the principles of compatibility and completeness, the trial function $\overline{U}(x, y)$ must have a constant term and constant first-order derivatives in x and y. Therefore, we choose a trial function of the form

$$\overline{U}(x, y) = a + bx + cy \tag{7.22}$$

Notice that the trial function is linear and has three terms only. This is dictated by the geometry of the element; it has three nodes, therefore three nodal values F_1, F_2, and F_3, and its sides are linear.

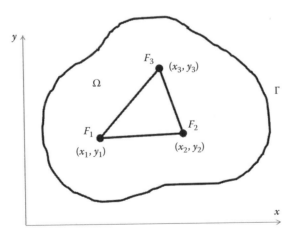

FIGURE 7.5 Linear triangle.

The trial function, expression (7.22), can be rewritten in the form

$$\overline{U}(x, y) = \begin{bmatrix} 1 & x & y \end{bmatrix} \begin{Bmatrix} a \\ b \\ c \end{Bmatrix} \tag{7.23}$$

At the nodes $1(x_1, y_1)$, $2(x_2, y_2)$, and $3(x_3, y_3)$, we make the trial function $\overline{U}(x, y)$ coincide with the unknown function $F(x, y)$, which leads to

$$\overline{U}(x_1, y_1) = a + bx_1 + cy_1 = F_1 \tag{7.24}$$

$$\overline{U}(x_2, y_2) = a + bx_2 + cy_2 = F_2 \tag{7.25}$$

$$\overline{U}(x_3, y_3) = a + bx_3 + cy_3 = F_3 \tag{7.26}$$

which may be again rewritten in matrix form as

$$\begin{bmatrix} 1 & x_1 & y_1 \\ 1 & x_2 & y_2 \\ 1 & x_3 & y_3 \end{bmatrix} \begin{Bmatrix} a \\ b \\ c \end{Bmatrix} = \begin{Bmatrix} F_1 \\ F_2 \\ F_3 \end{Bmatrix} \tag{7.27}$$

If the matrix of the system (7.27) is not singular, that is, the three nodes of the triangle are distinct and not aligned, the system can be solved for the constants a, b, and c, which are the general parameters of the approximation

$$\begin{Bmatrix} a \\ b \\ c \end{Bmatrix} = \begin{bmatrix} 1 & x_1 & y_1 \\ 1 & x_2 & y_2 \\ 1 & x_3 & y_3 \end{bmatrix}^{-1} \begin{Bmatrix} F_1 \\ F_2 \\ F_3 \end{Bmatrix} \tag{7.28}$$

Substituting for a, b, and c in Equation (7.23) yields

$$\overline{U}(x, y) = \begin{bmatrix} 1 & x & y \end{bmatrix} \begin{bmatrix} 1 & x_1 & y_1 \\ 1 & x_2 & y_2 \\ 1 & x_3 & y_3 \end{bmatrix}^{-1} \begin{Bmatrix} F_1 \\ F_2 \\ F_3 \end{Bmatrix} \tag{7.29}$$

which may be rewritten in the form

$$\overline{U}(x, y) = \begin{bmatrix} N_1(x, y) & N_2(x, y) & N_3(x, y) \end{bmatrix} \begin{Bmatrix} F_1 \\ F_2 \\ F_3 \end{Bmatrix} \tag{7.30}$$

Expression (7.30) is a nodal approximation as opposed to (7.22), which is a general approximation. The shape functions $N_i(x, y), i = 1, 2, 3$ are obtained as

$$N_1(x, y) = \frac{1}{2A}((y_3 - y_2)(x_2 - x) - (x_3 - x_2)(y_2 - y)) \tag{7.31}$$

$$N_2(x, y) = \frac{1}{2A}((y_1 - y_3)(x_3 - x) - (x_1 - x_3)(y_3 - y)) \tag{7.32}$$

$$N_3(x, y) = \frac{1}{2A}((y_2 - y_1)(x_1 - x) - (x_2 - x_1)(y_1 - y)) \tag{7.33}$$

with

$$A = \frac{1}{2} det \begin{bmatrix} 1 & x_1 & y_1 \\ 1 & x_2 & y_2 \\ 1 & x_3 & y_3 \end{bmatrix} \qquad (7.34)$$

The shape functions may also be rewritten as

$$N_1(x, y) = m_{11} + m_{12}x + m_{13}y$$
$$N_2(x, y) = m_{21} + m_{22}x + m_{23}y \qquad (7.35)$$
$$N_3(x, y) = m_{31} + m_{32}x + m_{33}y$$

and in turn

$$m_{11} = \frac{x_2 y_3 - x_3 y_2}{2A} \qquad m_{12} = \frac{y_2 - y_3}{2A} \qquad m_{13} = \frac{x_3 - x_2}{2A}$$

$$m_{21} = \frac{x_3 y_1 - x_1 y_3}{2A} \qquad m_{22} = \frac{y_3 - y_1}{2A} \qquad m_{23} = \frac{x_1 - x_3}{2A} \qquad (7.36)$$

$$m_{31} = \frac{x_1 y_2 - x_2 y_1}{2A} \qquad m_{32} = \frac{y_1 - y_2}{2A} \qquad m_{33} = \frac{x_2 - x_1}{2A}$$

The shape functions $N_i(x, y)$ satisfy the following conditions:

$$N_i(x_j, y_j) = \begin{cases} 1 \\ 0 \end{cases} \quad if \quad \begin{matrix} i = j \\ i \neq j \end{matrix} \qquad (7.37)$$

At node 1	$N_1(x_1, y_1) = 1$	$N_2(x_1, y_1) = 0$	$N_3(x_1, y_1) = 0$
At node 2	$N_1(x_2, y_2) = 0$	$N_2(x_2, y_2) = 1$	$N_3(x_2, y_2) = 0$
At node 3	$N_1(x_3, y_3) = 0$	$N_2(x_3, y_3) = 0$	$N_3(x_3, y_3) = 1$

Furthermore, if the shape functions are evaluated at any point (x, y) belonging to the triangle, they satisfy the relation

$$\sum_{i=1}^{3} N_i(x, y) = 1 \qquad (7.38)$$

7.5.1.2 Reference Element

A different way of constructing the trial function $\overline{U}(x, y)$, Equation (7.22), is to construct it over a reference element, then transform it to the parent element using a geometrical transformation τ as represented in Figure 7.6. The geometrical transformation τ represented in Figure 7.6 defines the coordinates (x, y) of each point of the parent element from the coordinates (ξ, η) of the corresponding point of the reference element

$$\tau : \quad (\xi, \eta) \longmapsto \quad (x, y) = \tau(\xi, \eta) \qquad (7.39)$$

The transformation is chosen in such a way that

- Each point of the parent element corresponds to one and only one point of the reference element, and inversely

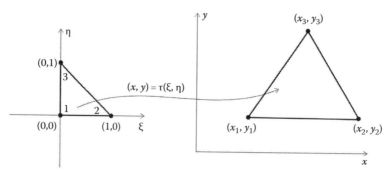

FIGURE 7.6   Geometrical transformation for a triangular element.

- The nodes of the parent element correspond to the nodes of the reference element and inversely
- Each portion of the boundary of the parent element defined by the nodes attached to it correspond to the portion of the boundary of the reference element defined by the corresponding nodes attached to it

To define the geometrical transformation, we assume that the coordinates (x, y) of an arbitrary point of the parent element are the unknown functions defined over the domain represented by the reference element in its local coordinate system (ξ, η). Notice that both the variables x and y belong to the C^0 class of functions since they are continuous and their first derivatives are constant equal to 1. Therefore, we start by constructing a general approximation for x in terms of ξ and η

$$x = \alpha_1 + \alpha_2\xi + \alpha_3\eta \tag{7.40}$$

or in a matrix form as

$$x = \begin{bmatrix} 1 & \xi & \eta \end{bmatrix} \begin{bmatrix} \alpha_1 \\ \alpha_2 \\ \alpha_3 \end{bmatrix} \tag{7.41}$$

As it is now familiar, we will transform the general approximation, Equation (7.40), to a nodal approximation by using the nodal values x_1, x_2, and x_3 respectively at nodes 1, 2, and 3. Notice also that the couple (ξ, η) takes on the values of $(0, 0)$, $(1, 0)$, and $(0, 1)$ respectively at nodes 1, 2, and 3. It follows

$$x_1 = \alpha_1$$
$$x_2 = \alpha_1 + \alpha_2 \tag{7.42}$$
$$x_3 = \alpha_1 + \alpha_3$$

which, when rewritten in a matrix form, yields

$$\begin{Bmatrix} x_1 \\ x_2 \\ x_3 \end{Bmatrix} = \begin{bmatrix} 1 & 0 & 0 \\ 1 & 1 & 0 \\ 1 & 0 & 1 \end{bmatrix} \begin{Bmatrix} \alpha_1 \\ \alpha_2 \\ \alpha_3 \end{Bmatrix} \tag{7.43}$$

or in a more compact form as

$$\{X\} = [A]\{\alpha\} \tag{7.44}$$

The parameters α_i can be easily obtained by solving the system (7.44). The inverse of the matrix $[A]$ is obtained as

$$[A]^{-1} = \begin{bmatrix} 1 & 0 & 0 \\ -1 & 1 & 0 \\ -1 & 0 & 1 \end{bmatrix} \tag{7.45}$$

and the parameters α_i as

$$\begin{Bmatrix} \alpha_1 \\ \alpha_2 \\ \alpha_3 \end{Bmatrix} = \begin{bmatrix} 1 & 0 & 0 \\ -1 & 1 & 0 \\ -1 & 0 & 1 \end{bmatrix} \begin{Bmatrix} x_1 \\ x_2 \\ x_3 \end{Bmatrix} \tag{7.46}$$

Substituting for the parameters α_i in Equation (7.41) yields

$$x(\xi, \eta) = \begin{bmatrix} 1 & \xi & \eta \end{bmatrix} \begin{bmatrix} 1 & 0 & 0 \\ -1 & 1 & 0 \\ -1 & 0 & 1 \end{bmatrix} \begin{Bmatrix} x_1 \\ x_2 \\ x_3 \end{Bmatrix} \tag{7.47}$$

Expanding and rearranging Equation (7.47) yields

$$x(\xi, \eta) = \tau_1(\xi, \eta) x_1 + \tau_2(\xi, \eta) x_2 + \tau_3(\xi, \eta) x_3 \tag{7.48}$$

with

$$\begin{aligned} \tau_1(\xi, \eta) &= 1 - \xi - \eta \\ \tau_2(\xi, \eta) &= \xi \\ \tau_3(\xi, \eta) &= \eta \end{aligned} \tag{7.49}$$

Following exactly the same process for the variable y yields

$$y(\xi, \eta) = \tau_1(\xi, \eta) y_1 + \tau_2(\xi, \eta) y_2 + \tau_3(\xi, \eta) y_3 \tag{7.50}$$

Expressions (7.48) and (7.50) represent well and truly a linear geometrical transformation. This can be easily checked. The x coordinate of the midpoint between node 1 and node 2 of the parent element is given as $x = (x_1 + x_2)/2$. The (ξ, η) coordinates of the corresponding point on the reference element are given as $(1/2, 0)$. Substituting these values in expression (7.49) and then in expressions (7.48) yields

$$x = (1 - 0.5 - 0)x_1 + 0.5x_2 + 0x_3 = \frac{(x_1 + x_2)}{2}$$

The Jacobian of the transformation is given by

$$[J] = \begin{bmatrix} \dfrac{\partial x}{\partial \xi} & \dfrac{\partial y}{\partial \xi} \\ \dfrac{\partial x}{\partial \eta} & \dfrac{\partial y}{\partial \eta} \end{bmatrix} = \begin{bmatrix} \sum_{i=1}^{3} \dfrac{\partial \tau_i}{\partial \xi} x_i & \sum_{i=1}^{3} \dfrac{\partial \tau_i}{\partial \xi} y_i \\ \sum_{i=1}^{3} \dfrac{\partial \tau_i}{\partial \eta} x_i & \sum_{i=1}^{3} \dfrac{\partial \tau_i}{\partial \eta} y_i \end{bmatrix} \tag{7.51}$$

After deriving and rearranging, the Jacobian is written in the form of a product of two matrices:

$$[J] = \begin{bmatrix} -1 & 1 & 0 \\ -1 & 0 & 1 \end{bmatrix} \begin{bmatrix} x_1 & y_1 \\ x_2 & y_2 \\ x_3 & y_3 \end{bmatrix} \tag{7.52}$$

Since the geometrical transformation is well defined, we will construct the trial function $\overline{U}(x, y)$ for an unknown function $F(x, y)$ over the reference element. The unknown function, defined over the parent element, is of class C^0 with nodal values F_1, F_2, and F_3. Since it is of the same class as the coordinates x and y, we will reuse the same trial function; that is,

$$\overline{U}(\xi, \eta) = \alpha_1 + \alpha_2 \xi + \alpha_3 \eta \tag{7.53}$$

Following exactly the same procedure as previously, and replacing x_1, x_2, and x_3 respectively with the nodal values F_1, F_2, and F_3, we end up with

$$\overline{U}(\xi, \eta) = N_1(\xi, \eta)F_1 + N_2(\xi, \eta)F_2 + N_3(\xi, \eta)F_3 \tag{7.54}$$

with

$$
\begin{aligned}
N_1(\xi, \eta) &= 1 - \xi - \eta \\
N_2(\xi, \eta) &= \xi \\
N_3(\xi, \eta) &= \eta
\end{aligned}
\tag{7.55}
$$

Remark: The shape functions $N_i(\xi, \eta)$ are exactly the same as the functions $\tau_i(\xi, \eta)$ of the geometrical transformation. This is due to the fact that the function $\overline{U}(\xi, \eta)$ is of the same class as the coordinates x and y, and most importantly the geometrical nodes (the nodes used to define the geometry of the element) are the same as the interpolation nodes (the nodes used to define the nodal values of the unknown function). Such an element is called an isoparametric (same parameters) element since it uses the same nodes to define both the geometry and interpolate the function.

7.5.1.3 Area Coordinates

Let us consider an arbitrary point O of the triangular element shown in Figure 7.7. The area coordinates L_1, L_2, and L_3 are defined as

$$L_1 = \frac{\text{Area}_{O23}}{\text{Area}_{123}} \tag{7.56}$$

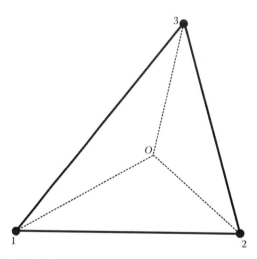

FIGURE 7.7   Three-node triangular element with an arbitrary point O.

$$L_2 = \frac{\text{Area}_{O13}}{\text{Area}_{123}} \tag{7.57}$$

$$L_3 = \frac{\text{Area}_{O12}}{\text{Area}_{123}} \tag{7.58}$$

From these definitions, it follows that

$$L_1 + L_2 + L_3 = 1 \tag{7.59}$$

It is also obvious that

- When point O coincides with node 1, $L_1 = 1$, $L_2 = 0$, and $L_3 = 0$.
- When point O coincides with node 2, $L_1 = 0$, $L_2 = 1$, and $L_3 = 0$.
- When point O coincides with node 3, $L_1 = 0$, $L_2 = 0$, and $L_3 = 1$.

In addition, moving point O in any direction will result in a linear variation of the area coordinates L_1, L_2, and L_3 in terms of x and y. Therefore, it should be clear to the reader that the area coordinates L_1, L_2, and L_3 are indeed the same as the shape functions N_1, N_2, and N_3 given in Equation (7.31); that is,

$$
\begin{aligned}
L_1 &= N_1(x, y) \\
L_2 &= N_2(x, y) \\
L_3 &= N_3(x, y)
\end{aligned}
\tag{7.60}
$$

In the case of a reference triangular element as shown in Figure 7.8, the area coordinates are expressed in terms of the coordinates (ξ, η) as follows:

$$
\begin{aligned}
L_1 &= N_1(\xi, \eta) = 1 - \xi - \eta \\
L_2 &= N_2(\xi, \eta) = \xi \\
L_3 &= N_3(\xi, \eta) = \eta
\end{aligned}
\tag{7.61}
$$

7.5.2 Linear Quadrilateral Element for C^0 Problems

7.5.2.1 Geometrical Transformation

In the quadrilateral family of elements, except for the square or the rectangle, it is impossible to construct the shape functions directly in terms of x and y as we did for the triangle. The only way to construct these functions is to use a reference element, which is a square of side 2 (units) as represented in Figure 7.9. To define the geometrical transformation, we will assume that the coordinates (x, y) of an arbitrary point of the parent element are the unknown functions defined over the domain represented by the reference element in its local coordinate system (ξ, η). Notice that both the variables x and y belong to the C^0 class of functions since they are continuous and their first derivatives are constant equal to 1. Therefore, we start by constructing a general approximation for x in terms of ξ and η

$$x = \alpha_1 + \alpha_2 \xi + \alpha_3 \eta + \alpha_4 \xi \eta \tag{7.62}$$

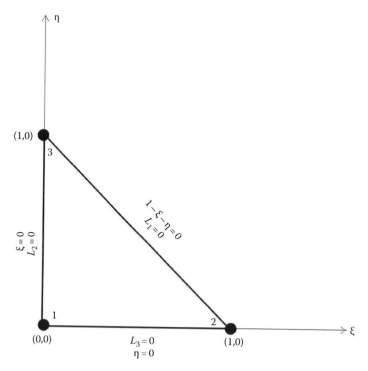

FIGURE 7.8   Three-node triangular reference element.

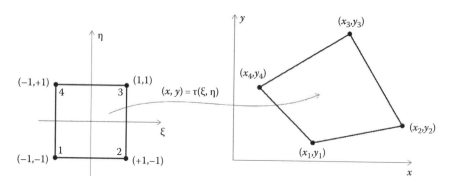

FIGURE 7.9   Geometrical transformation.

or in a matrix form as

$$x = \begin{bmatrix} 1 & \xi & \eta & \xi\eta \end{bmatrix} \begin{bmatrix} \alpha_1 \\ \alpha_2 \\ \alpha_3 \\ \alpha_4 \end{bmatrix} \tag{7.63}$$

Then, we will transform the general approximation, Equation (7.62), to a nodal approximation by using the nodal values x_1, x_2, x_3, and x_4 respectively at nodes 1, 2, 3, and 4. Notice also that the couple (ξ, η) takes on the values of $(-1, -1)$, $(+1, -1)$, $(1, 1)$, and $(-1, +1)$ respectively at nodes 1, 2, 3, and 4. It follows

$$x_1 = \alpha_1 - \alpha_2 - \alpha_3 + \alpha_4$$
$$x_2 = \alpha_1 + \alpha_2 - \alpha_3 - \alpha_4$$
$$x_3 = \alpha_1 + \alpha_2 + \alpha_3 + \alpha_4$$
$$x_4 = \alpha_1 - \alpha_2 + \alpha_3 - \alpha_4$$

(7.64)

which, when rewritten in a matrix form, yields

$$\begin{Bmatrix} x_1 \\ x_2 \\ x_3 \\ x_4 \end{Bmatrix} = \begin{bmatrix} 1 & -1 & -1 & 1 \\ 1 & 1 & -1 & -1 \\ 1 & 1 & 1 & 1 \\ 1 & -1 & 1 & -1 \end{bmatrix} \begin{Bmatrix} \alpha_1 \\ \alpha_2 \\ \alpha_3 \\ \alpha_4 \end{Bmatrix}$$

(7.65)

or in a more compact form as

$$\{X\} = [A]\{\alpha\}$$

(7.66)

The parameters α_i can be obtained easily by solving the system (7.65). It can be noticed that the columns of the matrix $[A]$ are actually orthogonal vectors of norm 4. Hence, the inverse of the matrix $[A]$ is obtained as

$$[A]^{-1} = \frac{1}{4}[A]^T = \frac{1}{4} \begin{bmatrix} 1 & 1 & 1 & 1 \\ -1 & 1 & 1 & -1 \\ -1 & -1 & 1 & 1 \\ 1 & -1 & 1 & -1 \end{bmatrix}$$

(7.67)

and the parameters α_i as

$$\begin{Bmatrix} \alpha_1 \\ \alpha_2 \\ \alpha_3 \\ \alpha_4 \end{Bmatrix} = \frac{1}{4} \begin{bmatrix} 1 & 1 & 1 & 1 \\ -1 & 1 & 1 & -1 \\ -1 & -1 & 1 & 1 \\ 1 & -1 & 1 & -1 \end{bmatrix} \begin{Bmatrix} x_1 \\ x_2 \\ x_3 \\ x_4 \end{Bmatrix}$$

(7.68)

Substituting for the parameters α_i in Equation (7.63) yields

$$x(\xi, \eta) = \begin{bmatrix} 1 & \xi & \eta & \xi\eta \end{bmatrix} \frac{1}{4} \begin{bmatrix} 1 & 1 & 1 & 1 \\ -1 & 1 & 1 & -1 \\ -1 & -1 & 1 & 1 \\ 1 & -1 & 1 & -1 \end{bmatrix} \begin{Bmatrix} x_1 \\ x_2 \\ x_3 \\ x_4 \end{Bmatrix}$$

(7.69)

Expanding and rearranging Equation (7.69) leads to

$$x(\xi, \eta) = \tau_1(\xi, \eta)x_1 + \tau_2(\xi, \eta)x_2 + \tau_3(\xi, \eta)x_3 + \tau_4(\xi, \eta)x_4$$

(7.70)

with

$$\tau_1(\xi, \eta) = 0.25(1 - \xi - \eta + \xi\eta)$$
$$\tau_2(\xi, \eta) = 0.25(1 + \xi - \eta - \xi\eta)$$
$$\tau_3(\xi, \eta) = 0.25(1 + \xi + \eta + \xi\eta)$$
$$\tau_4(\xi, \eta) = 0.25(1 - \xi + \eta - \xi\eta)$$

(7.71)

Following exactly the same process for the variable y, we obtain

$$y(\xi, \eta) = \tau_1(\xi, \eta)y_1 + \tau_2(\xi, \eta)y_2 + \tau_3(\xi, \eta)y_3 + \tau_4(\xi, \eta)y_4 \tag{7.72}$$

Expressions (7.70) and (7.72) represent well and truly a linear geometrical transformation. This can be easily checked as follows. The center of the reference square is given by $(\xi, \eta) = (0, 0)$. Substituting these values in expression (7.71) and then in expressions (7.70) and (7.72) yields

$$x = \frac{1}{4}(x_1 + x_2 + x_3 + x_4)$$

$$y = \frac{1}{4}(y_1 + y_2 + y_3 + y_4)$$

which are the coordinates of the center of the parent element in the (x, y) coordinate system. The Jacobian of the transformation is given by

$$[J] = \begin{bmatrix} \dfrac{\partial x}{\partial \xi} & \dfrac{\partial y}{\partial \xi} \\ \dfrac{\partial x}{\partial \eta} & \dfrac{\partial y}{\partial \eta} \end{bmatrix} = \begin{bmatrix} \sum_{i=1}^{4} \dfrac{\partial \tau_i}{\partial \xi} x_i & \sum_{i=1}^{4} \dfrac{\partial \tau_i}{\partial \xi} y_i \\ \sum_{i=1}^{4} \dfrac{\partial \tau_i}{\partial \eta} x_i & \sum_{i=1}^{4} \dfrac{\partial \tau_i}{\partial \eta} y_i \end{bmatrix} \tag{7.73}$$

After deriving and rearranging, the Jacobian is written in the form of a product of two matrices

$$[J] = \frac{1}{4} \begin{bmatrix} -(1-\eta) & (1-\eta) & (1+\eta) & -(1+\eta) \\ -(1-\xi) & -(1+\xi) & (1+\xi) & (1-\xi) \end{bmatrix} \begin{bmatrix} x_1 & y_1 \\ x_2 & y_2 \\ x_3 & y_3 \\ x_4 & y_4 \end{bmatrix} \tag{7.74}$$

7.5.2.2 Construction of a Trial Function over a Linear Quadrilateral Element

Now, let us construct a trial function $\overline{U}(x, y)$ for an unknown function $F(x, y)$ of class C^0 with nodal values F_1, F_2, F_3, and F_4 defined over the parent element. Since the geometrical transformation is well defined, we will construct the trial function over the reference element. The function $F(x, y)$ is of the same class as the coordinates x and y, we will use the same trial function; that is,

$$\overline{U}(\xi, \eta) = \alpha_1 + \alpha_2 \xi + \alpha_3 \eta + \alpha_4 \xi \eta \tag{7.75}$$

Following exactly the same procedure as previously described, we end up with

$$\overline{U}(\xi, \eta) = \begin{bmatrix} 1 & \xi & \eta & \xi\eta \end{bmatrix} \frac{1}{4} \begin{bmatrix} 1 & 1 & 1 & 1 \\ -1 & 1 & 1 & -1 \\ -1 & -1 & 1 & 1 \\ 1 & -1 & 1 & -1 \end{bmatrix} \begin{Bmatrix} F_1 \\ F_2 \\ F_3 \\ F_4 \end{Bmatrix} \tag{7.76}$$

which, after expanding and rearranging, becomes

$$\overline{U}(\xi, \eta) = N_1(\xi, \eta)F_1 + N_2(\xi, \eta)F_2 + N_3(\xi, \eta)F_3 + N_4(\xi, \eta)F_4 \tag{7.77}$$

with

$$
\begin{aligned}
N_1(\xi, \eta) &= 0.25(1 - \xi - \eta + \xi\eta) \\
N_2(\xi, \eta) &= 0.25(1 + \xi - \eta - \xi\eta) \\
N_3(\xi, \eta) &= 0.25(1 + \xi + \eta + \xi\eta) \\
N_4(\xi, \eta) &= 0.25(1 - \xi + \eta - \xi\eta)
\end{aligned}
\tag{7.78}
$$

The bilinear quadrilateral element is also isoparametric since the shape functions are the same as the functions $\tau_i(\xi, \eta)$ of the geometrical transformation.

7.6 SHAPE FUNCTIONS OF SOME CLASSICAL ELEMENTS FOR C⁰ PROBLEMS

7.6.1 One-Dimensional Elements

7.6.1.1 Two-Nodded Linear Element (Figure 7.10)

$$\begin{Bmatrix} N_1(\xi) \\ N_2(\xi) \end{Bmatrix} = \begin{Bmatrix} \dfrac{1}{2}(1 - \xi) \\ \dfrac{1}{2}(1 + \xi) \end{Bmatrix} \tag{7.79}$$

7.6.1.2 Three-Nodded Quadratic Element

$$\begin{Bmatrix} N_1(\xi) \\ N_2(\xi) \\ N_3(\xi) \end{Bmatrix} = \begin{Bmatrix} \dfrac{1}{2}\xi(1 - \xi) \\ (1 - \xi^2) \\ \dfrac{1}{2}\xi(1 + \xi) \end{Bmatrix} \tag{7.80}$$

7.6.2 Two-Dimensional Elements

7.6.2.1 Four-Nodded Bilinear Quadrilateral (Figure 7.11)

$$\begin{Bmatrix} N_1(\xi, \eta) \\ N_2(\xi, \eta) \\ N_3(\xi, \eta) \\ N_4(\xi, \eta) \end{Bmatrix} = \begin{Bmatrix} 0.25(1 - \xi - \eta + \xi\eta) \\ 0.25(1 + \xi - \eta - \xi\eta) \\ 0.25(1 + \xi + \eta + \xi\eta) \\ 0.25(1 - \xi + \eta - \xi\eta) \end{Bmatrix} \tag{7.81}$$

FIGURE 7.10 One-dimensional elements.

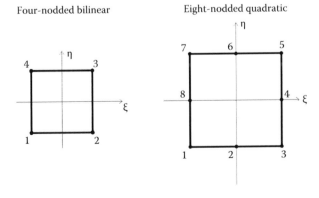

FIGURE 7.11 Two-dimensional quadrilateral elements.

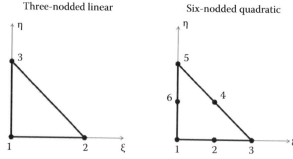

FIGURE 7.12 Two-dimensional triangular elements.

7.6.2.2 Eight-Nodded Quadratic Quadrilateral

$$
\begin{Bmatrix}
N_1(\xi,\eta) \\
N_2(\xi,\eta) \\
N_3(\xi,\eta) \\
N_4(\xi,\eta) \\
N_5(\xi,\eta) \\
N_6(\xi,\eta) \\
N_7(\xi,\eta) \\
N_8(\xi,\eta)
\end{Bmatrix}
=
\begin{Bmatrix}
-0.25(1-\xi)(1-\eta)(1+\xi+\eta) \\
0.50(1-\xi^2)(1-\eta) \\
-0.25(1+\xi)(1-\eta)(1-\xi+\eta) \\
0.50(1+\xi)(1-\eta^2) \\
-0.25(1+\xi)(1+\eta)(1-\xi-\eta) \\
0.50(1-\xi^2)(1+\eta) \\
-0.25(1-\xi)(1+\eta)(1+\xi-\eta) \\
0.50(1-\xi)(1-\eta^2)
\end{Bmatrix}
\tag{7.82}
$$

7.6.2.3 Three-Nodded Linear Triangle (Figure 7.12)

$$
\begin{Bmatrix}
N_1(\xi,\eta) \\
N_2(\xi,\eta) \\
N_3(\xi,\eta)
\end{Bmatrix}
=
\begin{Bmatrix}
1-\xi-\eta \\
\xi \\
\eta
\end{Bmatrix}
\tag{7.83}
$$

7.6.2.4 Six-Nodded Quadratic Triangle

$$
\begin{Bmatrix}
N_1(\xi,\eta) \\
N_2(\xi,\eta) \\
N_3(\xi,\eta) \\
N_4(\xi,\eta) \\
N_5(\xi,\eta) \\
N_6(\xi,\eta)
\end{Bmatrix}
=
\begin{Bmatrix}
-(1-\xi-\eta)(1-2(1-\xi-\eta)) \\
4\xi(1-\xi-\eta) \\
-\xi(1-2\xi) \\
4\xi\eta \\
-\eta(1-2\eta) \\
4\eta(1-\xi-\eta)
\end{Bmatrix}
\tag{7.84}
$$

7.6.3 Three-Dimensional Elements

7.6.3.1 Four-Nodded Linear Tetrahedra

$$
\begin{Bmatrix}
N_1(\xi,\eta,\zeta) \\
N_2(\xi,\eta,\zeta) \\
N_3(\xi,\eta,\zeta) \\
N_4(\xi,\eta,\zeta)
\end{Bmatrix}
=
\begin{Bmatrix}
1-\xi-\eta-\zeta \\
\xi \\
-\eta \\
\zeta
\end{Bmatrix}
\tag{7.85}
$$

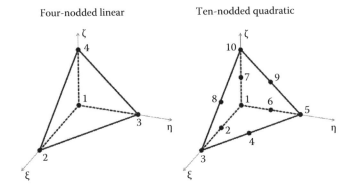

FIGURE 7.13 Three-dimensional tetrahedric elements.

7.6.3.2 Ten-Nodded Quadratic Tetrahedra (Figure 7.13)

$$
\begin{Bmatrix}
N_1(\xi,\eta,\zeta) \\
N_2(\xi,\eta,\zeta) \\
N_3(\xi,\eta,\zeta) \\
N_4(\xi,\eta,\zeta) \\
N_5(\xi,\eta,\zeta) \\
N_6(\xi,\eta,\zeta) \\
N_7(\xi,\eta,\zeta) \\
N_8(\xi,\eta,\zeta) \\
N_9(\xi,\eta,\zeta) \\
N_{10}(\xi,\eta,\zeta)
\end{Bmatrix}
=
\begin{Bmatrix}
-(1-\xi-\eta-\zeta)(1-2(1-\xi-\eta-\zeta)) \\
4\xi(1-\xi-\eta-\zeta) \\
-\xi(1-2\xi) \\
4\xi\eta \\
-\eta(1-2\eta) \\
4\eta(1-\xi-\eta-\zeta) \\
4\zeta(1-\xi-\eta-\zeta) \\
4\xi\zeta \\
4\eta\zeta \\
-\zeta(1-2\zeta)
\end{Bmatrix}
\tag{7.86}
$$

7.6.3.3 Eight-Nodded Linear Brick Element

$$
\begin{Bmatrix}
N_1(\xi,\eta,\zeta) \\
N_2(\xi,\eta,\zeta) \\
N_3(\xi,\eta,\zeta) \\
N_4(\xi,\eta,\zeta) \\
N_5(\xi,\eta,\zeta) \\
N_6(\xi,\eta,\zeta) \\
N_7(\xi,\eta,\zeta) \\
N_8(\xi,\eta,\zeta)
\end{Bmatrix}
=
\frac{1}{8}
\begin{Bmatrix}
(1-\xi)(1-\eta)(1-\zeta) \\
(1+\xi)(1-\eta)(1-\zeta) \\
(1+\xi)(1+\eta)(1-\zeta) \\
(1-\xi)(1+\eta)(1-\zeta) \\
(1-\xi)(1-\eta)(1+\zeta) \\
(1+\xi)(1-\eta)(1+\zeta) \\
(1+\xi)(1+\eta)(1+\zeta) \\
(1-\xi)(1+\eta)(1+\zeta)
\end{Bmatrix}
\tag{7.87}
$$

7.6.3.4 Twenty-Nodded Quadratic Brick Element (Figure 7.14)

$$
\begin{Bmatrix}
N_1(\xi, \eta, \zeta) \\
N_2(\xi, \eta, \zeta) \\
N_3(\xi, \eta, \zeta) \\
N_4(\xi, \eta, \zeta) \\
N_5(\xi, \eta, \zeta) \\
N_6(\xi, \eta, \zeta) \\
N_7(\xi, \eta, \zeta) \\
N_8(\xi, \eta, \zeta) \\
N_9(\xi, \eta, \zeta) \\
N_{10}(\xi, \eta, \zeta) \\
N_{11}(\xi, \eta, \zeta) \\
N_{12}(\xi, \eta, \zeta) \\
N_{13}(\xi, \eta, \zeta) \\
N_{14}(\xi, \eta, \zeta) \\
N_{15}(\xi, \eta, \zeta) \\
N_{16}(\xi, \eta, \zeta) \\
N_{17}(\xi, \eta, \zeta) \\
N_{18}(\xi, \eta, \zeta) \\
N_{19}(\xi, \eta, \zeta) \\
N_{20}(\xi, \eta, \zeta)
\end{Bmatrix}
=
\begin{Bmatrix}
\frac{1}{8}(1 - \xi)(1 - \eta)(1 - \zeta)(-2 - \xi - \eta - \zeta) \\
\frac{1}{4}(1 - \xi^2)(1 - \eta)(1 - \zeta) \\
\frac{1}{8}(1 + \xi)(1 - \eta)(1 - \zeta)(-2 + \xi - \eta - \zeta) \\
\frac{1}{4}(1 + \xi)(1 - \eta^2)(1 - \zeta) \\
\frac{1}{8}(1 + \xi)(1 + \eta)(1 - \zeta)(-2 + \xi + \eta - \zeta) \\
\frac{1}{4}(1 - \xi^2)(1 + \eta)(1 - \zeta) \\
\frac{1}{8}(1 - \xi)(1 + \eta)(1 - \zeta)(-2 - \xi + \eta - \zeta) \\
\frac{1}{4}(1 - \xi)(1 - \eta^2)(1 - \zeta) \\
\frac{1}{4}(1 - \xi)(1 - \eta)(1 - \zeta^2) \\
\frac{1}{4}(1 + \xi)(1 - \eta)(1 - \zeta^2) \\
\frac{1}{4}(1 + \xi)(1 + \eta)(1 - \zeta^2) \\
\frac{1}{4}(1 - \xi)(1 + \eta)(1 - \zeta^2) \\
\frac{1}{8}(1 - \xi)(1 - \eta)(1 + \zeta)(-2 - \xi - \eta + \zeta) \\
\frac{1}{4}(1 - \xi^2)(1 - \eta)(1 + \zeta) \\
\frac{1}{8}(1 + \xi)(1 - \eta)(1 + \zeta)(-2 + \xi - \eta + \zeta) \\
\frac{1}{4}(1 + \xi)(1 - \eta^2)(1 + \zeta) \\
\frac{1}{8}(1 + \xi)(1 + \eta)(1 + \zeta)(-2 + \xi + \eta + \zeta) \\
\frac{1}{4}(1 - \xi^2)(1 + \eta)(1 + \zeta) \\
\frac{1}{8}(1 - \xi)(1 + \eta)(1 + \zeta)(-2 - \xi + \eta + \zeta) \\
\frac{1}{4}(1 - \xi)(1 - \eta^2)(1 + \zeta)
\end{Bmatrix}
\tag{7.88}
$$

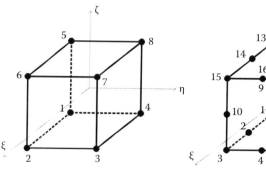

FIGURE 7.14 Three-dimensional brick elements.

8 Numerical Integration

8.1 INTRODUCTION

In Section 6.6.4, analytical integration was used to integrate the expression of the theorem of virtual work during the evaluation of the stiffness matrix of the beam element. That was relatively easy because a beam element is unidimensional. However, when the number of elements is large, and/or their geometrical shape is general, as is the case in most finite element applications, the use of analytical integration is quite cumbersome and ill-suited for computer coding. The alternative is to use numerical integration.

There exist many numerical methods for evaluating a definite integral. Simpson's rule, Newton–Cotes, and Gauss quadrature are examples of such methods. The basic idea of numerical integration is to replace the continuous integral with a series of finite sums:

$$\int_a^b f(x)\, dx = \sum_{i=1}^n A_i f(x_i) + \text{error} \tag{8.1}$$

The parameters A_i are called the weights of the integration.

In finite element application, Gauss quadrature, also called the Gauss–Legendre method, is the most widely used as it is the most precise.

8.2 GAUSS QUADRATURE

To begin the explanation of Gauss quadrature, we consider a one-dimensional problem without reference to the finite element method. Given a polynomial function of degree $m \le 2r - 1$, we assume that we can evaluate exactly the following integral with the method of Gauss quadrature on the interval $[-1, +1]$:

$$\int_{-1}^{+1} f(\xi)\, d\xi = \sum_{i=1}^r W_i f(\xi_i) \tag{8.2}$$

Based on our assumption, it follows that Equation (8.2) is verified for any polynomial function of the form

$$f(\xi) = \alpha_1 + \alpha_2 \xi + \alpha_3 \xi^2 + \cdots + \alpha_{2r}\xi^{2r-1} \tag{8.3}$$

To obtain the weights W_i and the abscissa ξ_i, which are the unknowns, we substitute Equation (8.3) for $f(\xi)$ in Equation (8.2), which yields

$$\alpha_1 \int_{-1}^{+1} d\xi + \alpha_2 \int_{-1}^{+1} \xi\, d\xi + \cdots + \alpha_{2r} \int_{-1}^{+1} \xi^{2r-1} d\xi = \alpha_1 (W_1 + W_2 + \cdots + W_r)$$

$$+ \alpha_2 (W_1 \xi_1 + W_2 \xi_2 + \cdots + W_r \xi_r) + \cdots \alpha_{2r}(W_1 \xi_1^{2r-1} + W_2 \xi_2^{2r-1} + \cdots + W_r \xi_r^{2r-1}) \tag{8.4}$$

For Equation (8.4) to be identically satisfied for all α_i, we must have the following equalities:

$$\int_{-1}^{+1} \xi^\alpha d\xi = \frac{2}{\alpha+1} = \sum_{i=1}^{r} W_i f(\xi_i^\alpha) \quad \alpha = 0, 2, 4, \ldots, 2r \tag{8.5}$$

$$\int_{-1}^{+1} \xi^\alpha d\xi = 0 = \sum_{i=1}^{r} W_i f(\xi_i^\alpha) \quad \alpha = 1, 3, 5, \ldots, 2r - 1 \tag{8.6}$$

which gives

$$2 = W_1 + W_2 + \cdots + W_r$$
$$0 = W_1 \xi_1 + W_2 \xi_2 + \cdots + W_r \xi_r$$
$$2/3 = W_1 \xi_1^2 + W_2 \xi_2^2 + \cdots + W_r \xi_r^2 \tag{8.7}$$
$$\cdots$$
$$0 = W_1 \xi_1^{2r-1} + W_2 \xi_2^{2r-1} + \cdots + W_r \xi_r^{2r-1}$$

The system (8.7) is linear in W_i but nonlinear in ξ_i, and determines the parameters of (8.2) under the conditions

$$\left. \begin{array}{c} W_i > 0 \\ -1 \leq \xi \leq +1 \end{array} \right\} \quad i = 1, 2, \ldots, r \tag{8.8}$$

However, there is no need to solve the system (8.7) to obtain the abscissa ξ_i and the weights W_i. The abscissa ξ_i are the roots of Legendre polynomials of order r, which are defined, for $k = 1, 2, \ldots, r$, as

$$P_0(\xi) = 1$$
$$P_1(\xi) = \xi$$
$$\cdots = \cdots \tag{8.9}$$
$$P_k(\xi) = \frac{2k-1}{k} \xi P_{k-1}(\xi) - \frac{k-1}{k} \xi P_{k-2}(\xi)$$

and the weights W_i are obtained as

$$W_i = \frac{2(1 - \xi_i^2)}{(r(P_{r-1}(\xi_i)))^2} \tag{8.10}$$

Example 1: Weights and abscissa for $r = 2$

Find the abscissas ξ_i and the weights W_i for $r = 2$.
   The Legendre polynomials up to order 2 are written as

$$P_0(\xi) = 1$$
$$P_1(\xi) = \xi$$
$$P_2(\xi) = \frac{3}{2} \xi^2 - \frac{1}{2}$$

TABLE 8.1
Abscissa and Weights for Gauss Quadrature

r	ξ	W
1	0.000000 000000 000000	2.000000 000000 000000
2	0.577350 269189 635764	1.000000 000000 000000
3	0.774596 669241 483377	0.555555 555555 555555
	0.000000 000000 000000	0.888888 888888 888888
4	0.861136 371594 052575	0.347854 845137 453857
	0.339981 043584 856264	0.652145 154862 546142
5	0.906179 845938 663992	0.236926 885056 189087
	0.538469 310105 683091	0.478628 670499 366468
	0.000000 000000 000000	0.568888 888888 888888

The roots of $P_2(\xi) = 0$ are given as

$$\xi_i = \pm \frac{1}{\sqrt{3}}$$

The weights W_1 and W_2 can be obtained from the system (8.7) as

$$2 = W_1 + W_2$$

$$0 = -\frac{1}{\sqrt{3}}W_1 + \frac{1}{\sqrt{3}}W_2$$

or directly from Equation (8.10). In both cases, we obtain

$$W_1 = W_2 = 1$$

Table 8.1 gives the abscissa ξ_i and the weights W_i for $r = 1, \ldots, 5$
The abscissae are symmetrical with respect to 0, and the corresponding weights are equal; for example for $r = 5$, we get

$$\xi_1 = -0.906179845938663992 \quad W_1 = 0.538469310105683091$$
$$\xi_2 = -0.478628670499366468 \quad W_2 = 0.568888888888888888$$
$$\xi_3 = 0.000000000000000000 \quad W_2 = 0.236926885056189087$$
$$\xi_4 = 0.478628670499366468 \quad W_4 = 0.568888888888888888$$
$$\xi_5 = 0.906179845938663992 \quad W_5 = 0.538469310105683091$$

Example 2: Integral Evaluation

Evaluate the integral $\int_{-1}^{+1}(\xi^2 + \sin(\xi/2))d\xi$ using three Gauss points, $r = 3$.
    Using Table 8.1, the abscissa and weights for three Gauss points are

$$\xi_1 = -0.774596669241483377 \quad W_1 = 0.555555555555555555$$
$$\xi_2 = 0.000000000000000000 \quad W_2 = 0.888888888888888888$$
$$\xi_3 = 0.774596669241483377 \quad W_3 = 0.555555555555555555$$

Using only six significant digits for the abscissa and the weights, the integral becomes

$$\xi_1, \quad W_1: \quad [(-0.774596)^2 + \sin(-0.774596/2)]0.555555 +$$

$$\xi_2, \quad W_2: \quad [0.000000^2 + \sin(0.000000/2)]0.888888 +$$

$$\xi_3, \quad W_3: \quad [0.774596^2 + \sin(0.774596/2)]0.555555 = 0.666664$$

Compared to the analytical solution, we have

$$I_{exact} = \left[\frac{\xi^3}{3} - 2\cos(\xi/2)\right]_{-1}^{+1} = 0.666666$$

With only six significant figures, the integration is exact up to five digits after the decimal point.

8.2.1 Integration over an Arbitrary Interval [a, b]

Up to now, the method of Gauss quadrature has been presented only for evaluating integrals in the domain $[-1, +1]$. What about if the interval of integration is of the general form such as $[a, b]$? That is, evaluating an integral of the form

$$\int_a^b f(x)\, dx \tag{8.11}$$

In this case, we transform the interval $[-1, +1]$ to the interval $[a, b]$ through a change of variable. In other words, we define a linear transformation between $[-1, +1]$ and $[a, b]$. The analytical expression of the linear transformation between the two intervals is given by

$$x = \frac{b-a}{2}\xi + \frac{b+a}{2} \tag{8.12}$$

Differentiating yields

$$dx = \frac{b-a}{2}d\xi \tag{8.13}$$

Substituting Equations (8.13) and (8.12) in Equation (8.11) yields

$$\int_a^b f(x)\, dx = \frac{b-a}{2}\int_{-1}^{+1} f(x(\xi))\, d\xi = \frac{b-a}{2}\sum_{i=1}^r W_i f(x(\xi_i)) \tag{8.14}$$

Example 3: Evaluation of a General Integral

Evaluate the integral $\int_3^7 \frac{1}{1.1+x}\, dx$ with two ($r = 2$) and three ($r = 3$) Gauss points. First, we operate the following variable change given by Equation (8.12):

$$x = \frac{7-3}{2}\xi + \frac{7+3}{2} = 2\xi + 5$$

a. Two Gauss points r = 2

Using Table 8.1, we obtain the abscissa and the weights for two Gauss points:

$$\xi_1 = -0.577350269189635764 \quad W_1 = 1.000000000000000000$$

$$\xi_2 = 0.577350269189635764 \quad\quad W_2 = 1.000000000000000000$$

Using only six significant digits, the integral becomes

$$\xi_1, \quad W_1: \quad \frac{7-3}{2} \left[\frac{1}{1.1 + (2(-0.577350) + 5)} \right] 1.000000 +$$

$$\xi_2, \quad W_2: \quad \frac{7-3}{2} \left[\frac{1}{1.1 + (2(+0.577350) + 5)} \right] 1.000000 = 0.680107$$

b. Three Gauss points $I = 3$

The abscissa and weights for three Gauss points are

$$\xi_1 = -0.774596669241483377 \quad W_1 = 0.555555555555555555$$

$$\xi_2 = 0.000000000000000000 \quad W_2 = 0.888888888888888888$$

$$\xi_3 = 0.774596669241483377 \quad W_3 = 0.555555555555555555$$

Using only six significant digits, the integral becomes

$$\xi_1, \quad W_1: \quad \frac{7-3}{2} \left[\frac{1}{1.1 + (2(-0.774596) + 5)} \right] 0.555555 +$$

$$\xi_2, \quad W_2: \quad \frac{7-3}{2} \left[\frac{1}{1.1 + (2(+0.000000) + 5)} \right] 0.888888 +$$

$$\xi_3, \quad W_3: \quad \frac{7-3}{2} \left[\frac{1}{1.1 + (2(+0.774596) + 5)} \right] 0.555555 = 0.68085$$

Compared to the analytical solution, we have

$$I_{exact} = \left[\ln(1.1 + x) \right]_3^7 = 6.80877$$

8.2.2   Integration in Two and Three Dimensions

Integrating in two and three dimensions consists of using a single integral in each dimension. For instance, the evaluation of $\int_{-1}^{+1} \int_{-1}^{+1} f(\xi, \eta) \, d\xi \, d\eta$ is carried out as follows:

$$\int_{-1}^{+1} \int_{-1}^{+1} f(\xi, \eta) \, d\xi \, d\eta = \sum_{i=1}^{r_1} \sum_{j=1}^{r_2} W_i W_j f(\xi_i, \eta_j)) \qquad (8.15)$$

Notice that different number of Gauss points can be used in each direction. The method integrates exactly the product of a polynome of degree $2r_1 - 1$ in ξ and a polynome of degree $2r_2 - 1$ in η.

In three dimensions, Equation (8.15) becomes

$$\int_{-1}^{+1} \int_{-1}^{+1} \int_{-1}^{+1} f(\xi, \eta, \zeta) \, d\xi \, d\eta \, d\zeta = \sum_{i=1}^{r_1} \sum_{j=1}^{r_2} \sum_{k=1}^{r_3} W_i W_j W_k f(\xi_i, \eta_j, \zeta_j) \qquad (8.16)$$

Example 4: Evaluation of a Double Integral

Using Gauss quadrature, evaluate the following integral using three Gauss points in each direction:

$$I = \int_0^\pi \int_0^3 (x^2 - x) \sin y \, dx \, dy$$

In this case, it is necessary to operate two variable changes to evaluate numerically this integral. The variable changes are

In x-direction: $x = \dfrac{3}{2}\xi + \dfrac{3}{2}$

In y-direction: $y = \dfrac{\pi}{2}\eta + \dfrac{\pi}{2}$

The integral is written as

$$I = \frac{3\pi}{4}\int_{-1}^{+1}\int_{-1}^{+1}(x(\xi)^2 - x(\xi))\sin y(\eta)\,d\xi\,d\eta$$

and can be replaced by the following series:

$$I = \frac{3\pi}{4}\sum_{i=1}^{3}\sum_{j=1}^{3}W_i W_j(x(\xi_i)^2 - x(\xi_i))\sin y(\eta_j)$$

Using Table 8.1 for $r = 3$, we have

$$x(\xi_1) = 0.3381 \quad y(\eta_1) = 0.3541 \quad W_1 = 0.5555$$
$$x(\xi_2) = 1.5000 \quad y(\eta_2) = 1.5708 \quad W_2 = 0.8888$$
$$x(\xi_3) = 2.6619 \quad y(\eta_3) = 2.7875 \quad W_3 = 0.5555$$

Developing the series yields

$i = 1 \quad j = 1:\quad I = \dfrac{3\pi}{4}\Big[0.5555((0.3381)^2 - 0.3381)0.5555\sin(0.3541)$

$\quad\quad\ j = 2:\quad +0.5555((0.3381)^2 - 0.3381)0.8888\sin(1.5708)$

$\quad\quad\ j = 3:\quad +0.5555((0.3381)^2 - 0.3381)0.5555\sin(2.7875)\Big]$

$i = 2 \quad j = 1:\quad +\dfrac{3\pi}{4}\Big[0.8888((1.5000)^2 - 1.5000)0.5555\sin(0.3541)$

$\quad\quad\ j = 2:\quad +0.8888((1.5000)^2 - 1.5000)0.8888\sin(1.5708)$

$\quad\quad\ j = 3:\quad +0.8888((1.5000)^2 - 1.5000)0.5555\sin(2.7875)\Big]$

$i = 3 \quad j = 1:\quad +\dfrac{3\pi}{4}\Big[0.5555((2.6619)^2 - 2.6619)0.5555\sin(0.3541)$

$\quad\quad\ j = 2:\quad +0.5555((2.6619)^2 - 2.6619)0.8888\sin(1.5708)$

$\quad\quad\ j = 3:\quad +0.5555((2.6619)^2 - 2.6619)0.5555\sin(2.7875)\Big] = 9.0047$

The analytical solution is obtained as

$$I = \left[\frac{x^3}{3} - \frac{x^2}{2}\right]_0^3\left[-\cos(y)\right]_0^\pi = 9$$

8.3 INTEGRATION OVER A REFERENCE ELEMENT

As we have seen in Section 8.2, Gauss quadrature evaluates single integrals between $[-1, +1]$, double integrals over a square of side 2, and triple integrals over a cube of side 2. For instance, to evaluate an integral over a quadrilateral, it is necessary to transform the quadrilateral into a reference element over which the integration can be carried out. For example, the evaluation of the integral $\int_A f(x, y) dA$ over a quadrilateral area is carried out as follows:

- Since the bilinear quadrilateral is isoparametric, we write the coordinates x and y in terms of the reference coordinates ξ and η as

$$x(\xi, \eta) = N_1(\xi, \eta)x_1 + N_2(\xi, \eta)x_2 + N_3(\xi, \eta)x_3 + N_4(\xi, \eta)x_4$$
$$y(\xi, \eta) = N_1(\xi, \eta)y_1 + N_2(\xi, \eta)y_2 + N_3(\xi, \eta)y_3 + N_4(\xi, \eta)y_4$$

the shape functions $N_i(\xi, \eta)$ are as given by Equations (7.78)
- Use Equation (7.74) of the Jacobian of the transformation to express the elementary area $dA = dxdy$ in terms of the corresponding elementary area $d\xi d\eta$ of the reference element

$$dxdy = det[J] d\xi\, d\eta$$

- Construct a nodal approximation for the function using its nodal values

$$\bar{f}(\xi, \eta) = \sum_{i=1}^{n} N_i(\xi, \eta)f_i$$

- Finally, the integral becomes

$$I = \int_{-1}^{+1}\int_{-1}^{+1} \left(\sum_{i=1}^{n} N_i(\xi, \eta)f_i \right) det[J]\, d\xi\, d\eta \qquad (8.17)$$

8.4   INTEGRATION OVER A TRIANGULAR ELEMENT

The main reason for introducing the area coordinates in Section 7.5.1.3 was to allow the evaluation of simple integrals that arise in the finite element method when the linear triangular element is used.

8.4.1   Simple Formulas

The following simple formulas can be used to evaluate integrals over the side or the area of a triangular element:

- Integrals over length

$$\int_l L_i^\alpha L_j^\beta\, dl = \frac{\alpha!\beta!}{(\alpha + \beta + 1)!} l_{ij} \qquad (8.18)$$

where dl represents an element of length between nodes i and j
- Integrals over area

$$\int_A L_i^\alpha L_j^\beta L_k^\gamma\, dA = \frac{\alpha!\beta!\gamma!}{(\alpha + \beta + \gamma + 2)!} 2A \qquad (8.19)$$

where dA represents an element of area.

If the shape functions, $N_i(x, y)$, of the triangular element are defined directly in terms of the coordinates x and y, then they can be directly substituted for the area coordinates $L_i(x, y)$.

8.4.2   Numerical Integration over a Triangular Element

The simple formulas given by expressions (8.18) and (8.19) are only useful when the linear triangular element is used since the shape functions $N_i(x, y)$ are the same as the area coordinates $L_i(x, y)$. However, when higher order triangular elements are used, the simple formulas described earlier become quite cumbersome and numerical integration over a reference triangular element is the most indicated. Expression (8.20) gives the formulas for integrating over a triangular reference element:

$$I = \int_0^{+1} \int_0^{1-\xi} f(\xi, \eta)\, d\eta\, d\xi = \sum_{i=1}^{r} W_i f(\xi_i, \eta_i) \tag{8.20}$$

These formulas integrate exactly monomes $\xi^\alpha \eta^\beta$ such that $\alpha + \beta \le m$. They are referred to as Hammer formulas. The abscissa ξ_i, η_i, and the weights W_i are different from those used by Gauss quadrature. Table 8.2 from [1,2] gives the abscissa and weights for integration over a triangular reference element. Notice that there are two sets of abscissa and weights for $m = 2$. Figure 8.1 shows the positions of the sampling points for orders 1, 2, and 3.

TABLE 8.2
Abscissae and Weights for a Triangle

Order m	Number of Points r	ξ	η	W
1	1	0.333333333333	0.333333333333	0.5
2	3	0.5	0.5	0.166666666666
		0	0.5	0.166666666666
		0.5	0	0.166666666666
2	3	0.166666666666	0.166666666666	0.166666666666
		0.666666666666	0.166666666666	0.166666666666
		0.166666666666	0.666666666666	0.166666666666
3	4	0.333333333333	0.333333333333	−0.28125
		0.2	0.2	0.260416666666
		0.6	0.2	0.260416666666
		0.2	0.6	0.260416666666
4	6	0.44594849092	0.44594849092	0.111690794839
		0.10810301817	0.44594849092	0.111690794839
		0.44594849092	0.10810301817	0.111690794839
		0.09157621351	0.09157621351	0.054975871827
		0.81684757289	0.09157621351	0.054975871827
		0.09157621351	0.81684757289	0.054975871827
5	7	0.33333333333	0.33333333333	0.1125
		0.470142064105	0.470142064105	0.066197076394
		0.05971587179	0.470142064105	0.066197076394
		0.470142064105	0.05971587179	0.066197076394
		0.101286507324	0.101286507324	0.708802923606
		0.898713492676	0.101286507324	0.708802923606
		0.101286507324	0.898713492676	0.708802923606

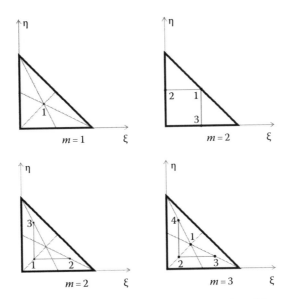

FIGURE 8.1   Positions of the sampling points for a triangle: Orders 1, 2, and 3.

8.5   SOLVED PROBLEMS

8.5.1   PROBLEM 8.1

Use Gauss quadrature to evaluate the second moment of area of the quarter annulus shown in Figure 8.2 with respect to the axis x.

Solution

To evaluate the integral $I_{xx} = \iint y^2 dA$, we introduce a double change of variables. First, we express x and y in terms of the polar coordinates r and θ, then we express the polar coordinates in terms of the reference coordinates ξ and η as depicted in Figure 8.3.

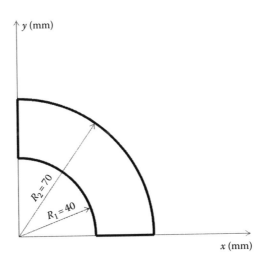

FIGURE 8.2   Gauss quadrature over an arbitrary area.

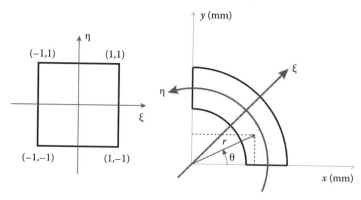

FIGURE 8.3   Double change of variables.

$$x = r \cos \theta$$

$$y = r \sin \theta$$

$$\theta = \frac{\pi}{4}\eta + \frac{\pi}{4}$$

$$r = \frac{R_2 - R_1}{2}\xi + \frac{R_2 + R_1}{2}$$

In terms of polar coordinates, the infinitesimal area $dA = dxdy$ is written as

$$dA = r \, dr \, d\theta$$

Substituting in the expression of the second moment of area, the latter can be written as

$$I_{xx} = \int\limits_{0}^{\pi/2} \int\limits_{R_1}^{R_2} (r \sin \theta)^2 r \, dr \, d\theta$$

Such an integral can be easily evaluated analytically; the result is obtained as

$$I_{xx} = \int\limits_{R_1}^{R_2} r^3 dr \int\limits_{0}^{\pi/2} (\sin \theta)^2 d\theta = \left[\frac{r^4}{4}\right]_{R_1}^{R_2} \left[\frac{\theta}{2} - \frac{1}{4}\sin(2\theta)\right]_{0}^{\pi/2} = \frac{\pi(R_2^4 - R_1^4)}{16}$$

Using numerical values, $R_1 = 40$ mm and $R_2 = 70$ mm, we obtain

$$I_{xx} = 4{,}211{,}700 \ \text{mm}^4$$

In terms of the reference coordinates ξ and η, the integral is written as

$$I_{xx} = \frac{\pi}{4}\left(\frac{R_2 - R_1}{2}\right)\int\limits_{-1}^{+1}\int\limits_{-1}^{+1}\left(\frac{R_2 - R_1}{2}\xi + \frac{R_2 + R_1}{2}\right)^3 \left(\sin\left(\frac{\pi}{4}\eta + \frac{\pi}{4}\right)\right)^2 d\xi \, d\eta$$

Introducing the method of Gauss quadrature, we obtain

$$I_{xx} = \frac{\pi}{4}\left(\frac{R_2 - R_1}{2}\right)\sum_{i=1}^{n_1}\sum_{j=1}^{n_2}\left(\frac{R_2 - R_1}{2}\xi_i + \frac{R_2 + R_1}{2}\right)^3 \left(\sin\left(\frac{\pi}{4}\eta_j + \frac{\pi}{4}\right)\right)^2 W_i W_j$$

Using two Gauss points in the direction of ξ, three in the direction of η, and introducing the same numerical values, $R_1 = 40$ mm and $R_2 = 70$ mm, we obtain

$$
\begin{aligned}
\xi_1 &= -0.577350 & r_1 &= 15(-0.577350) + 55 = 46.3397 & W_1 &= 1 \\
\xi_2 &= 0.577350 & r_2 &= 15(0.577350) + 55 = 63.6603 & W_2 &= 1 \\
\eta_1 &= -0.774596 & \theta_1 &= (\pi/4)(-0.774596) + (\pi/4) = 0.1770 & W_1 &= 0.55555 \\
\eta_2 &= 0.0000000 & \theta_2 &= (\pi/4)(0.000000) + (\pi/4) = 0.7854 & W_2 &= 0.88888 \\
\eta_3 &= 0.774596 & \theta_3 &= (\pi/4)(0.774596) + (\pi/4) = 1.3938 & W_3 &= 0.55555
\end{aligned}
$$

After substitution, the sum equation becomes

$$
\begin{aligned}
I_{xx} = 11.7810\Big[&(46.3397^3)(\sin(0.1770))^2 \times 1 \times 0.55555 + (46.3397^3)(\sin(0.7854))^2 \\
&\times 1 \times 0.88888 + (46.3397^3)(\sin(1.3938))^2 \times 1 \times 0.55555 \\
&+ (63.6603^3)(\sin(0.1770))^2 \times 1 \times 0.55555 + (63.6603^3)(\sin(0.7854))^2 \\
&\times 1 \times 0.88888 + (63.6603^3)(\sin(1.3938))^2 \times 1 \times 0.55555 \Big] \\
&= 4{,}211{,}700 \text{ mm}^4
\end{aligned}
$$

8.5.2 Problem 8.2

Use coarse and fine meshes of respectively 2 and 8 quadratic isoparametric 8-nodded elements as shown in Figures 8.4 and 8.5 to compute the second moment of area I_{xx} of the annulus in Worked Example 8.1.

Solution

The second moment of the area of the annulus is obtained as the sum of the second moments of area of the two elements; that is,

$$
I_{xx} = I_{xx}^{(1)} + I_{xx}^{(2)}
$$

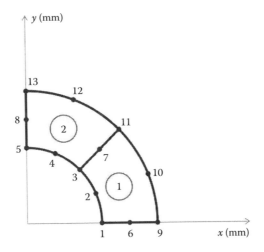

FIGURE 8.4   Coarse mesh of two 8-nodded elements.

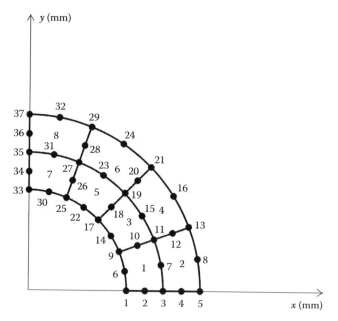

FIGURE 8.5   Eight elements finite element approximation with two 8-nodded elements.

The second moment of area of elements 1 and 2 are obtained respectively as

$$I_{xx}^{(1)} = \int_{A1} y^2 dx\, dy$$

$$I_{xx}^{(2)} = \int_{A2} y^2 dx\, dy$$

To be able to evaluate the aforementioned integrals, we introduce the reference coordinates ξ and η. For each element, the y ordinate is approximated in terms of the nodal coordinates of the element as

$$y = N_1(\xi, \eta)y_1 + N_2(\xi, \eta)y_2 + \cdots + N_8(\xi, \eta)y_8$$

The infinitesimal element of area $dxdy$ is obtained as

$$dxdy = [J(\xi, \eta)]d\xi d\eta = \begin{bmatrix} \dfrac{\partial x}{\partial \xi} & \dfrac{\partial y}{\partial \xi} \\[2mm] \dfrac{\partial x}{\partial \eta} & \dfrac{\partial y}{\partial \eta} \end{bmatrix} d\xi d\eta$$

After substitution, the aforementioned integrals become

$$I_{xx}^{(1)} = \int_{-1}^{+1}\int_{-1}^{+1} \left(\sum_{k=1}^{8} N_k(\xi, \eta)y_k^{(1)} \right)^2 det[J^{(1)}(\xi, \eta)]\, d\xi\, d\eta$$

$$I_{xx}^{(2)} = \int_{-1}^{+1}\int_{-1}^{+1} \left(\sum_{k=1}^{8} N_k(\xi, \eta)y_k^{(2)} \right)^2 det[J^{(2)}(\xi, \eta)]\, d\xi\, d\eta$$

Notice that the only difference between the two equations are the coordinates of the nodes. Introducing Gauss quadrature, the integrals become

$$I_{xx}^{(1)} = \sum_{i=1}^{ngp} \sum_{j=1}^{ngp} \left(\sum_{k=1}^{8} N_k(\xi_i, \eta_j) y_k^{(1)} \right)^2 W_i W_j det[J^{(1)}(\xi_i, \eta_j)]$$

$$I_{xx}^{(2)} = \sum_{i=1}^{ngp} \sum_{j=1}^{ngp} \left(\sum_{k=1}^{8} N_k(\xi_i, \eta_j) y_k^{(2)} \right)^2 W_i W_j det[J^{(2)}(\xi_i, \eta_j)]$$

The aforementioned sums involve matrix multiplication, and their evaluation by hand is very tedious. Therefore, it is better and quicker to evaluate them with a MATLAB® code. In addition, to choose the required number of Gauss points ngp, we need to investigate the order of the polynomials involved in the aforementioned equations. The functions $N_k(\xi, \eta)$ are of degree 2 in ξ and η. When they are squared they become of order 4. The determinant of the Jacobian matrix is linear in both ξ and η. Therefore, the order of the polynomials involved is 5. As such, we need three Gauss points in each direction, $ngp = 3$.

The code named **IXX.m**, listed next, begins with the input data. It uses either a mesh of two or eight elements. The input data, which consist of the number of nodes nnd, their coordinates stored in the matrix $geom(nnd, 2)$, the number of elements nel, the number of nodes per element nne, and the connectivity matrix $connec(nel, nne)$, are given respectively in the scripts **Two_Q8.m** and **Eight_Q8.m** listed next.

IXX.m

```
% Evaluation of the second moment of area of a geometrical domain
% Using finite element approximation with an 8 Nodes
% isoparametric element elements.
%
clc
clear
%
global geom connec nel nne nnd RI RE
%
RI = 40;      % Internal radius
RE = 70;      % External radius
%
Eight_Q8          % Load input for fine mesh
%
%  Number of Gauss points
%
ngp = 3           % The polynomials involved are of degree 5
%
samp = gauss(ngp)   % Gauss abscissae and weights
%
%
Ixx = 0.; % Initialize the second moment of area to zero
%
for k=1:nel
    coord = coord_q8(k,nne, geom, connec);    % Retrieve the coordinates of
                                              % the nodes of element k
    X = coord(:,1);                           % X coordinates of element k
    Y = coord(:,2)                            % Y coordinates of element k
    X = coord(:,1);                           % X coordinates of element k
    Y = coord(:,2)                            % Y coordinates of element k
%
    for i=1:ngp
        xi = samp(i,1);
        WI = samp(i,2);
        for j =1:ngp
```

```
        eta = samp(j,1);
        WJ = samp(j,2);
        [der,fun] = fmquad(samp, i,j);   % Form the vector of the shape functions
                                         % and the matrix of their derivatives
        JAC = der*coord;                 % Evaluate the Jacobian
        DET =det(JAC)                    % Evaluate determinant of Jacobian matrix
        Ixx =Ixx+ (dot(fun,Y))^2*WI*WJ*DET;
        end
    end
end
Ixx
```

Two_Q8

```
% Input module Two_Q8.m
% Two elements mesh
%
global geom connec nel nne nnd RI RE
nnd = 13          % Number of nodes
%
% The matrix geom contains the x and y coordinates of the nodes
%
geom =  ...
[RI                0.;                           ... % node 1
RI*cos(pi/8)      RI*sin(pi/8);                  ... % node 2
RI*cos(pi/4)      RI*sin(pi/4);                  ... % node 3
RI*cos(3*pi/8)    RI*sin(3*pi/8);               ... % node 4
RI*cos(pi/2)      RI*sin(pi/2);                  ... % node 5
(RI+RE)/2         0.;                            ... % node 6
((RI+RE)/2)*cos(pi/4)   ((RI+RE)/2)*sin(pi/4);... % node 7
((RI+RE)/2)*cos(pi/2)   ((RI+RE)/2)*sin(pi/2);... % node 8
RE                0.;                            ... % node 9
RE*cos(pi/8)      RE*sin(pi/8);                  ... % node 10
RE*cos(pi/4)      RE*sin(pi/4);                  ... % node 11
RE*cos(3*pi/8)    RE*sin(3*pi/8);               ... % node 12
RE*cos(pi/2)      RE*sin(pi/2)]                      % node 13

nel = 2           % Number of elements
nne = 8           % Number of nodes per element
%
% The matrix connec contains the connectivity of the elements
%
connec = [1    6    9    10    11    7    3    2; ...   % Element 1
          3    7    11   12    13    8    5    4]       % Element 2
%
% End of input module Two_Q8.m
```

Eight_Q8.m

```
% Eight elements mesh
%
global geom connec nel nne nnd  RI RE
nnd = 37          % Number of nodes
%
% The matrix geom contains the x and y coordinates of the nodes
%
geom = ...
[RI                0.;                                    ... % node 1
RI+(RE-RI)/4      0.;                                     ... % node 2
RI+(RE-RI)/2      0.;                                     ... % node 3
RI+3*(RE-RI)/4    0.;                                     ... % node 4
RE                0.;                                     ... % node 5
RI*cos(pi/16)     RI*sin(pi/16);                          ... % node 6
(RI+(RE-RI)/2)*cos(pi/16)   (RI+(RE-RI)/2)*sin(pi/16);    ... % node 7
RE*cos(pi/16)     RE*sin(pi/16);                          ... % node 8
```

```
RI*cos(pi/8)    RI*sin(pi/8);                                        ...  % node 9
(RI+(RE-RI)/4)*cos(pi/8)    (RI+(RE-RI)/4)*sin(pi/8);                ...  % node 10
(RI+(RE-RI)/2)*cos(pi/8)    (RI+(RE-RI)/2)*sin(pi/8);                ...  % node 11
(RI+3*(RE-RI)/4)*cos(pi/8)  (RI+3*(RE-RI)/4)*sin(pi/8);             ...  % node 12
RE*cos(pi/8)    RE*sin(pi/8);                                        ...  % node 13
RI*cos(3*pi/16)    RI*sin(3*pi/16);                                  ...  % node 14
(RI+(RE-RI)/2)*cos(3*pi/16)    (RI+(RE-RI)/2)*sin(3*pi/16);          ...  % node 15
RE*cos(3*pi/16)    RE*sin(3*pi/16);                                  ...  % node 16
RI*cos(pi/4)    RI*sin(pi/4);                                        ...  % node 17
(RI+(RE-RI)/4)*cos(pi/4)    (RI+(RE-RI)/4)*sin(pi/4);                ...  % node 18
(RI+(RE-RI)/2)*cos(pi/4)    (RI+(RE-RI)/2)*sin(pi/4);                ...  % node 19
(RI+3*(RE-RI)/4)*cos(pi/4)  (RI+3*(RE-RI)/4)*sin(pi/4);             ...  % node 20
RE*cos(pi/4)    RE*sin(pi/4);                                        ...  % node 21
RI*cos(5*pi/16)    RI*sin(5*pi/16);                                  ...  % node 22
(RI+(RE-RI)/2)*cos(5*pi/16)    (RI+(RE-RI)/2)*sin(5*pi/16);          ...  % node 23
RE*cos(5*pi/16)    RE*sin(5*pi/16);                                  ...  % node 24
RI*cos(6*pi/16)    RI*sin(6*pi/16);                                  ...  % node 25
(RI+(RE-RI)/4)*cos(6*pi/16)    (RI+(RE-RI)/4)*sin(6*pi/16);          ...  % node 26
(RI+(RE-RI)/2)*cos(6*pi/16)    (RI+(RE-RI)/2)*sin(6*pi/16);          ...  % node 27
(RI+3*(RE-RI)/4)*cos(6*pi/16)  (RI+3*(RE-RI)/4)*sin(6*pi/16); ...  % node 28
RE*cos(6*pi/16)    RE*sin(6*pi/16);                                  ...  % node 29
RI*cos(7*pi/16)    RI*sin(7*pi/16);                                  ...  % node 30
(RI+(RE-RI)/2)*cos(7*pi/16)    (RI+(RE-RI)/2)*sin(7*pi/16);          ...  % node 31
RE*cos(7*pi/16)    RE*sin(7*pi/16);                                  ...  % node 32
RI*cos(pi/2)    RI*sin(pi/2);                                        ...  % node 33
(RI+(RE-RI)/4)*cos(pi/2)    (RI+(RE-RI)/4)*sin(pi/2);                ...  % node 34
(RI+(RE-RI)/2)*cos(pi/2)    (RI+(RE-RI)/2)*sin(pi/2);                ...  % node 35
(RI+3*(RE-RI)/4)*cos(pi/2)  (RI+3*(RE-RI)/4)*sin(pi/2);             ...  % node 36
RE*cos(pi/2)    RE*sin(pi/2)]                                        % node 37
%
nel = 8              % Number of elements
nne = 8              % Number of nodes per element
%
% The matrix connec contains the connectivity of the elements
%
connec = [1    2    3    7    11    10    9    6; ...      % Element 1
          3    4    5    8    13    12    11    7; ...      % Element 2
          9    10    11    15    19    18    17    14; ...   % Element 3
          11    12    13    16    21    20    19    15; ...   % Element 4
          17    18    19    23    27    26    25    22; ...   % Element 5
          19    20    21    24    29    28    27    23; ...   % Element 6
          25    26    27    31    35    34    33    30; ...   % Element 7
          27    28    29    32    37    36    35    31]; ...  % Element 8
%
% End script Eight_Q8.m
```

Next, we provide the abscissae and weights necessary to perform a gauss quadrature. These are given in the script **gauss.m**, listed in Appendix A, which is a function that returns the matrix **samp(ngp, 2)**. The first column contains the abscissa and the second column the weights.

For each element, we retrieve the coordinates of its nodes using the script **coord_q8.m** also listed in Appendix A.

The double sum is evaluated using three Gauss points $ngp = 3$. The shape functions $N_i(\xi_i, \eta_j)$, given in the vector $fun(nne)$, as well as their derivatives, returned in the matrix $der(2, nne)$, are all evaluated at the Gauss points using the script **fmquad.m** listed in Appendix A.

The Jacobian is simply evaluated as $jac = der * coord$. The second moment of area is obtained as a sum of all the terms $Ixx = Ixx + (dot(fun, Y))^2 * WI * WJ * DET$ with Ixx being previously initialized to zero.

After execution of the code, the second moment of area obtained with the coarse mesh is

$$I_{xx} = 4,205,104 \text{ mm}^4$$

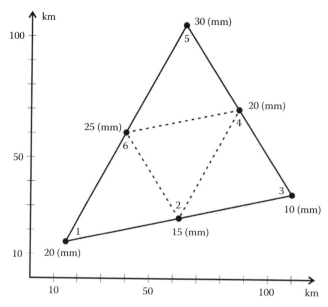

FIGURE 8.6   Estimation of rainfall using finite element approximation.

This is not quite the result anticipated. Indeed, it is not possible to approximate a circle with a quadratic polynomial. Now let us increase the number of elements by using the fine mesh of eight elements. The new result is

$$I_{xx} = 4,211,281 \text{ mm}^4$$

It can be seen that the precision of the second moment of area has greatly improved. Obviously if we keep refining the mesh, the computed value will converge to the exact one.

8.5.3 PROBLEM 8.3

Six rain gages are placed in a triangular shape, as shown in Figure 8.6. Estimate the total rainfall on the region as well as its area by using

- Four linear triangular elements
- One quadratic triangular element

The coordinates of the rain gauges and the precipitations recorded are as given in Table 8.3.

Solution

If $Q(x, y)$ is the unknown function for rainfall, then the total quantity of rainfall over the area A is given as

$$Q_T = \int_A Q(x, y) \, dA$$

To be able to estimate Q_T, we need first to construct a trial function for the unknown function $Q(x, y)$.

TABLE 8.3
Coordinates of Rain Gages and Precipitations

Gage	x (km)	y (km)	Precipitation (mm)
1	15	15	20
2	62.5	25	15
3	110	35	10
4	87.5	70	20
5	65	105	30
6	40	60	25

Four linear triangular elements

The total area A is divided into four triangular elements having nodes $1-2-6, 2-3-4, 2-4-6$, and $6-4-5$. We will use the same nodal approximation as a trial function for all the elements. For an arbitrary linear triangular element with nodes $i-j-k$, we have

$$\overline{Q}(x, y) = N_i(x, y)Q_i + N_j(x, y)Q_j + N_k(x, y)Q_k$$

The shape functions $N_i(x, y)$, $N_j(x, y)$, and $N_k(x, y)$ are given by expressions (7.31). The quantity of rainfall over the element is given as

$$Q_e = \int_{A_e} \overline{Q}(x, y) \, dA = Q_1 \int_{A_e} N_1(x, y) \, dA + Q_2 \int_{A_e} N_2(x, y) \, dA + Q_3 \int_{A_e} N_3(x, y) \, dA$$

Using the formulas for integration over a triangle, expression (8.19), the previous expression becomes

$$\overline{Q}_e = (Q_1 + Q_2 + Q_3)\frac{A_e}{3}$$

The area A_e is given in terms of the nodal coordinates by Equation (7.34).

The aforementioned computations can be easily coded in a MATLAB code, which we will name **precipitation_T3.m** listed next.

```
%    PROGRAM  precipitation_T3.M
%
%    This program estimates the quantity of rainfall over
%    an area discretized with linear triangular elements
%
nel = 4      % Total number of elements
nnd = 6      % Total number of nodes
nne = 3      % Number of nodes per element
%
% Coordinates of the rain gauges (nodes) in km
%
geom = [15.      15.    ; ... %  Node 1
        62.5     25.    ; ... %  Node 2
        110.     35.    ; ... %  Node 3
        87.5     70.    ; ... %  Node 4
        65.      105.   ; ... %  Node 5
        40.      60.]   ;     %  Node 6
%
% Precipitations recorded by the rain gauges
%
q = [20.; 15.; 10.; 20.; 30.; 25.] ;
%
%  Connectivity
```

```
%
connec = [1    2    6; ... % Element 1
          2    3    4; ... % Element 2
          2    4    6; ... % Element 3
          6    4    5];    % Element 4
%
AT = 0. ; % Initialize total area to zero
QT = 0. ; % Initialize total rainfall to zero
%
for i=1:nel
%
%    for each element retrieve the x and y coordinates of its nodes
%
     xi = geom(connec(i,1),1); yi = geom(connec(i,1),2);
     xj = geom(connec(i,2),1); yj = geom(connec(i,2),2);
     xk = geom(connec(i,3),1); yk = geom(connec(i,3),2);
%
%    Retrieve the precipitations recorded at its nodes
%
     qi = q(connec(i,1)); qj =q(connec(i,2)); qk =q(connec(i,3));
%
%    calculate its area
%
     A = (0.5)*det([1   xi    yi;...
                    1   xj    yj;...
                    1   xk    yk]);
     AT = AT + A;
%
%    Estimate quantity of rain over its area
%
     Q = (qi+qj+qk)*A/3;
     QT = QT + Q;
end
%
AT
QT
```

After execution of the code, the area and the total quantity of rainfall are obtained respectively as

$$A = 3{,}775 \text{ km}^2$$

$$Q = 75{,}500 \text{ mm km}^2$$

One quadratic triangular element

In this case, we will construct the trial function on a triangular reference element with local coordinates ξ and η:

$$\overline{Q}(x, y) = \sum_{i=1}^{6} N_i(\xi, \eta) Q_i$$

The shape functions $N_i(\xi, \eta)$, $i = 1$ to 6 are given by Equation (7.84). The element is isoparametric, and the coordinates x and y of any point of the parent element are given as

$$x = \sum_{i=1}^{6} N_i(\xi, \eta) x_i$$

$$y = \sum_{i=1}^{6} N_i(\xi, \eta) y_i$$

The total area of the element is given as

$$A = \int_A dA = \int_0^{+1} \int_0^{1-\xi} \det(J(\xi, \eta))\, d\xi\, d\eta = \sum_k^{npt} W_k \det(J(\xi_k, \eta_k))$$

and total rainfall over the element is obtained as

$$Q_e = \int_A \overline{Q}(x, y)\, dA$$

$$= \int_o^{+1} \int_0^{1-\xi} \left(\sum_{i=1}^{6} N_i(\xi, \eta) Q_i \right) det[J(\xi, \eta)]\, d\xi\, d\eta$$

$$= \sum_k^{npt} W_k \left(\sum_{i=1}^{6} N_i(\xi_k, \eta_k) Q_i \right) det[J(\xi_k, \eta_k)]$$

The number of integration points npt, the weights W_k, and the abscissae ξ_k and η_k are given in Table 8.2. The matrix $[J]$ is the Jacobian and is given as

$$[J] = \begin{bmatrix} \dfrac{\partial x}{\partial \xi} & \dfrac{\partial y}{\partial \xi} \\[2mm] \dfrac{\partial x}{\partial \eta} & \dfrac{\partial y}{\partial \eta} \end{bmatrix} = \begin{bmatrix} \sum_{i=1}^{6} \dfrac{\partial N_i}{\partial \xi} x_i & \sum_{i=1}^{6} \dfrac{\partial N_i}{\partial \xi} y_i \\[3mm] \sum_{i=1}^{6} \dfrac{\partial N_i}{\partial \eta} x_i & \sum_{i=1}^{6} \dfrac{\partial N_i}{\partial \eta} y_i \end{bmatrix}$$

The partial derivatives $\frac{\partial N_i}{\partial \xi}$ and $\frac{\partial N_i}{\partial \eta}$ are obtained by deriving the shape functions $N_i(\xi, \eta)$, $i = 1$ to 6 with respect to ξ and η.

The calculations in this example are quite elaborate as they involve numerical integration, matrix multiplication, and evaluation of determinants. Therefore, it is better to write a MATLAB code named **precipitation_T6.m** containing the functions **hammer.m** and **fmT6_quad.m** both listed in Appendix A. The function **hammer.m** returns the weight and abscissa listed in Table 8.2. The function **fmT6_quad.m** returns the shape functions stored in the vector **fun** and their derivatives with respect to ξ and η stored in the array **der**. The Jacobian is simply evaluated as **jac = der ∗ coord**. The array **coord** contains the coordinates x and y of the nodes of the element.

precipitation_T6.m

```
%     PROGRAM  precipitation_T6.m
%
%     This program estimates the quantity of rainfall over
%     an area discretized with linear triangular elements
%
clear
clc
nel = 1      % Total number of elements
nnd = 6      % Total number of nodes
nne = 6      % Number of nodes per element
npt=4;
samp=hammer(npt);
%
% Coordinates of the rain gauges (nodes)in km
%
geom = [15.      15.    ; ... %  Node 1
          62.5    25.    ; ... %  Node 2
          110.    35.    ; ... %  Node 3
          87.5    70.    ; ... %  Node 4
```

```
        65.      105.   ; ... %  Node 5
        40.      60.]   ;      %  Node 6
%
% Precipitations recorded by the rain gauges
%
q = [20.; 15.; 10.; 20.; 30.; 25.] ;
%
%  Connectivity
%
connec = [1   2   3    4    5    6]; ... % Element 1
%
AT = 0. ; % Initialize total area to zero
QT = 0. ; % Initialize total rainfall to zero
%
for i=1:nel
%
%    for each element retrieve the vector qe containing the
%    precipitations  at its nodes as well as the matrix coord
%    containing  the x and y coordinates of the nodes
%
    for k=1: nne
        qe(k) = q(connec(i,k));
        for j=1:2
        coord(k,j)=geom(connec(i,k),j);
        end
    end
%
    for ig = 1:npt
        WI = samp(ig,3);
        [der,fun] = fmT6_quad(samp, ig);
        JAC = der*coord;
        DET = det(JAC);
%
%   calculate its area
%
    AT = AT+ WI*DET;
%
%  Estimate quantity of rain over its area
%
   QT = QT + WI*dot(fun,qe)*DET;
    end
 end
%
AT QT
```

After execution of the code, we obtain exactly the same results as with the linear triangular elements; that is,

$$A = 3,775 \ \text{km}^2$$

$$Q = 75,500 \ \text{mm}\,\text{km}^2$$

9 Plane Problems

9.1 INTRODUCTION

By now, it should have become clear to the reader that in any finite element analysis we are not analyzing the actual physical problem, but a mathematical model of it. As a result, we introduce some simplifications, and hence some modeling errors. In reality all solids are three-dimensional. Fortunately, for many problems which are of practical interest, some simplifying assumptions can be made regarding the stress or strain distributions. For example, in Chapters 2 through 4 dealing with skeletal structures, line-type elements were used because of the predominance of the longitudinal stress. In Section 5.4.4, we have also seen that when the loading and/or geometry permit it, a solid can be analyzed as a plane stress or plane strain problem. There are also other simplifications for solids that posses a symmetry of revolution in both geometry and loading, and for flat solids loaded perpendicular to their plane. These will be dealt respectively in Chapters 10 and 11. However, unlike skeletal structures, whose discretization into an assembly of elements is relatively easy, the connecting joints naturally constitute the nodes, such an intuitive approach does not exist for a two- or three-dimensional continuum. There are no joints to be used as nodes or cleavage lines to be used as elements' edges. Hence, the discretization becomes a process that requires an understanding of the physical problem at hand. It should be also added that the more physical details one tries to capture, the more complex the model becomes. In particular, the user has to decide on the choice of element type and size. These depend on the physical make-up of the body, the loading, and on how close to the actual behavior the user wants the results to be. He/she also has to decide whether the model can be simplified? And how could the results be checked? There are, of course, no definite answers to these questions. In this chapter dealing with plane problems, and in Chapters 10 and 11, we will formulate the finite element method, and in the process attempt to answer some of these questions. The user, however, is reminded that only practice makes perfect.

9.2 FINITE ELEMENT FORMULATION FOR PLANE PROBLEMS

The stress–strain relationships for a plane problem, see Section 5.4.4, are given for plane stress as

$$\begin{Bmatrix} \sigma_{xx} \\ \sigma_{yy} \\ \tau_{xy} \end{Bmatrix} = \frac{E}{1-\nu^2} \begin{bmatrix} 1 & \nu & 0 \\ \nu & 1 & 0 \\ 0 & 0 & \dfrac{(1-\nu)}{2} \end{bmatrix} \begin{Bmatrix} \epsilon_{xx} \\ \epsilon_{yy} \\ \gamma_{xz} \end{Bmatrix} \tag{9.1}$$

and for plane strain as

$$\begin{Bmatrix} \sigma_{xx} \\ \sigma_{yy} \\ \tau_{xy} \end{Bmatrix} = \frac{E}{(1+\nu)(1-2\nu)} \begin{bmatrix} 1-\nu & -\nu & 0 \\ -\nu & 1-\nu & 0 \\ 0 & 0 & \dfrac{(1-2\nu)}{2} \end{bmatrix} \begin{Bmatrix} \epsilon_{xx} \\ \epsilon_{yy} \\ \gamma_{xy} \end{Bmatrix} \tag{9.2}$$

Whether it is a state of plane stress or plane strain, a material point can only move in the directions x and y. Therefore, the two displacement variables that play a role are $u(x,y)$ and $v(x,y)$.

231

The infinitesimal strain displacements relations for both theories are the same (refer to Section 5.3.6), and they are given as

$$\epsilon_{xx} = \frac{\partial u}{\partial x} \tag{9.3}$$

$$\epsilon_{yy} = \frac{\partial v}{\partial y} \tag{9.4}$$

$$\gamma_{xy} = \frac{\partial u}{\partial y} + \frac{\partial v}{\partial x} \tag{9.5}$$

These relations can be written in a matrix form as

$$\left\{ \begin{array}{c} \epsilon_{xx} \\ \epsilon_{yy} \\ \gamma_{xy} \end{array} \right\} = \begin{bmatrix} \dfrac{\partial}{\partial x} & 0 \\ 0 & \dfrac{\partial}{\partial y} \\ \dfrac{\partial}{\partial y} & \dfrac{\partial}{\partial x} \end{bmatrix} \left\{ \begin{array}{c} u \\ v \end{array} \right\} \tag{9.6}$$

or in a more compact form as

$$\{\epsilon\} = [L]U \tag{9.7}$$

where $[L]$ is a linear differential operator.

The only unknowns in Equations (9.1) through (9.7) are actually the displacements u and v. If these are known, then the strains and the stresses can be obtained in a unique fashion, provided of course that the compatibility equations (5.89) are satisfied.

Let us consider a finite element approximation for the unknown functions u and v. For an element having n nodes, the unknown displacements are interpolated using nodal approximations as

$$u = N_1 u_1 + N_2 u_2 + \cdots + N_n u_n \tag{9.8}$$
$$v = N_1 v_1 + N_2 v_2 + \cdots + N_n v_n \tag{9.9}$$

which, when written in a matrix form, yields

$$\left\{ \begin{array}{c} u \\ v \end{array} \right\} = \begin{bmatrix} N_1 & 0 & | & N_2 & 0 & | & \ldots & | & N_n & 0 \\ 0 & N_1 & | & 0 & N_2 & | & \ldots & | & 0 & N_n \end{bmatrix} \left\{ \begin{array}{c} u_1 \\ v_1 \\ u_2 \\ v_2 \\ \vdots \\ u_n \\ v_n \end{array} \right\} \tag{9.10}$$

or simply as

$$\{U\} = [N]a \tag{9.11}$$

with $\{a\} = \{u_1, v_1, u_2, v_2, \ldots, u_n, v_n\}$ being the vector of nodal displacements. The number and the form of the shape functions depend on the element used.

Substituting for $\{U\}$ using Equation (9.10), the strain displacement Equation (9.6) become

$$\{\epsilon\} = [B]\{a\} \tag{9.12}$$

with

$$[B] = \begin{bmatrix} \dfrac{\partial N_1}{\partial x} & 0 & | & \dfrac{\partial N_2}{\partial x} & 0 & | & \cdots & | & \dfrac{\partial N_n}{\partial x} & 0 \\[2ex] 0 & \dfrac{\partial N_1}{\partial y} & | & 0 & \dfrac{\partial N_2}{\partial y} & | & \cdots & | & 0 & \dfrac{\partial N_n}{\partial y} \\[2ex] \dfrac{\partial N_1}{\partial y} & \dfrac{\partial N_1}{\partial x} & | & \dfrac{\partial N_2}{\partial y} & \dfrac{\partial N_2}{\partial x} & | & \cdots & | & \dfrac{\partial N_n}{\partial y} & \dfrac{\partial N_n}{\partial x} \end{bmatrix} \tag{9.13}$$

The matrix $[B]$ is called the strain matrix; it relates the nodal displacements to the strains. It is formed by the partial derivatives of the shape functions $N_i(x, y)$.

To derive the matrix relationship between the loads acting on the element and its nodal displacements, we will make use of the principle of virtual work, which has already been introduced in Section 6.4. For a single finite element, the principle of virtual work is written as

$$\int_{V_e} \delta\{\epsilon\}^T \{\sigma\}\, dV = \int_{V_e} \delta\{U\}^T \{b\}\, dV + \int_{\Gamma_e} \delta\{U\}^T \{t\}\, d\Gamma + \sum_i \delta\{U\}^T_{(\{x\}=\{\bar{x}\})} \{P\}_i \tag{9.14}$$

where

$\{\epsilon\}$ represents the strain vector
$\{\sigma\}$ is the stress vector
$\{U\}$ is the displacements vector
$\{b\}$ is the body forces vector
$\{t\}$ is the traction forces vector
$\{P\}_i$ is the vector of concentrated forces applied at $\{x\} = \{\bar{x}\}$
dv is an element of volume
$d\Gamma$ is an element of the boundary of the element on which the traction forces $\{t\}$ are applied

The variation in the strains $\{\delta\epsilon\}$ and in the displacements $\{\delta U\}$ can now be respectively expressed as

$$\{\delta\epsilon\} = \delta([B]\{a\}) = [B]\{\delta a\} \tag{9.15}$$

$$\{\delta U\} = \delta([N]\{a\}) = [N]\{\delta a\} \tag{9.16}$$

Substituting for $\{\epsilon\}$ using Equation (9.12), the stress–strain relationship is written as

$$\{\sigma\} = [D]\{\epsilon\} = [D][B]\{a\} \tag{9.17}$$

Substituting for $\delta\{U\}$, $\delta\{\epsilon\}$, and $\{\sigma\}$ in Equation (9.14), the principle of virtual work is written as

$$\int_{V_e} \delta\{a\}^T [B]^T [D][B]\{a\}\, dV = \int_{V_e} \delta\{a\}^T [N]^T \{b\}\, dV + \int_{\Gamma_e} \delta\{a\}^T [N]^T \{t\}\, d\Gamma + \sum_i \delta\{a\}^T [N_{(\{x\}=\{\bar{x}\})}]^T \{P\}_i \tag{9.18}$$

Note that for a plane element the element of volume dv and the element of boundary $d\Gamma$ can be written respectively as $dv = t\,dA$ and $d\Gamma = t\,dl$, where t represents the thickness of the element, dA an infinitesimal element of its area, and dl an infinitesimal element of its boundary.

Since $\delta\{a\}$ is a variation in the nodal values, therefore independent of the spatial coordinates, it can be taken out of the integral signs and completely eliminated from the earlier equation, which becomes

$$\left[\int_{A_e}[B]^T[D][B]tdA\right]\{a\} = \int_{A_e}[N]^T\{b\}tdA + \int_{L_e}[N]^T\{t\}tdl + \sum_i[N_{(\{x\}=\{\bar{x}\})}]^T\{P\}_i \tag{9.19}$$

Equation (9.19) can be rewritten in a matrix form as

$$[K_e]\{a\} = f_e \tag{9.20}$$

$$[K_e] = \left[\int_{A_e}[B]^T[D][B]tdA\right] \tag{9.21}$$

is the element stiffness matrix, and

$$\{f_e\} = \int_{A_e}[N]^T\{b\}tdA + \int_{L_e}[N]^T\{t\}tdl + \sum_i[N_{(\{x\}=\{\bar{x}\})}]^T\{P\}_i$$

is the element force vector.

9.3 SPATIAL DISCRETIZATION

The first step in any finite element analysis is the partition of the domain into a suitable mesh of elements. There is of course no unique way in achieving a mesh. However, the foregoing considerations must be addressed.

- Two distinct elements can only have in common nodes situated along their common boundary if the latter exists. This condition excludes any overlapping between two or more elements. Figure 9.1 shows one of the most common discretization errors involving overlapping between elements.
- The meshed domain should resemble as much as possible the original domain. Holes between elements as shown in Figure 9.2 are not permitted unless the holes physically exist in the original domain.
- Elongated or highly skewed elements as shown in Figure 9.3 should be avoided as they result in decreased accuracy.
- When meshing domains with curved boundaries, a geometrical discretization error is unavoidable. However, it can be reduced by refining the mesh or using higher-order elements, as shown in Figure 9.4.

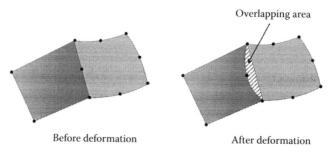

Before deformation   After deformation

FIGURE 9.1   Discretization error involving overlapping.

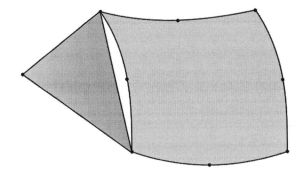

FIGURE 9.2 Discretization error involving holes between elements.

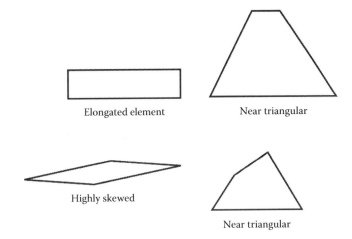

Elongated element

Near triangular

Highly skewed

Near triangular

FIGURE 9.3 Plane elements with shape distortions.

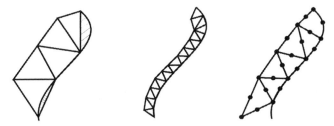

FIGURE 9.4 Geometrical discretization error.

9.4 CONSTANT STRAIN TRIANGLE

The linear triangular element shown in Figure 9.5 is perhaps the earliest finite element. It has three nodes, and each node has two degrees of freedom. Its shape functions have already been obtained in Chapter 7, and they are given as

$$
\begin{aligned}
N_1(x, y) &= m_{11} + m_{12}x + m_{13}y \\
N_2(x, y) &= m_{21} + m_{22}x + m_{23}y \\
N_3(x, y) &= m_{31} + m_{32}x + m_{33}y
\end{aligned}
\tag{9.22}
$$

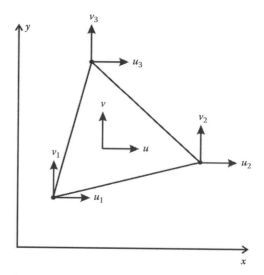

FIGURE 9.5 Linear triangular element.

with

$$m_{11} = \frac{x_2 y_3 - x_3 y_2}{2A} \qquad m_{12} = \frac{y_2 - y_3}{2A} \qquad m_{13} = \frac{x_3 - x_2}{2A}$$

$$m_{21} = \frac{x_3 y_1 - x_1 y_3}{2A} \qquad m_{22} = \frac{y_3 - y_1}{2A} \qquad m_{23} = \frac{x_1 - x_3}{2A} \qquad (9.23)$$

$$m_{31} = \frac{x_1 y_2 - x_2 y_1}{2A} \qquad m_{32} = \frac{y_1 - y_2}{2A} \qquad m_{33} = \frac{x_2 - x_1}{2A}$$

and

$$A = \frac{1}{2}\det \begin{bmatrix} 1 & x_1 & y_1 \\ 1 & x_2 & y_2 \\ 1 & x_3 & y_3 \end{bmatrix} \qquad (9.24)$$

9.4.1 Displacement Field

The displacement field over the element is approximated as

$$u = N_1 u_1 + N_2 u_2 + N_3 u_3 \qquad (9.25)$$

$$v = N_1 v_1 + N_2 v_2 + N_3 v_3 \qquad (9.26)$$

or in a matrix form as

$$\begin{Bmatrix} u \\ v \end{Bmatrix} = \begin{bmatrix} N_1 & 0 & | & N_2 & 0 & | & N_3 & 0 \\ 0 & N_1 & | & 0 & N_2 & | & 0 & N_3 \end{bmatrix} \begin{Bmatrix} u_1 \\ v_1 \\ u_2 \\ v_2 \\ u_3 \\ v_3 \end{Bmatrix} \qquad (9.27)$$

or more compactly as

$$\{U\} = [N]\{a\} \tag{9.28}$$

9.4.2 Strain Matrix

Substituting for the displacements u and v in Equation (9.6) using Equation (9.27), the strain vector is obtained as

$$\{\epsilon\} = [B]\{a\} \tag{9.29}$$

with

$$[B] = \begin{bmatrix} \dfrac{\partial N_1}{\partial x} & 0 & \bigg| & \dfrac{\partial N_2}{\partial x} & 0 & \bigg| & \dfrac{\partial N_3}{\partial x} & 0 \\[3mm] 0 & \dfrac{\partial N_1}{\partial y} & \bigg| & 0 & \dfrac{\partial N_2}{\partial y} & \bigg| & 0 & \dfrac{\partial N_3}{\partial y} \\[3mm] \dfrac{\partial N_1}{\partial y} & \dfrac{\partial N_1}{\partial x} & \bigg| & \dfrac{\partial N_2}{\partial y} & \dfrac{\partial N_2}{\partial x} & \bigg| & \dfrac{\partial N_3}{\partial y} & \dfrac{\partial N_3}{\partial x} \end{bmatrix} \tag{9.30}$$

Substituting Equations (9.22) and (9.23) in (9.30), the matrix $[B]$ becomes

$$[B] = \begin{bmatrix} m_{12} & 0 & | & m_{22} & 0 & | & m_{32} & 0 \\ 0 & m_{13} & | & 0 & m_{23} & | & 0 & m_{33} \\ m_{13} & m_{12} & | & m_{23} & m_{22} & | & m_{33} & m_{32} \end{bmatrix} \tag{9.31}$$

Remark: The matrix $[B]$ is independent of the Cartesian coordinates x and y. It is a function of the nodal coordinates only, and it is constant all over the element. It follows therefore that the strain vector is constant over the element. That is the reason why the element is termed "constant strain triangle."

9.4.3 Stiffness Matrix

The stiffness matrix of the element is given by Equation (9.21). Since both the matrices $[B]$ and $[D]$ are constant, the stiffness matrix becomes

$$[K_e] = [B]^T[D][B]tA_e \tag{9.32}$$

where A_e represents the area of the element and is given by Equation (9.24).

9.4.4 Element Force Vector

The element force vector is given by Equation (9.22).

9.4.4.1 Body Forces

Considering that the body forces $\{b\}$ are due to gravity, the first term of Equation (9.22) is evaluated as

$$
\int_{A_e} [N]^T \{b\} t \, dA = t \int_{A_e}
\begin{bmatrix}
N_1 & 0 \\
0 & N_1 \\
N_2 & 0 \\
0 & N_2 \\
N_3 & 0 \\
0 & N_3
\end{bmatrix}
\left\{
\begin{array}{c}
0 \\
-\rho g
\end{array}
\right\} dA = t
\begin{bmatrix}
0 \\
-\int_{A_e} N_1 \rho g \, dA \\
0 \\
-\int_{A_e} N_2 \rho g \, dA \\
0 \\
-\int_{A_e} N_3 \rho g \, dA
\end{bmatrix}
\tag{9.33}
$$

The individual integrals over the area involving the shape functions are evaluated using the integration formulas over a triangle presented in Equations (8.18) and (8.19). Applying these formulas, the individual integrals are evaluated as follows:

$$
\int_{A_e} N_1 \rho g \, dA = \rho g \int_{A_e} N_1^1 N_2^0 N_3^0 \, dA = \rho g \frac{1!0!0!}{(1+0+0+2)!} 2A_e = \rho g \frac{A_e}{3}
\tag{9.34}
$$

$$
\int_{A_e} N_2 \rho g \, dA = \rho g \int_{A_e} N_1^0 N_2^1 N_3^0 \, dA = \rho g \frac{0!1!0!}{(0+1+0+2)!} 2A_e = \rho g \frac{A_e}{3}
\tag{9.35}
$$

$$
\int_{A_e} N_3 \rho g \, dA = \rho g \int_{A_e} N_1^0 N_2^0 N_3^1 \, dA = \rho g \frac{0!0!1!}{(0+0+1+2)!} 2A_e = \rho g \frac{A_e}{3}
\tag{9.36}
$$

Substituting back in Equation (9.33), we obtain

$$
\int_{A_e} [N]^T \{b\} t \, dA = -\frac{t}{3}
\left\{
\begin{array}{c}
0 \\
\rho g A_e \\
0 \\
\rho g A_e \\
0 \\
\rho g A_e
\end{array}
\right\}
\tag{9.37}
$$

It can be noticed that the self-weight of the element is shared equally between the nodes.

9.4.4.2 Traction Forces

Consider the element shown in Figure 9.6 subject to a uniformly distributed load of magnitude q normal to the side 2–3 and at an angle θ with the global axis x. The vector of the traction forces can therefore be written as $\{t\} = \{-q\cos\theta, -q\sin\theta\}^T$.

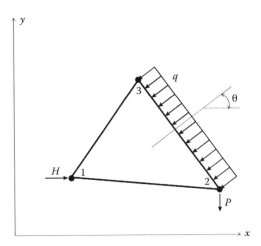

FIGURE 9.6   Element nodal forces.

The second term of Equation (9.22) is evaluated as

$$\int_{L_e} [N]^T \{t\} t\, dl = \int_{L_{2-3}} \begin{bmatrix} 0 & 0 \\ 0 & 0 \\ N_2 & 0 \\ 0 & N_2 \\ N_3 & 0 \\ 0 & N_3 \end{bmatrix} \begin{Bmatrix} -q\cos\theta \\ -q\sin\theta \end{Bmatrix} t\, dl = t \int_{L_{2-3}} \begin{Bmatrix} 0 \\ 0 \\ -N_2 q\cos\theta \\ -N_2 q\sin\theta \\ -N_3 q\cos\theta \\ -N_3 q\sin\theta \end{Bmatrix} dl \qquad (9.38)$$

Notice that $N_1 = 0$ on side 2–3.

The integrals over the length are evaluated using the integration formula over a side of a triangle given by Equation (8.18). Applying this formula, the aforementioned integral becomes

$$\int_{L_e} [N]^T \{t\} t\, dl = t \begin{Bmatrix} 0 \\ 0 \\ -q\cos\theta L_{2-3}/2 \\ -q\sin\theta L_{2-3}/2 \\ -q\cos\theta L_{2-3}/2 \\ -q\sin\theta L_{2-3}/2 \end{Bmatrix} \qquad (9.39)$$

It can be noticed that the nodes 2 and 3 share the applied load qL_{2-3} equally between them.

9.4.4.3 Concentrated Forces

Finally, considering the element shown in Figure 9.6 subject to a horizontal force H and a vertical force P applied respectively at nodes 1 and 2, the third term of Equation (9.22) is evaluated as follows:

$$\sum_i [N_{(\{x\}=\{\bar{x}\})}]^T \{P\}_i = \begin{bmatrix} N_1 = 1 & 0 \\ 0 & N_1 = 1 \\ 0 & 0 \\ 0 & 0 \\ 0 & 0 \\ 0 & 0 \end{bmatrix} \begin{Bmatrix} H \\ 0 \end{Bmatrix} + \begin{bmatrix} 0 & 0 \\ 0 & 0 \\ N_2 = 1 & 0 \\ 0 & N_2 = 1 \\ 0 & 0 \\ 0 & 0 \end{bmatrix} \begin{Bmatrix} 0 \\ -P \end{Bmatrix} = \begin{Bmatrix} 0 \\ 0 \\ H \\ 0 \\ 0 \\ -P \end{Bmatrix} \tag{9.40}$$

Notice that $N_1 = 1$ when evaluated at node 1 and equal to zero when evaluated at nodes 2 and 3. In a similar fashion, $N_2 = 1$ when evaluated at node 2 and equal to zero when evaluated at nodes 1 and 3. $N_3 = 0$ when evaluated at nodes 1 and 2 where the loads are applied.

9.4.5 COMPUTER CODES USING THE CONSTANT STRAIN TRIANGLE

Writing a finite element code using any type of element follows exactly the same principles as those we used in Chapters 2 through 4 for writing the codes **Truss.m**, **Beam.m**, and **Frame.m**. Therefore, in the development of the codes **CST_PLANE_STRESS.m** and **CST_PLANE_STRESS_MESH.m** to follow, we will not only use the same style, but we will also borrow some functions from the codes **Truss.m**, **Beam.m**, and **Frame.m**.

Let us consider the cantilever beam shown in Figure 9.7, which has an exact analytical solution. The vertical displacement of any point is given as

$$v = \frac{\nu P x y^2}{2EI} + \frac{P x^3}{6EI} - \frac{P L^2 x}{2EI} + \frac{P L^3}{3EI} \tag{9.41}$$

where I represents the second moment of area of the section with respect to the axis z. Note that in the axis y is oriented from top to the bottom. As a result, Equation (9.41) yields positive values for the vertical displacement v.

To carry out a finite element analysis of the cantilever, it is necessary to introduce some numerical values for the dimensions, the elastic constants and the loading. Let us consider $C = 10$ mm, $L = 60$ mm, $t = 5$ mm for the geometrical properties, a Young's modulus of 200000 MPa and a Poisson's ratio of 0.3 for the material properties, as well as a concentrated force P of 1000 N. We will use 24 elements to discretize the domain as shown in Figure 9.8. The nodes numbered 19, 20, and 21 represent the fixed end.

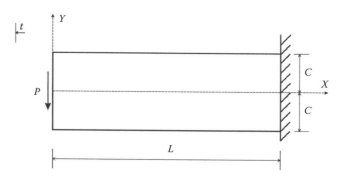

FIGURE 9.7   Analysis of a cantilever beam in plane stress.

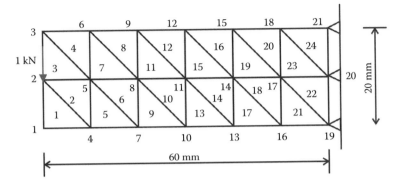

FIGURE 9.8   Finite element discretization with linear triangular elements.

9.4.5.1   Data Preparation

To read the data, we will use the M-file **CST_COARSE_MESH_DATA.m** listed next:

FILE: CST_COARSE_MESH_DATA.m

```
%               File:   CST_COARSE_MESH_DATA.m
%
% The following variables are declared as global in order
% to be used by all the functions (M-files) constituting
% the program
%
%
global nnd nel nne nodof eldof n
global geom connec dee nf Nodal_loads
%
format short e
%
nnd = 21 ;                 % Number of nodes:
nel = 24 ;                 % Number of elements:
nne = 3 ;                  % Number of nodes per element:
nodof =2;                  % Number of degrees of freedom per node
eldof = nne*nodof;         % Number of degrees of freedom per element
%
% Nodes coordinates x and y
%
geom = zeros(nnd,2);
%
geom = [ 0,    -10;  ... % Node 1
         0,     0;   ... % Node 2
         0,     10;  ... % Node 3
         10,    -10;  ... % Node 4
         10,    0;   ... % Node 5
         10,    10;  ... % Node 6
         20,    -10;  ... % Node 7
         20,    0;   ... % Node 8
         20,    10;  ... % Node 9
         30,    -10;  ... % Node 10
         30,    0;   ... % Node 11
         30,    10;  ... % Node 12
         40,    -10;  ... % Node 13
         40,    0;   ... % Node 14
         40,    10;  ... % Node 15
         50,    -10;  ... % Node 16
         50,    0;   ... % Node 17
         50,    10;  ... % Node 18
         60,    -10;  ... % Node 19
```

```
            60,    0;   ... % Node 20
            60,   10];      % Node 21
%
% Element connectivity
%
connec=zeros(nel,3);
connec = [ 1,    4,   2;  ...% Element 1
           4,    5,   2;  ...% Element 2
           2,    5,   3;  ...% Element 3
           5,    6,   3;  ...% Element 4
           4,    7,   5;  ...% Element 5
           7,    8,   5;  ...% Element 6
           5,    8,   6;  ...% Element 7
           8,    9,   6;  ...% Element 8
           7,   10,   8;  ...% Element 9
          10,   11,   8;  ...% Element 10
           8,   11,   9;  ...% Element 11
          11,   12,   9;  ...% Element 12
          10,   13,  11;  ...% Element 13
          13,   14,  11;  ...% Element 14
          11,   14,  12;  ...% Element 15
          14,   15,  12;  ...% Element 16
          13,   16,  14;  ...% Element 17
          16,   17,  14;  ...% Element 18
          14,   17,  15;  ...% Element 19
          17,   18,  15;  ...% Element 20
          16,   19,  17;  ...% Element 21
          19,   20,  17;  ...% Element 22
          17,   20,  18;  ...% Element 23
          20,   21,  18];     % Element 24
%
% Material
%
E = 200000.;     % Elastic modulus in MPa
vu = 0.3;        % Poisson's ratio
thick = 5.;      % Beam thickness in mm
%
% Form the elastic matrix for plane stress
%
dee = formdsig(E,vu);
%
% Boundary conditions
%
nf = ones(nnd, nodof);     % Initialize the matrix nf to 1
nf(19,1) = 0; nf(19,2) = 0;  % Prescribed nodal freedom of node 19
nf(20,1) = 0; nf(20,2) = 0;  % Prescribed nodal freedom of node 20
nf(21,1) = 0; nf(21,2) = 0;  % Prescribed nodal freedom of node 21
%
% Counting of the free degrees of freedom
%
n=0;
for i=1:nnd
    for j=1:nodof
        if nf(i,j) ~= 0
            n=n+1;
            nf(i,j)=n;
        end
    end
end
%
% loading
%
Nodal_loads= zeros(nnd, 2);
%
Nodal_loads(2,1) = 0.; Nodal_loads(2,2) = -1000.;    % Node 2
%
%%%%%%%%%%%   End of input        %%%%%%%%%%%
```

The input data for this beam consist of

- **nnd = 21**; number of nodes
- **nel = 24**; number of elements
- **nne = 3**; number of nodes per element
- **nodof = 2**; number of degrees of freedom per node

The thickness of the beam, which is a geometrical property, is given as **thick = 5**.

9.4.5.2 Nodes Coordinates

The coordinates x and y of the nodes are given in the form of a matrix **geom(nnd, 2)**.

9.4.5.3 Element Connectivity

The element connectivity is given in the matrix **connec(nel, 3)**. Note that the internal numbering of the nodes is anticlockwise.

9.4.5.4 Material Properties

The material properties, namely, elastic modulus and Poisson's ratio, are given in the variables **E = 200000** and **vu = 0.3**. With these properties we form the elastic matrix for plane stress using the function **formdsig.m** listed in Appendix A, which returns the matrix **dee**.

9.4.5.5 Boundary Conditions

In the same fashion as for a truss or a beam, a restrained degree of freedom is assigned the digit 0, while a free degree of freedom is assigned the digit 1. As previously explained, a node has two degrees of freedom: a horizontal translation along the axis X and a vertical translation along the axis Y. As shown in Figure 9.8, nodes 19, 20, and 21 represent the fixed end of the cantilever, which is fully fixed. The prescribed degrees of freedom of these nodes are assigned the digit 0. All the degrees of freedom of all the other nodes, which are free, are assigned the digits 1. The information on the boundary conditions is given in the matrix **nf(nnd, nodof)**.

9.4.5.6 Loading

The concentrated force of 1000 N is applied at node 2. The force will be assembled into the global force vector **fg** in the main program.

9.4.5.7 Main Program

The main program **CST_PLANE_STRESS.m** is listed next:

```
% THIS PROGRAM USES AN 3-NODE LINEAR TRIANGULAR ELEMENT FOR THE
% LINEAR ELASTIC STATIC ANALYSIS OF A TWO DIMENSIONAL PROBLEM
%
clear all
clc
%
% Make these variables global so they can be shared by other functions
%
global nnd nel nne nodof eldof n
global geom connec dee nf Nodal_loads
%
format long g
%
% ALTER NEXT LINES TO CHOOSE THE NAME OF THE OUTPUT FILE
%
fid =fopen('CST_COARSE_MESH_RESULTS.txt','w');
```

```
%
% To change the size of the problem or change elastic properties
% supply another input file
%
CST_COARSE_MESH_DATA;
%
%%%%%%%%%%%%%%%%%%%%%%%%%% End of input%%%%%%%%%%%%%%%%%%%%%%%%%%%%%%%%%%
%
% Assemble the global force vector
%
fg=zeros(n,1);
for i=1: nnd
if nf(i,1) ~= 0
fg(nf(i,1))= Nodal_loads(i,1);
end
if nf(i,2) ~= 0;
fg(nf(i,2))= Nodal_loads(i,2);
end
end
%
% Assembly of the global stiffness matrix
%
% initialize the global stiffness matrix to zero
%
kk = zeros(n, n);
%
for i=1:nel
[bee,g,A] = elem_T3(i); % Form strain matrix, and steering vector
ke=thick*A*bee'*dee*bee; % Compute stiffness matrix
kk=form_kk(kk,ke, g); % assemble global stiffness matrix
end
%
%
%%%%%%%%%%%%%%%%%%%%%%%%%% End of assembly %%%%%%%%%%%%%%%%%%%%%%%%%%%%%%%%
%
%
delta = kk\fg ; % solve for unknown displacements
%
node_disp=zeros(nnd,2);
%
for i=1: nnd %
if nf(i,1) == 0 %
x_disp =0.; %
else
x_disp = delta(nf(i,1)); %
end
%
if nf(i,2) == 0 %
y_disp = 0.; %
else
y_disp = delta(nf(i,2)); %
end
node_disp(i,:) =[x_disp y_disp];
end
%
% Retrieve the x_coord and y_disp of the nodes located on the neutral axis
%
k = 0;
for i=1:nnd;

    if geom(i,2)== 0.
        k=k+1;
        x_coord(k) = geom(i,1);
        vertical_disp(k)=node_disp(i,2);
    end
end
```

```
%
%
for i=1:nel
[bee,g,A] = elem_T3(i); % Form strain matrix, and steering vector
eld=zeros(eldof,1); % Initialize element displacement to zero
for m=1:eldof
if g(m)==0 eld(m)=0.;
else %
eld(m)=delta(g(m)); % Retrieve element displacement
end
end
%
eps=bee*eld; % Compute strains
EPS(i,:)=eps ; % Store strains for all elements
sigma=dee*eps; % Compute stresses
SIGMA(i,:)=sigma ; % Store strains for all elements
end
%
% Print results to file
%
print_CST_results;
%
% Plot the stresses in the x_direction
%
x_stress = SIGMA(:,1);
cmin = min(x_stress);
cmax = max(x_stress);
caxis([cmin cmax])
patch('Faces', connec, 'Vertices', geom, 'FaceVertexCData',x_stress,  ...
'Facecolor','flat','Marker','o')
colorbar
%
plottools
```

After declaring the global variables that will be used by the functions, and the naming of the output results file **'CST_COARSE_MESH_RESULTS.txt'**, the program starts by uploading the data file and assembling the global force vector **fg**. The elements' stiffness matrices, the assembly of the global stiffness matrix, the solution of the global equations, and the computation of stresses and strains are obtained as follows.

9.4.5.8 Element Stiffness Matrix

For each element, from 1 to *nel*, we set up its strain matrix **bee**, its steering vector **g**, and calculate its area **A**. This is achieved in the function **elem_T3.m**, which can be found in Appendix A.

- For any element **i**, retrieve the coordinates *x* and *y* of its nodes

$$\mathbf{x1} = \mathbf{geom(connec(i, 1), 1)}; \qquad \mathbf{y1} = \mathbf{geom(connec(i, 1), 2)}$$

$$\mathbf{x2} = \mathbf{geom(connec(i, 2), 1)}; \qquad \mathbf{y2} = \mathbf{geom(connec(i, 2), 2)}$$

$$\mathbf{x3} = \mathbf{geom(connec(i, 3), 1)}; \qquad \mathbf{y3} = \mathbf{geom(connec(i, 3), 2)}$$

- Calculate the area of the element using Equation (7.34), and the coefficients $\mathbf{m_{jk}}$, $\mathbf{j}$, $\mathbf{k} = $ **1, 2, 3** using Equation (7.36)
- Using the coefficients $\mathbf{m_{jk}}$, assemble the matrix **bee** using Equation (9.31)

- Using the matrix of nodal freedom **nf** in combination with the connectivity matrix, retrieve the steering vector **g** for the element

$$
\mathbf{g} = \begin{Bmatrix} \mathbf{nf}(\mathbf{connec}(1,1),1) \\ \mathbf{nf}(\mathbf{connec}(1,1),2) \\ \mathbf{nf}(\mathbf{connec}(2,1),1) \\ \mathbf{nf}(\mathbf{connec}(2,1),2) \\ \mathbf{nf}(\mathbf{connec}(3,1),1) \\ \mathbf{nf}(\mathbf{connec}(3,1),2) \end{Bmatrix}
$$

Once the matrix **bee** is formed, the element stiffness matrix **ke** is obtained as

$$
\mathbf{ke} = \mathbf{thick} \times \mathbf{A} \times \mathbf{bee}^{\mathrm{T}} \times \mathbf{dee} \times \mathbf{bee}
$$

9.4.5.9 Assembly of the Global Stiffness Matrix

As shown in Figure 9.5, a linear triangular element has in total 6 degrees of freedom. The global stiffness matrix [**KK**] is assembled using a double loop over the components of the vector **g**. The script is exactly the same as the one used in the codes **Truss.m**, **Beam.m**, and **Frame.m**. It is given in the function **form_KK.m** listed in Appendix A.

9.4.5.10 Solution of the Global System of Equations

The solution of the global system of equations is obtained with one statement:

$$
\mathbf{delta} = \mathbf{KK}\backslash\mathbf{fg}
$$

9.4.5.11 Nodal Displacements

Once the global displacements vector **delta** is obtained, it is possible to retrieve any nodal displacements. A loop is carried over all the nodes. If a degree of freedom j of a node i is free; that is, $\mathbf{nf}(\mathbf{i},\mathbf{j}) \neq \mathbf{0}$, then it could have a displacement different from zero. The value of the displacement is extracted from the global displacements vector **delta**:

$$
\mathbf{node\_disp}(\mathbf{i},\mathbf{j}) = \mathbf{delta}(\mathbf{nf}(\mathbf{i},\mathbf{j}))
$$

9.4.5.12 Element Stresses and Strains

To obtain the element stresses and strains, a loop is carried over all the elements:

1. Form element strain matrix **bee** and "steering" vector **g**
   a. Loop over the degrees of freedom of the element to obtain element displacements vector **edg**
   b. If $\mathbf{g}(\mathbf{j}) = \mathbf{0}$, then the degree of freedom is restrained; $\mathbf{edg}(\mathbf{j}) = \mathbf{0}$
   c. Otherwise $\mathbf{edg}(\mathbf{j}) = \mathbf{delta}(\mathbf{g}(\mathbf{j}))$
2. Obtain element strain vector **eps** = **bee** × **edg**
3. Obtain element stress vector **sigma** = **dee** × **bee** × **edg**
4. Store the strains for all the elements **EPS**($\mathbf{i}$, :) = **eps** for printing to result file
5. Store the stresses for all the elements **SIGMA**($\mathbf{i}$, :) = **sigma** for printing to result file

9.4.5.13 Results and Discussion

After running the program **CST_PLANE_STRESS.m**, the results are written to the text file **CST_COARSE_MESH_RESULTS.txt** listed next:

CST_COARSE_MESH_RESULTS.txt

```
--------------------------------------------------------

******* PRINTING ANALYSIS RESULTS ************

--------------------------------------------------------
Nodal displacements
Node      disp_x            disp_y
  1,     1.45081e-002,     -6.49329e-002
  2,     3.28049e-004,     -6.52078e-002
  3,    -1.42385e-002,     -6.47141e-002
  4,     1.42332e-002,     -4.97317e-002
  5,     1.82950e-004,     -4.94530e-002
  6,    -1.38358e-002,     -4.94091e-002
  7,     1.29745e-002,     -3.50495e-002
  8,     1.37982e-004,     -3.46630e-002
  9,    -1.26721e-002,     -3.47556e-002
 10,     1.09224e-002,     -2.19922e-002
 11,     8.95233e-005,     -2.14870e-002
 12,    -1.07002e-002,     -2.16958e-002
 13,     8.08085e-003,     -1.13485e-002
 14,     2.56420e-005,     -1.07261e-002
 15,    -7.90991e-003,     -1.10480e-002
 16,     4.46383e-003,     -3.88383e-003
 17,    -6.63586e-005,     -3.19069e-003
 18,    -4.26507e-003,     -3.66370e-003
 19,     0.00000e+000,      0.00000e+000
 20,     0.00000e+000,      0.00000e+000
 21,     0.00000e+000,      0.00000e+000


--------------------------------------------------------
                   Element stresses
element    sigma_(xx)         sigma_(yy)          tau_(xy)
  1,     -7.8546e+000,      -7.8546e+000,       7.8546e+000
  2,     -1.3515e+000,       5.1683e+000,       1.3112e+001
  3,      6.6118e-002,       9.8937e+000,       9.1400e+000
  4,      9.1400e+000,       3.6192e+000,       9.8937e+000
  5,     -2.5827e+001,      -2.1744e+000,       4.8607e+000
  6,      1.5601e+000,       8.1980e+000,       1.5027e+001
  7,     -6.9913e-001,       6.6741e-001,       5.9323e+000
  8,      2.4966e+001,       5.6374e+000,       1.4180e+001
  9,     -4.2552e+001,      -5.0356e+000,       1.6983e+000
 10,      2.2662e+000,       1.0785e+001,       1.8024e+001
 11,     -1.6757e+000,      -2.3552e+000,       2.8152e+000
 12,      4.1961e+001,       8.4119e+000,       1.7462e+001
 13,     -5.9121e+001,      -7.6315e+000,      -1.4550e+000
 14,      2.6997e+000,       1.3258e+001,       2.0813e+001
 15,     -2.7809e+000,      -5.0108e+000,      -2.2163e-001
 16,      5.9202e+001,       1.1322e+001,       2.0864e+001
 17,     -7.5391e+001,      -1.0170e+001,      -4.5429e+000
 18,      2.5481e+000,       1.4627e+001,       2.3117e+001
 19,     -4.1445e+000,      -7.6816e+000,      -3.0783e+000
 20,      7.6988e+001,       1.3636e+001,       2.4504e+001
 21,     -9.3536e+001,      -1.4198e+001,      -4.9720e+000
 22,      1.4584e+000,       4.3753e-001,       2.4544e+001
 23,     -1.6603e+000,      -9.9582e+000,      -7.7540e+000
 24,      9.3738e+001,       2.8121e+001,       2.8182e+001

--------------------------------------------------------
                   Element strains
element   epsilon_(xx)       epsilon_(yy)         gamma_(xy)
  1,     -2.7491e-005,      -2.7491e-005,       1.0211e-004
```

```
2,      -1.4510e-005,    2.7869e-005,     1.7045e-004
3,      -1.4510e-005,    4.9369e-005,     1.1882e-004
4,       4.0271e-005,    4.3858e-006,     1.2862e-004
5,      -1.2587e-004,    2.7869e-005,     6.3189e-005
6,      -4.4967e-006,    3.8650e-005,     1.9535e-004
7,      -4.4967e-006,    4.3858e-006,     7.7120e-005
8,       1.1637e-004,   -9.2623e-006,     1.8434e-004
9,      -2.0521e-004,    3.8650e-005,     2.2078e-005
10,     -4.8459e-006,    5.0524e-005,     2.3431e-004
11,     -4.8459e-006,   -9.2623e-006,     3.6597e-005
12,      1.9719e-004,   -2.0883e-005,     2.2701e-004
13,     -2.8416e-004,    5.0524e-005,    -1.8915e-005
14,     -6.3881e-006,    6.2239e-005,     2.7057e-004
15,     -6.3881e-006,   -2.0883e-005,    -2.8812e-006
16,      2.7903e-004,   -3.2191e-005,     2.7123e-004
17,     -3.6170e-004,    6.2239e-005,    -5.9058e-005
18,     -9.2001e-006,    6.9314e-005,     3.0052e-004
19,     -9.2001e-006,   -3.2191e-005,    -4.0018e-005
20,      3.6448e-004,   -4.7301e-005,     3.1856e-004
21,     -4.4638e-004,    6.9314e-005,    -6.4636e-005
22,      6.6359e-006,    0.0000e+000,     3.1907e-004
23,      6.6359e-006,   -4.7301e-005,    -1.0080e-004
24,      4.2651e-004,    0.0000e+000,     3.6637e-004
```

Once the calculations are done, the first thing that needs to be checked is whether the results are reasonable or not. This task is even more difficult when "in-house" software is used as is the case here. The results, as shown earlier, are in the form of numbers, hence difficult to interpret.

The first thing we can do is to check whether the deflected shape is correct. For this, we plot the vertical displacement of the nodes situated along the neutral axis of the cantilever, as shown in Figure 9.9. As it appears, the shape is acceptable; however, the computed values are just over half those obtained with the analytical solution, Equation (9.41).

Next we plot a contour of the longitudinal stress σ_{xx} using the MATLAB® patch function, as shown in Figure 9.10. The elements above the neutral axis are in tension, while those below the neutral axis are in compression, which is obviously correct. Most importantly, the stress value is constant over each element. However, the neutral axis should be stress free, and that is not the case. As they are, the results are not satisfactory. Indeed, we are asking too much of the constant strain (stress) triangle; that is to model a stress gradient, when evidently it cannot do so. We have also used a coarse mesh without sufficient refinement to model the stress gradient.

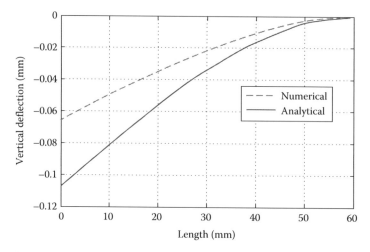

FIGURE 9.9 Deflection of the cantilever beam.

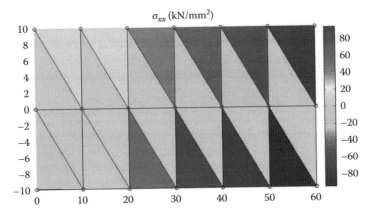

FIGURE 9.10   Stresses along the *x*-axis.

9.4.5.14   Program with Automatic Mesh Generation

To better model the stress gradient with a triangular element, we need to refine the mesh. However, this will require many elements and nodes, which is not easy to prepare by hand as we did for the coarse mesh. In the new program named **CST_PLANE_STRESS_MESH.m**, listed next, the mesh is automatically created by calling the function **T3_mesh.m**. This function prepares the elements' connectivity and nodal geometry matrices and is listed after the main program.

CST_PLANE_STRESS_MESH.m

```
% THIS PROGRAM USES AN 3-NODE LINEAR TRIANGULAR ELEMENT FOR THE
% LINEAR ELASTIC STATIC ANALYSIS OF A TWO DIMENSIONAL PROBLEM
% IT INCLUDES AN AUTOMATIC MESH GENERATION
%
% Make these variables global so they can be shared by other functions
%
clear all
clc
global nnd nel nne  nodof eldof  n
global geom  dee nf Nodal_loads
global Length Width NXE NYE X_origin Y_origin
%
 format long g
%
%
% To change the size of the problem or change elastic properties
%     supply another input file
%
Length = 60.; % Length of the model
Width =20.;    % Width
NXE = 24;      % Number of rows in the x direction
NYE = 10;      % Number of rows in the y direction
dhx = Length/NXE; % Element size in the x direction
dhy = Width/NYE;  % Element size in the x direction
X_origin = 0. ;  % X origin of the global coordinate system
Y_origin = Width/2. ;   % Y origin of the global coordinate system
%
nne = 3;
nodof = 2;
eldof = nne*nodof;
%
T3_mesh ;          % Generate the mesh
%
% Material
%
```

```
E = 200000.;        % Elastic modulus in MPa
vu = 0.3;           % Poisson's ratio
thick = 5.;         % Beam thickness in mm
%
% Form the elastic matrix for plane stress
%
dee = formdsig(E,vu);
%
%
% Boundary conditions
%
nf = ones(nnd, nodof);      % Initialize the matrix nf to 1
%
% Restrain in all directions the nodes situated @
% (x = Length)
%
for i=1:nnd
    if geom(i,1) == Length;
        nf(i,:) = [0 0];
    end
end
%
% Counting of the free degrees of freedom
%
n=0; for i=1:nnd
    for j=1:nodof
        if nf(i,j) ~= 0
            n=n+1;
            nf(i,j)=n;
        end
    end
end
%
% loading
%
Nodal_loads= zeros(nnd, 2);   % Initialize the matrix of nodal loads to 0
%
% Apply the load as a concentrated load on the node having coordinate X = Y =0.
%
Force = 1000.;  % N
%
for i=1:nnd
    if geom(i,1) == 0. && geom(i,2) == 0.
        Nodal_loads(i,:) = [0.  -Force];
    end
end
%
%%%%%%%%%%%%%%%%%%%%%%%%%%% End of input%%%%%%%%%%%%%%%%%%%%%%%%%%%%%%%%%%%%
%
% Assemble the global force vector
%
fg=zeros(n,1);
for i=1: nnd
    if nf(i,1) ~= 0
        fg(nf(i,1))= Nodal_loads(i,1);
    end
    if nf(i,2) ~= 0
        fg(nf(i,2))= Nodal_loads(i,2);
    end
end
%
% Assembly of the global stiffness matrix
%
%  initialize the global stiffness matrix to zero
%
kk = zeros(n, n);
%
```

```
for i=1:nel
    [bee,g,A] = elem_T3(i);      % Form strain matrix, and steering vector
    ke=thick*A*bee'*dee*bee;     % Compute stiffness matrix
    kk=form_kk(kk,ke, g);        % assemble global stiffness matrix
end
%
%
%%%%%%%%%%%%%%%%%%%%%%%%%   End of assembly  %%%%%%%%%%%%%%%%%%%%%%%%%%%%%%%%%%%
%
%
delta = kk\fg ;               % solve for unknown displacements
%
for i=1: nnd                                %
    if nf(i,1) == 0                         %
        x_disp =0.;                         %
    else
        x_disp = delta(nf(i,1));            %
    end
%
    if nf(i,2) == 0                         %
        y_disp = 0.;                        %
    else
        y_disp = delta(nf(i,2));            %
    end
    node_disp(i,:) =[x_disp  y_disp];
end
%
%
% Retrieve the x_coord and y_disp of the nodes located on the neutral axis
%
k = 0;
vertical_disp=zeros(1,NXE+1);
for i=1:nnd;

    if geom(i,2)== 0.
        k=k+1;
        x_coord(k) = geom(i,1);
        vertical_disp(k)=node_disp(i,2);
    end
end
%
for i=1:nel
    [bee,g,A] = elem_T3(i);      % Form strain matrix, and steering vector
    eld=zeros(eldof,1);          % Initialize element displacement to zero
    for m=1:eldof
        if g(m)==0
         eld(m)=0.;
        else                         %
         eld(m)=delta(g(m));     % Retrieve element displacement
        end
    end
%
   eps=bee*eld;                 % Compute strains
   EPS(i,:)=eps ;               % Store strains for all elements
   sigma=dee*eps;              % Compute stresses
   SIGMA(i,:)=sigma ;          % Store stresses for all elements
end
%
%
% Plot stresses in the x_direction
%
x_stress = SIGMA(:,1);
cmin = min(x_stress);
cmax = max(x_stress);
caxis([cmin cmax]);
patch('Faces', connec, 'Vertices', geom, 'FaceVertexCData',x_stress, ...
      'Facecolor','flat','Marker','o');
```

```
colorbar;
%
plottools;
```

T3_mesh.m

```
% This function generates a mesh of triangular elements
%
global nnd nel nne  nodof eldof  n
global geom connec dee nf Nodal_loads
global Length Width NXE NYE X_origin Y_origin dhx dhy
%
nnd = 0;
k = 0;
for i = 1:NXE
    for j=1:NYE
        k = k + 1;
        n1 = j + (i-1)*(NYE + 1);
        geom(n1,:) = [(i-1)*dhx - X_origin    (j-1)*dhy - Y_origin ];
        n2 = j + i*(NYE+1);
        geom(n2,:) = [i*dhx - X_origin        (j-1)*dhy - Y_origin ];
        n3 = n1 + 1;
        geom(n3,:) = [(i-1)*dhx - X_origin       j*dhy - Y_origin ];
        n4 = n2 + 1;
        geom(n4,:) = [i*dhx- X_origin        j*dhy - Y_origin     ];
        nel = 2*k;
        m = nel -1;
        connec(m,:) = [n1  n2  n3];
        connec(nel,:) = [n2  n4  n3];
        nnd = n4;
    end
end
%
```

The variables **NXE** and **NYE** represent respectively the number of intervals along the x and y directions, as shown in Figure 9.11. For each interval i and j, four nodes n_1, n_2, n_3, and n_4 and two elements are created. The first element has nodes n_1, n_2, n_3, while the second element has nodes n_2, n_4, n_3. In total the number of elements and nodes created are respectively equal to

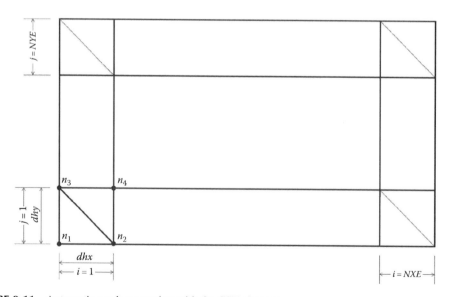

FIGURE 9.11 Automatic mesh generation with the CST element.

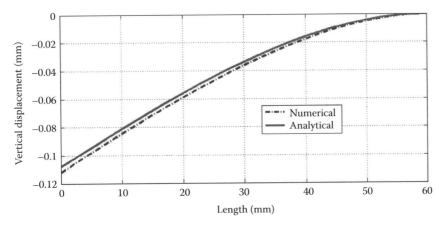

FIGURE 9.12   Deflection of the cantilever beam obtained with the fine mesh.

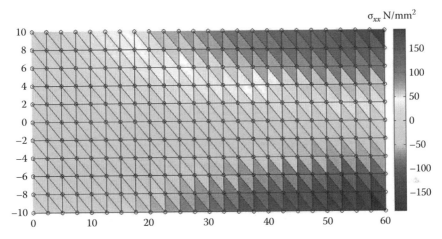

FIGURE 9.13   Stresses along the x-axis obtained with the fine mesh.

$nel = 2 \times NXE \times NYE$, and $nnd = (NXE + 1) \times (NYE + 1)$. The module also returns the matrices $geom(nnd, 2)$ and $connec(nel, nne)$.

The results obtained with the fine mesh are displayed in Figures 9.12 and 9.13. Figure 9.12 shows the deflection of the nodes situated along the center line (neutral axis). It can be clearly seen that the solution matches closely the analytical solution. Figure 9.13 displays a contour of the stresses in the x-direction. The stress gradient can be clearly seen even though each element displays a constant stress. Those elements within the vicinity of the neutral axis display stress values close to zero.

9.4.6   ANALYSIS WITH ABAQUS USING THE CST

9.4.6.1   Interactive Edition

In this section, we will analyze the cantilever beam shown in Figure 9.7 with the Abaqus interactive edition. We keep the same geometrical properties, $C = 10$ mm, $L = 60$ mm, $t = 5$ mm, the same mechanical properties, a Young's modulus of 200000 MPa and a Poisson's ratio of 0.3 and the same loading; a concentrated force P of 1000 N.

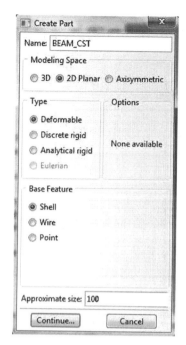

Start **Abaqus CAE**. Click on **Create Model Database**. On the main menu, click on **File** and set **Set Work Directory** to choose your working directory. Click on **Save As** and name the file **BEAM_CST.cae**. On the left-hand-side menu, click on **Part** to begin creating the model. Name the part **Beam_CST**, check **2D Planar**, check **Deformable** in the type. Choose **Shell** as the base feature. Enter an approximate size of 100 mm and click on **Continue** (Figure 9.14).

FIGURE 9.14   Creating the Beam_CST Part.

In the sketcher menu, choose the **Create-Lines Rectangle** icon to begin drawing the geometry of the beam. Click on **Done** in the bottom-left corner of the viewport window (Figure 9.15).

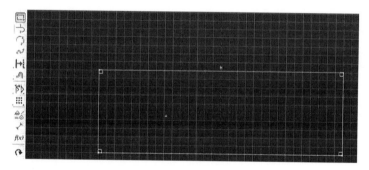

FIGURE 9.15   Drawing using the create-lines rectangle icon.

If we want to make sure that we will have nodes lying on the neutral axis of the beam, it is advisable to partition the beam along the neutral axis. On the main menu, click on **Tools** then on **Partition**. In the dialog box, check **Face** in **Type**, and **Use shortest path between 2 points** in **Method**. Select the two end points as shown in Figure 9.16, and in the prompt area, click on **Create partition**.

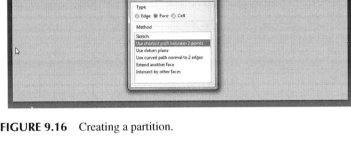

FIGURE 9.16   Creating a partition.

Define a material named steel with an elastic modulus of 200000 MPa and a Poisson's ratio of 0.3. Next, click on **Sections** to create a section named **Beam_section**. In the **Category** check **Solid**, and in the **Type**, check **Homogeneous**. Click on **Continue**. In the **Edit Section** dialog box, check **Plane stress/strain thickness** and enter 5 mm as the thickness. Click on **OK** (Figure 9.17).

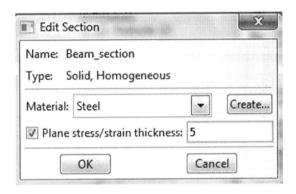

FIGURE 9.17   Creating a plane stress section.

Expand the menu under **Parts** and **BEAM_CST** and double click on **Section Assignments**. With the mouse select the whole part. In the **Edit Section Assignments** dialog box, select **Beam_section** and click on **OK** (Figure 9.18).

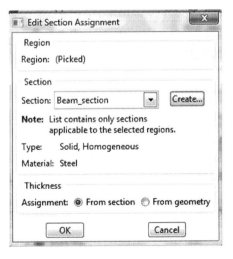

FIGURE 9.18   Editing section assignments.

In the model tree, double click on **Mesh** under the **BEAM_CST**. In the main menu, under **Mesh**, click on **Mesh Controls**. In the dialog box, check **Tri** for **Element shape** and **Structured** for **Technique**. Click on **OK** (Figure 9.19).

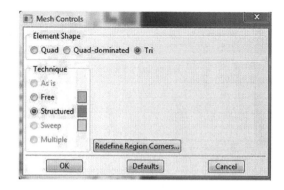

FIGURE 9.19 Mesh controls.

In the main menu, under **Mesh**, click on **Element Type**. With the mouse select all the part in the viewport. In the dialog box, select **Standard** for element library, **Linear** for geometric order. The description of the element **CPS3 A 3-node linear plane stress triangle** can be seen in the dialog box. Click on **OK** (Figure 9.20).

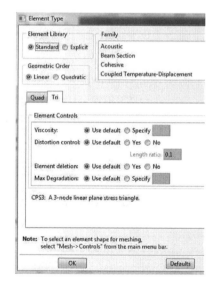

FIGURE 9.20 Selecting element type.

In the main menu, under **Seed**, click on **Part**. In the dialog box, enter 5 for **Approximate global size**. Click on **OK** and on **Done** (Figure 9.21).

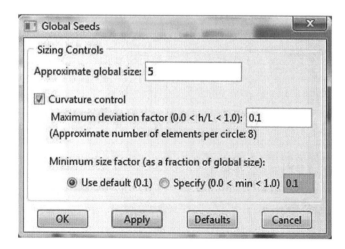

FIGURE 9.21 Seeding part by size.

In the main menu, under **Mesh**, click on **Part**. In the prompt area, click on **Yes**. In the main menu, select **View**, then **Part Display Options**. In the **Part Display Options**, under **Mesh**, check **Show node labels** and **Show element labels**. Click **Apply**. The element and node labels will appear in the viewport (Figure 9.22).

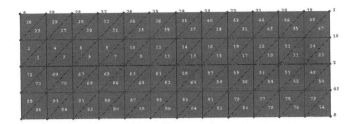

FIGURE 9.22 Mesh.

In the model tree, expand the **Assembly** and double click on **Instances**. Select **BEAM_CST** for **Parts** and click **OK**. In the model tree, expand **Steps** and **Initial** and double click on **BC**. Name the boundary condition **FIXED**, select **Symmetry/Antisymmetry/Encastre** for the type, and click on **Continue**. Keep the shift key down, and with the mouse select the right edge and click on **Done** in the prompt area. In the **Edit Boundary Condition** check **ENCASTRE**, Click **OK** (Figure 9.23).

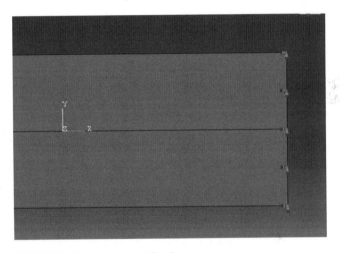

FIGURE 9.23 Imposing BC using geometry.

In the model tree, double click on **Steps**. Name the step **Apply_loads**. Set the procedure to **General** and select **Static, General**. Click on **Continue**. Give the step a description and click **OK**. In the model tree, under **steps**, and under **Apply_loads**, click on **Loads**. Name the load **Point_Load** and select **Concentrated Force** as the type. Click on **Continue**. Using the mouse click on the middle of the left edge, and click on **Done** in the prompt area. In the **Edit Load** dialog box, enter −1000 for **CF2**. Click **OK** (Figure 9.24).

FIGURE 9.24 Imposing a concentrated force using geometry.

In the model tree, expand the **Field Output Requests** and then double click on **F-Output-1**. **F-Output-1** is the default and is automatically generated when creating the step. Uncheck the variables **Contact** and select any other variable you wish to add to the field output. Click on **OK**. Under **Analysis**, right click on **Jobs** and then click on **Create**.

In the **Create Job** dialog box, name the job **BEAM_CST** and click on **Continue**. In the **Edit Job** dialog box, enter a description for the job. Check **Full analysis**, select to run the job in **Background**, and check to start it **immediately**. Click **OK**. Expand the tree under **Jobs**, right click on **BEAM_CST**. Then, click on **Submit**. If you get the following message **BEAM_CST completed successfully** in the bottom window, then your job is free of errors and was executed properly (Figure 9.25). Notice that Abaqus has generated an input file for the job **BEAM_CST.inp**, which you can open with your preferred text editor.

Under the top menu, in the **Module** scroll to **Visualization**, and click to load **Abaqus Viewer**. On the main menu, under **File**, click **Open**, navigate to your working directory, and open the file **BEAM_CST.odb**. It should have the same name as the job you submitted. Click on the **Common options** icon to display the **Common Plot options** dialog box. Under **labels**, check **Show Element labels** and **Show Node labels** to display elements and nodes' numbering. Click on the icon **Plot Contours on both shapes** to display the deformed shape of the beam. Under the main menu, select **U** and **U2** to plot the vertical displacement. It can be seen that the displacement of the left edge is equal to -0.965 mm, which is almost similar with the analytical solution and the results obtained with the MATLAB code (Figure 9.26). In the menu bar, click on **Report** and **Field Output**. In the **Report Field Output** dialog box, for **Position** select **Unique nodal**, check **U1**, and **U2** under **U: Spatial**

```
The job "BEAM_CST" has been created.
The job input file "BEAM_CST.inp" has been submitted for analysis.
Job BEAM_CST: Analysis Input File Processor completed successfully.
Job BEAM_CST: Abaqus/Standard completed successfully.
Job BEAM_CST completed successfully.
```

FIGURE 9.25   Analyzing a job in Abaqus CAE.

FIGURE 9.26   Plotting displacements on deformed and undeformed shapes.

displacement. Then click on click on **Set up**. Click on **Select** to navigate to your working directory. Name the file **BEAM_CST.rpt**. Uncheck **Append to file** and click **OK**. Use your favorite text editor and open the file **BEAM_CST.rpt**, which should be the same as the one listed next.

```
******************************************************************************
Field Output Report, written Wed May 11 01:15:14 2011

Source 1
---------

   ODB: C:/Abaqus_FILES/BEAM_CST.odb
   Step: Apply_loads
   Frame: Increment        1: Step Time =      1.000

Loc 1 : Nodal values from source 1

Output sorted by column "Node Label".

Field Output reported at nodes for part: BEAM_CST-1

          Node           U.U1            U.U2
          Label          @Loc 1          @Loc 1
     ------------------------------------------------
              1       -215.7E-06       -96.56E-03
              2       -26.59E-36        85.37E-36
              3       -2.141E-33       -262.8E-36
              4       -22.10E-03       -95.95E-03
              5        22.02E-03       -95.90E-03
              6        2.269E-33       -954.4E-36
              7       -77.91E-06       -84.47E-03
              8       -70.86E-06       -72.89E-03
              9       -68.46E-06       -61.63E-03
             10       -63.12E-06       -50.84E-03
             11       -55.86E-06       -40.66E-03
             12       -47.25E-06       -31.25E-03
             13       -36.83E-06       -22.76E-03
             14       -22.80E-06       -15.36E-03
             15       -2.085E-06       -9.187E-03
             16        25.90E-06       -4.408E-03
             17        35.64E-06       -1.232E-03
             18       -1.641E-33        27.34E-36
             19       -3.622E-03       -2.115E-03
             20       -6.811E-03       -5.371E-03
             21       -9.712E-03       -10.08E-03
             22       -12.32E-03       -16.16E-03
             23       -14.64E-03       -23.47E-03
             24       -16.65E-03       -31.86E-03
             25       -18.35E-03       -41.18E-03
             26       -19.75E-03       -51.26E-03
             27       -20.84E-03       -61.97E-03
             28       -21.61E-03       -73.14E-03
             29       -22.02E-03       -84.56E-03
             30       -10.99E-03       -96.03E-03
             31        10.85E-03       -96.00E-03
             32        21.87E-03       -84.45E-03
             33        21.44E-03       -73.01E-03
             34        20.68E-03       -61.85E-03
             35        19.62E-03       -51.15E-03
             36        18.24E-03       -41.06E-03
             37        16.55E-03       -31.75E-03
             38        14.56E-03       -23.35E-03
             39        12.25E-03       -16.03E-03
             40        9.620E-03       -9.948E-03
             41        6.673E-03       -5.248E-03
             42        3.418E-03       -2.033E-03
             43        1.539E-33        104.5E-36
```

44	-10.98E-03	-84.51E-03
45	-10.70E-03	-72.96E-03
46	-10.29E-03	-61.73E-03
47	-9.734E-03	-50.96E-03
48	-9.029E-03	-40.80E-03
49	-8.172E-03	-31.42E-03
50	-7.160E-03	-22.96E-03
51	-5.992E-03	-15.58E-03
52	-4.668E-03	-9.433E-03
53	-3.184E-03	-4.664E-03
54	-1.547E-03	-1.399E-03
55	1.601E-03	-1.416E-03
56	3.183E-03	-4.601E-03
57	4.635E-03	-9.356E-03
58	5.934E-03	-15.51E-03
59	7.081E-03	-22.89E-03
60	8.075E-03	-31.36E-03
61	8.916E-03	-40.75E-03
62	9.603E-03	-50.90E-03
63	10.14E-03	-61.67E-03
64	10.52E-03	-72.90E-03
65	10.77E-03	-84.42E-03

9.4.6.2   Keyword Edition

In this section, we will use a text editor to prepare an input file for the cantilever beam shown in Figure 9.7. The file is named **BEAM_CST_Keyword.inp** and is listed next:

```
*Heading
 Analysis of cantilever beam as a plane stress problem
*Preprint, echo=YES
**
**
** Node generation
**
**
*NODE
 1,    0.,       0.
 5,    0.,      20.
 61,  60.,       0.
 65,  60.,      20.
*NGEN,NSET=Left_Edge
 1,5
*NGEN,NSET=Right_Edge
 61,65
*NFILL
Left_Edge,Right_Edge,12,5
*NSET, NSET = Loaded_node
3
**
** Element generation
**
*ELEMENT,TYPE=CPS3
1, 1, 6, 7
*ELGEN, ELSET = ODD
1, 4, 1, 2, 12, 5, 8
**
*ELEMENT,TYPE=CPS3
2, 1, 7, 2
*ELGEN,ELSET = EVEN
2, 4, 1, 2, 12, 5, 8
*ELSET, ELSET = All_Elements
EVEN, ODD
*MATERIAL, NAME =STEEL
*ELASTIC
200000., 0.3
```

```
*SOLID SECTION, ELSET = All_Elements, MATERIAL = STEEL
5.
**
** BOUNDARY CONDITIONS
**
**
*Boundary
Right_Edge, encastre
**
** STEP: Apply_Loads
**
*Step, name=Apply_Loads
*Static
1., 1., 1e-05, 1.
**
** LOADS
**
*Cload
Loaded_node, 2, -1000.
**
**
** OUTPUT REQUESTS
**
**
*Output, field, variable=PRESELECT
**
*Output, history, variable=PRESELECT
*End Step
```

1. The input file always starts with the keyword ***HEADING**, which in this case is entered as **Analysis of cantilever beam as a plane stress problem**.
2. Using ***Preprint, echo=YES** will allow to print an echo of the input file to the file with an extension ***.dat**.
3. Using the keyword ***Node**, we define the four corner nodes 1, 5, 61, and 65 as shown in Figure 9.27.
4. Using the keyword ***NGEN** we generate the nodes located on the left edge. In the data line, we enter the number of the first end node 1, which has been previously defined, then the number of the second end node 5, which also must have been previously defined, followed by the increment in the numbers between each node along the line, which in this case is the default 1. We then group the nodes in a set named **Left_Edge**.
5. Using the keyword ***NGEN** again, we generate the nodes located on the right edge and group them in a set named **Right_Edge**.
6. Using the keyword ***NFILL**, we generate all the remaining nodes by filling in nodes between two bounds. In the data line, we enter first the node sets **Left_Edge** and

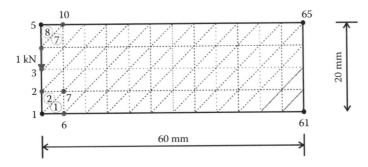

FIGURE 9.27   Generating a mesh manually in Abaqus.

Right_Edge followed by the number of intervals along each line between bounding nodes, in this case 12, and the increment in node numbers from the node number at the first bound set end, which in this case is 5 as shown in Figure 9.27.

7. Using the keyword ***NSET, NSET = Loaded_node**, we create a node set containing node 3. This will be used to apply the concentrated load of 1000 N.

8. Using the keyword ***ELEMENT** and **Type = CPS3**, which stands for a continuum plane stress three node triangle, we define elements 1 and 2 as well as their connectivity.

9. Using the keyword ***ELGEN**, we generate all the elements having an odd number which we group in the set **ODD**. The keyword ***ELGEN** requires in its data line:
   a. Master element number.
   b. Number of elements to be defined in the first row generated, including the master element.
   c. Increment in node numbers of corresponding nodes from element to element in the row. The default is 1.
   d. Increment in element numbers in the row. The default is 1.
   e. If necessary, copy this newly created master row to define a layer of elements.
   f. Number of rows to be defined, including the master row. The default is 1.
   g. Increment in node numbers of corresponding nodes from row to row.
   h. Increment in element numbers of corresponding elements from row to row.
   i. If necessary, copy this newly created master layer to define a block of elements (only necessary for a 3D mesh).
   j. Number of layers to be defined, including the master layer. The default is 1.
   k. Increment in node numbers of corresponding nodes from layer to layer.
   l. Increment in element numbers of corresponding elements from layer to layer.

10. Using the same procedure, we generate all the elements having an even number, which we group in the set **EVEN**.

11. Next, we use the keyword ***elset** to group all the elements in an element set named **All_Elements** consisting of element sets **ODD** and **EVEN** listed in the data line.

12. Using the keywords ***Material** and ***elastic**, we define a material named **steel** having an elastic modulus of 200,000 MPa and a Poisson's ratio of 0.3.

13. Using the keyword ***solid section**, we assign the material **steel** to all the elements, and in the data line we enter the thickness of the domain, which in this case is 5 mm.

14. Using the created node sets, we impose the boundary conditions with the keyword ***Boundary**. We fully fix the node set **Right_Edge** by using **encastre**.

15. Next using the keyword ***step**, we create a step named **Apply_Loads**. The keyword ***static** indicates that it will be a general static analysis.

16. Using the keyword ***cload**, we apply a concentrated load of −1000 N in the direction 2 to the node in node set **Loaded_node**.

17. Using the keywords ***Output, field, variable=PRESELECT**, and ***Output, history, variable=PRESELECT** we request the default variables for both field and history outputs.

18. Finally, we end the step and the file with ***End Step**.

At the command line type **Abaqus job=BEAM_CST_Keyword inter** followed by **Return**. If you get an error, open the file with extension ***.dat** to see what type of error. To load the visualization model, type **Abaqus Viewer** at the command line.

On the main menu, under **File**, click **Open**, navigate to your working directory, and open the file **BEAM_CST_Keyword.odb**. Click on the **Common options** icon to display the **Common Plot options** dialog box. Under **labels**, check **Show Element labels** and **Show Node labels** to display elements and nodes' numbering. Click on the icon **Plot Deformed Shape** to display the deformed shape of the beam. On the main menu, click on **Results** then on **Field Output** to open the **Field Output** dialog box. Choose **U Spatial displacements at nodes**. For component, choose $U2$ to plot

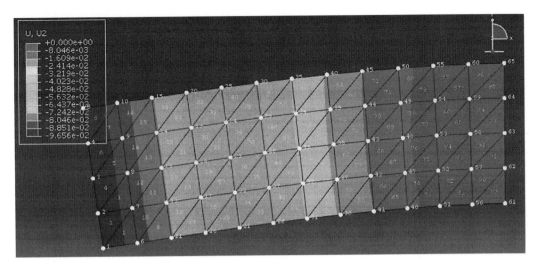

FIGURE 9.28 Displacement contour.

the vertical displacement. Notice that the displacements contour is exactly the same as obtained previously, except that the node and element numbering is different (Figure 9.28).

9.5 LINEAR STRAIN TRIANGLE

A more versatile element in the triangular family is the linear strain triangle shown in Figure 9.29. It has six nodes. The sides can be straight or curved. It can be used to mesh domains with curved boundaries. Its shape functions have already been defined in Chapter 7, and they are given as

$$
\begin{Bmatrix} N_1(\xi,\eta) \\ N_2(\xi,\eta) \\ N_3(\xi,\eta) \\ N_4(\xi,\eta) \\ N_5(\xi,\eta) \\ N_6(\xi,\eta) \end{Bmatrix} = \begin{Bmatrix} -\lambda(1-2\lambda) \\ 4\xi\lambda \\ -\xi(1-2\xi) \\ 4\xi\eta \\ -\eta(1-2\eta) \\ 4\eta\lambda \end{Bmatrix}
\tag{9.42}
$$

with $\lambda = 1 - \xi - \eta$

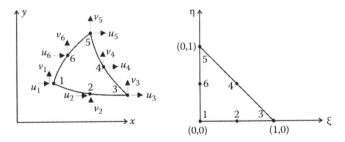

FIGURE 9.29 Linear strain triangular element.

9.5.1 Displacement Field

The displacement field over the element is approximated as

$$u = N_1 u_1 + N_2 u_2 + N_3 u_3 + N_4 u_4 + N_5 u_5 + N_6 u_6 \tag{9.43}$$

$$v = N_1 v_1 + N_2 v_2 + N_3 v_3 + N_4 v_4 + N_5 v_5 + N_6 v_6 \tag{9.44}$$

or in a matrix form as

$$\begin{Bmatrix} u \\ v \end{Bmatrix} = \begin{bmatrix} N_1 & 0 & | & N_2 & 0 & | & N_3 & 0 & | & N_4 & 0 & | & N_5 & 0 & | & N_6 & 0 \\ 0 & N_1 & | & 0 & N_2 & | & 0 & N_3 & | & 0 & N_4 & | & 0 & N_5 & | & 0 & N_6 \end{bmatrix} \begin{Bmatrix} u_1 \\ v_1 \\ u_2 \\ v_2 \\ u_3 \\ v_3 \\ u_4 \\ v_4 \\ u_5 \\ v_5 \\ u_6 \\ v_6 \end{Bmatrix} \tag{9.45}$$

or more compactly as

$$\{U\} = [N]\{a\} \tag{9.46}$$

The element is isoparametric, the coordinates x and y of any point of the parent element are given as

$$x = N_1 x_1 + N_2 x_2 + N_3 x_3 + N_4 x_4 + N_5 x_5 + N_6 x_6 \tag{9.47}$$

$$y = N_1 y_1 + N_2 y_2 + N_3 y_3 + N_4 y_4 + N_5 y_5 + N_6 y_6 \tag{9.48}$$

where the couple (x_i, y_i) represents the coordinates of the nodes.

The matrix $[J]$ is the Jacobian of the geometrical transformation and is given as

$$[J] = \begin{bmatrix} \dfrac{\partial x}{\partial \xi} & \dfrac{\partial y}{\partial \xi} \\ \dfrac{\partial x}{\partial \eta} & \dfrac{\partial y}{\partial \eta} \end{bmatrix} = \begin{bmatrix} \displaystyle\sum_{i=1}^{6} \dfrac{\partial N_i}{\partial \xi} x_i & \displaystyle\sum_{i=1}^{6} \dfrac{\partial N_i}{\partial \xi} y_i \\ \displaystyle\sum_{i=1}^{6} \dfrac{\partial N_i}{\partial \eta} x_i & \displaystyle\sum_{i=1}^{6} \dfrac{\partial N_i}{\partial \eta} y_i \end{bmatrix}$$

The partial derivatives $\frac{\partial N_i}{\partial \xi}$ and $\frac{\partial N_i}{\partial \eta}$ are obtained by deriving the shape functions $N_i(\xi, \eta)$, $i = 1$ to 6 with respect to ξ and η. The Jacobian is rewritten as

$$[J] = \frac{1}{4}\begin{bmatrix} 1 - 4\lambda & 4(\lambda - \xi) & -1 + 4\xi & 4\eta & 0 & -4\eta \\ 1 - 4\lambda & -4\xi & 0 & 4\xi & -1 + 4\eta & 4(\lambda - \eta) \end{bmatrix}\begin{bmatrix} x_1 & y_1 \\ x_2 & y_2 \\ x_3 & y_3 \\ x_4 & y_4 \\ x_5 & y_5 \\ x_6 & y_6 \end{bmatrix} \tag{9.49}$$

with $\lambda = 1 - \xi - \eta$.

9.5.2 STRAIN MATRIX

Substituting for the displacements u and v in Equation (9.6) using Equation (9.45), the strain vector is obtained as

$$\{\epsilon\} = [B]\{a\} \tag{9.50}$$

with

$$[B] = \begin{bmatrix} \frac{\partial N_1}{\partial x} & 0 & \Big| & \frac{\partial N_2}{\partial x} & 0 & \Big| & \frac{\partial N_3}{\partial x} & 0 & \Big| & \frac{\partial N_4}{\partial x} & 0 & \Big| & \frac{\partial N_5}{\partial x} & 0 & \Big| & \frac{\partial N_6}{\partial x} & 0 \\ 0 & \frac{\partial N_1}{\partial y} & \Big| & 0 & \frac{\partial N_2}{\partial y} & \Big| & 0 & \frac{\partial N_3}{\partial y} & \Big| & 0 & \frac{\partial N_4}{\partial y} & \Big| & 0 & \frac{\partial N_5}{\partial y} & \Big| & 0 & \frac{\partial N_6}{\partial y} \\ \frac{\partial N_1}{\partial y} & \frac{\partial N_1}{\partial x} & \Big| & \frac{\partial N_2}{\partial y} & \frac{\partial N_2}{\partial x} & \Big| & \frac{\partial N_3}{\partial y} & \frac{\partial N_3}{\partial x} & \Big| & \frac{\partial N_4}{\partial y} & \frac{\partial N_4}{\partial x} & \Big| & \frac{\partial N_5}{\partial y} & \frac{\partial N_5}{\partial x} & \Big| & \frac{\partial N_6}{\partial y} & \frac{\partial N_6}{\partial x} \end{bmatrix} \tag{9.51}$$

To evaluate the matrix $[B]$, it is necessary to relate the partial derivatives in the (x, y) coordinates to the local coordinates (ξ, η). This is achieved using the chain rule as

$$\frac{\partial N_i}{\partial \xi} = \frac{\partial N_i}{\partial x}\frac{\partial x}{\partial \xi} + \frac{\partial N_i}{\partial y}\frac{\partial y}{\partial \xi} \tag{9.52}$$

$$\frac{\partial N_i}{\partial \eta} = \frac{\partial N_i}{\partial x}\frac{\partial x}{\partial \eta} + \frac{\partial N_i}{\partial y}\frac{\partial y}{\partial \eta} \tag{9.53}$$

which can be rewritten in matrix form as

$$\begin{Bmatrix} \frac{\partial N_i}{\partial \xi} \\ \frac{\partial N_i}{\partial \eta} \end{Bmatrix} = \begin{bmatrix} \frac{\partial x}{\partial \xi} & \frac{\partial y}{\partial \xi} \\ \frac{\partial x}{\partial \eta} & \frac{\partial y}{\partial \eta} \end{bmatrix}\begin{Bmatrix} \frac{\partial N_i}{\partial x} \\ \frac{\partial N_i}{\partial y} \end{Bmatrix} \tag{9.54}$$

The derivatives of the shape functions in the (x, y) system are obtained by inversing the previous equation; that is,

$$
\begin{Bmatrix} \dfrac{\partial N_i}{\partial x} \\[2mm] \dfrac{\partial N_i}{\partial y} \end{Bmatrix} = [J]^{-1} \begin{Bmatrix} \dfrac{\partial N_i}{\partial \xi} \\[2mm] \dfrac{\partial N_i}{\partial \eta} \end{Bmatrix}
\tag{9.55}
$$

In practice, the matrix $[B]$ is not calculated but assembled from the values of $\frac{\partial N_i}{\partial x}$ and $\frac{\partial N_i}{\partial y}$ obtained with Equation (9.55).

9.5.3 STIFFNESS MATRIX

The stiffness matrix of the element is given as

$$
[K_e] = \left[\int_{A_e} [B]^T [D][B] t \, dA \right]
\tag{9.56}
$$

The integration over the volume is evaluated using the Hammer formula (see Chapter 7):

$$
[K_e] = t \int_0^{+1} \int_0^{1-\xi} [B(\xi, \eta)]^T [D][B(\xi, \eta)] \det[J(\xi, \eta)] d\eta \, d\xi
$$

$$
= t \sum_{i=1}^{nhp} W_i [B(\xi_i, \eta_i)]^T [D][B(\xi_i, \eta_i)] \det[J(\xi_i, \eta_i)]
\tag{9.57}
$$

where *nhp* represents the number of Hammer points.

9.5.4 COMPUTER CODE: LST_PLANE_STRESS_MESH.m

The program is virtually identical to its predecessor **CST_PLANE_STRESS_MESH.m,** except that the stiffness matrix is computed using numerical integration. The size of some of the arrays has increased to account for the extra degrees of freedom. In order to assess the performance of the element, we will analyze the cantilever beam shown in Figure 9.7.

The program is listed next and includes an automatic mesh generation, function *T6_mesh.m*, as well as another function, *prepare_contour_data.m*, that prepares the stress data for plotting using the MATLAB function **contourf**.

```
% THIS PROGRAM USES A 6-NODE LINEAR TRIANGULAR ELEMENT FOR THE
% LINEAR ELASTIC STATIC ANALYSIS OF A TWO DIMENSIONAL PROBLEM
% IT INCLUDES AN AUTOMATIC MESH GENERATION
%
% Make these variables global so they can be shared by other functions
%
clear all
clc
global nnd nel nne  nodof eldof  n
global connec geom  dee nf Nodal_loads XIG YIG
global Length Width NXE NYE X_origin Y_origin
%
 format long g
```

```
%
%
% To change the size of the problem or change elastic properties
%     supply another input file
%
Length = 60.; % Length of the model
Width =20.;    % Width
NXE = 12;      % Number of rows in the x direction
NYE = 5;       % Number of rows in the y direction
XIG = zeros(2*NXE+1,1); YIG=zeros(2*NYE+1,1); % Vectors holding grid coordinates
dhx = Length/NXE; % Element size in the x direction
dhy = Width/NYE;  % Element size in the x direction
X_origin = 0. ;   % X origin of the global coordinate system
Y_origin = Width/2. ;   % Y origin of the global coordinate system
%
nne = 6;
nodof = 2;
eldof = nne*nodof;
%
T6_mesh ;          % Generate the mesh
%
% Material
%
E = 200000.;       % Elastic modulus in MPa
vu = 0.3;          % Poisson's ratio
thick = 5.;        % Beam thickness in mm
nhp = 3;           % Number of sampling points
%
% Form the elastic matrix for plane stress
%
dee = formdsig(E,vu);
%
%
% Boundary conditions
%
nf = ones(nnd, nodof);     % Initialize the matrix nf to 1
%
% Restrain in all directions the nodes situated @
% (x = Length)
%
for i=1:nnd
    if geom(i,1) == Length;
        nf(i,:) = [0 0];
    end
end
%
% Counting of the free degrees of freedom
%
n=0;
for i=1:nnd
    for j=1:nodof
        if nf(i,j) ~= 0
            n=n+1;
            nf(i,j)=n;
        end
    end
end
%
% loading
%
Nodal_loads= zeros(nnd, 2);   % Initialize the matrix of nodal loads to 0
%
% Apply an equivalent nodal load of (Pressure*thick*dhx) to the central
% node located at x=0 and y = 0.
%
Force = 1000.;   % N
%
```

```
for i=1:nnd
    if geom(i,1) == 0. && geom(i,2) == 0.
        Nodal_loads(i,:) = [0.   -Force];
    end
end
%
%%%%%%%%%%%%%%%%%%%%%%%%%%%% End of input%%%%%%%%%%%%%%%%%%%%%%%%%%%%%%%%%%
%
% Assemble the global force vector
%
fg=zeros(n,1);
for i=1: nnd
    if nf(i,1) ~= 0
        fg(nf(i,1))= Nodal_loads(i,1);
    end
    if nf(i,2) ~= 0
        fg(nf(i,2))= Nodal_loads(i,2);
    end
end
%
% Assembly of the global stiffness matrix
%
%
%  Form the matrix containing the abscissas and the weights of Hammer points
%
samp=hammer(nhp);
%
%  initialize the global stiffness matrix to zero
%
kk = zeros(n, n);
%
for i=1:nel
    [coord,g] = elem_T6(i);     % Form strain matrix, and steering vector
    ke=zeros(eldof,eldof) ;         % Initialize the element stiffness matrix to zero
     for ig = 1:nhp
        wi = samp(ig,3);
        [der,fun] = fmT6_quad(samp, ig);
        jac = der*coord;
        d = det(jac);
        jac1=inv(jac);                     % Compute inverse of the Jacobian
        deriv=jac1*der;                    % Derivative of shape functions in global coordinates
        bee=formbee(deriv,nne,eldof);    % Form matrix [B]
        ke=ke + d*thick*wi*bee'*dee*bee; % Integrate stiffness matrix
     end
    kk=form_kk(kk,ke, g);                    % assemble global stiffness matrix
end
%
%
%%%%%%%%%%%%%%%%%%%%%%%%%  End of assembly %%%%%%%%%%%%%%%%%%%%%%%%%%%%%%%%%
%
%
delta = kk\fg ;              % solve for unknown displacements
%
for i=1: nnd                             %
    if nf(i,1) == 0                      %
        x_disp =0.;                      %
    else
        x_disp = delta(nf(i,1));         %
    end
%
    if nf(i,2) == 0                      %
        y_disp = 0.;                     %
    else
        y_disp = delta(nf(i,2));         %
    end
    node_disp(i,:) =[x_disp  y_disp];
```

```
end
%
%
% Retrieve the x_coord and y_disp of the nodes located on the neutral axis
%
k = 0;
for i=1:nnd;

    if geom(i,2)== 0.
        k=k+1;
        x_coord(k) = geom(i,1);
        vertical_disp(k)=node_disp(i,2);
    end
end
%
nhp = 1;    % Calculate stresses at the centroid of the element
samp=hammer(nhp);
%
for i=1:nel
    [coord,g] = elem_T6(i);          % Retrieve coordinates and steering vector
    eld=zeros(eldof,1);              % Initialize element displacement to zero
    for m=1:eldof                    %
        if g(m)==0                   %
            eld(m)=0.;               %
        else                         %
            eld(m)=delta(g(m));      % Retrieve element displacement from the
                                     % global displacement vector

        end
    end
%
    for ig=1: nhp
        [der,fun] = fmT6_quad(samp, ig);      % Derivative of shape functions in
                                              % local coordinates

        jac=der*coord;               % Compute Jacobian matrix
        jac1=inv(jac);               % Compute inverse of the Jacobian
        deriv=jac1*der;              % Derivative of shape functions
                                     %  in global coordinates

        bee=formbee(deriv,nne,eldof);  % Form matrix [B]
        eps=bee*eld;                   % Compute strains
        sigma=dee*eps ;                % Compute stresses
    end            % Compute stresses
    SIGMA(i,:)=sigma ;               % Store stresses for all elements
end
%
% Prepare stresses for plotting
%
[ZX, ZY, ZT, Z1, Z2]=prepare_contour_data(SIGMA);
%
%  Plot mesh using patches
%
% patch('Faces', connec, 'Vertices', geom, 'FaceVertexCData',hsv(nel), ...
        'Facecolor','none','Marker','o');
%
% Plot stresses in the x_direction
%
[C,h]= contourf(XIG,YIG,ZX,40);
%clabel(C,h);
colorbar plottools;
```

T6_mesh.m

```
% This function generates a mesh of the linear strain triangular element
%
global nnd nel geom connec XIG YIG
global Length Width NXE NYE X_origin Y_origin dhx dhy
%
```

```
%
nnd = 0;
k = 0;
for i = 1:NXE
    for j=1:NYE
            k = k + 1;
            n1 = (2*j-1) + (2*i-2)*(2*NYE+1) ;
            n2 = (2*j-1) + (2*i-1)*(2*NYE+1);
            n3 = (2*j-1) + (2*i)*(2*NYE+1);
            n4 = n1 + 1;
            n5 = n2 + 1;
            n6 = n3 + 1 ;
            n7 = n1 + 2;
            n8 = n2 + 2;
            n9 = n3 + 2;
            %
            geom(n1,:) = [(i-1)*dhx - X_origin              (j-1)*dhy - Y_origin];
            geom(n2,:) = [((2*i-1)/2)*dhx - X_origin        (j-1)*dhy - Y_origin ];
            geom(n3,:) = [i*dhx - X_origin                  (j-1)*dhy - Y_origin ];
            geom(n4,:) = [(i-1)*dhx - X_origin              ((2*j-1)/2)*dhy - Y_origin ];
            geom(n5,:) = [((2*i-1)/2)*dhx - X_origin        ((2*j-1)/2)*dhy - Y_origin ];
            geom(n6,:) = [i*dhx - X_origin                  ((2*j-1)/2)*dhy - Y_origin ];
            geom(n7,:) = [(i-1)*dhx - X_origin              j*dhy - Y_origin];
            geom(n8,:) = [((2*i-1)/2)*dhx - X_origin        j*dhy - Y_origin];
            geom(n9,:) = [i*dhx - X_origin                  j*dhy - Y_origin];
            %
            nel = 2*k;
            m = nel -1;
            connec(m,:) = [n1  n2   n3     n5     n7    n4];
            connec(nel,:) = [n3   n6    n9    n8    n7    n5];
            max_n = max([n1   n2   n3    n4   n5   n6   n7   n8     n9]);
            if(nnd <= max_n); nnd = max_n; end;
            %
            % XIN and YIN are two vectors that holds the coordinates X and Y
            % of the grid necessary for the function contourf (XIN,YIN, stress)
            %
            XIG(2*i-1) = geom(n1,1); XIG(2*i) = geom(n2,1); XIG(2*i+1) = geom(n3,1);
            YIG(2*j-1) = geom(n1,2); YIG(2*j) = geom(n4,2); YIG(2*j+1) = geom(n7,2);
    end
end
%
```

The variables **NXE** and **NYE** represent respectively the number of intervals along the x and y directions, as shown in Figure 9.30. For each interval i and j, nine nodes $n_1, n_2, n_3, n_4, n_5, n_6, n_7, n_8$ and n_9 and two elements are created. The first element has nodes $n_1, n_2, n_3, n_5, n_7, n_4$, while the second element has nodes $n_3, n_6, n_9, n_8, n_7, n_5$. In total the number of elements and nodes created are respectively equal to $NEL = 2 \times NXE \times NYE$, and $nnd = (2 \times NXE + 1) \times (2 \times NYE + 1)$. The module also returns the matrices $geom(nnd, 2)$ and $connec(nel, nne)$ as well as two vectors $XIG(2 \times NXE + 1)$ and $YIG(2 \times NYE + 1)$ holding the grid coordinates. These will be used for contour plotting using the MATLAB function *contour f*.

9.5.4.1 Numerical Integration of the Stiffness Matrix

The stiffness matrix is evaluated as

$$[K_e] = t \sum_{i=1}^{nhp} W_i [B(\xi_i, \eta_i]^T [D][B(\xi_i, \eta_i)] \det[J(\xi_i, \eta_i)] \qquad (9.58)$$

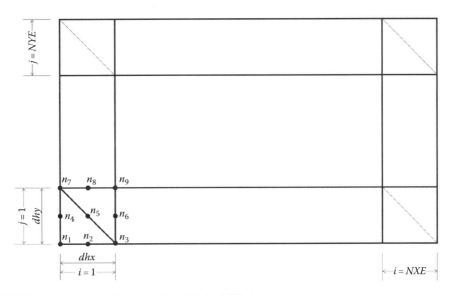

FIGURE 9.30 Automatic mesh generation with the LST element.

1. For every element $i = 1$ to *nel*
2. Retrieve the coordinates of its nodes **coord**(nne, 2) and its steering vector **g**(eldof) using the function *elem_t6.m*
3. Initialize the stiffness matrix to zero.
   a. Loop over the Hammer points $ig = 1$ to *nhp*
   b. Retrieve the weight **wi** as **samp**(ig, 3)
   c. Use the function *fmT6_quad.m* to compute the shape functions, vector **fun**, and their local derivatives, **der**, at the local coordinates $\xi = $ **samp**(ig, 1) and $\eta = $ **samp**(ig, 2)
   d. Evaluate the Jacobian **jac** = **der** $*$ **coord**
   e. Evaluate the determinant of the Jacobian as **d** = **det**(jac)
   f. Compute the inverse of the Jacobian as **jac1** = **inv**(jac)
   g. Compute the derivatives of the shape functions with respect to the global coordinates x and y as **deriv** = **jac1** $*$ **der**
   h. Use the function *formbee.m* to form the strain matrix **bee**
   i. Compute the stiffness matrix as **ke** = **ke** + **d** $*$ **thick** $*$ **wi** $*$ **bee**$'$ $*$ **dee** $*$ **bee**
4. Assemble the stiffness matrix **ke** into the global matrix **kk**

The abscissa and weights for the Hammer formula are listed in Table 8.2 and given by the function *hammer.m* listed in Appendix A.

9.5.4.2 Computation of the Stresses and Strains

Once the global system of equations is solved, we will compute the stresses at the centroid of the elements. For this we set *nhp* = 1. Then for each element:

1. Retrieve the coordinates of its nodes **coord**(nne, 2) and its steering vector **g**(eldof) using the function *elem_t6.m*
2. Retrieve its nodal displacements **eld**(eldof) from the global vector of displacements **delta**(n)

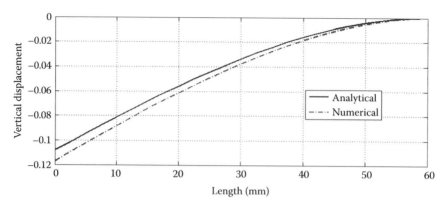

FIGURE 9.31   Deflection of the cantilever beam obtained with the LST element.

    a. Loop over the Hammer points $ig = 1$ to nhp
    b. Use the function *fmT6_quad.m* to compute the shape functions, vector **fun**, and their local derivatives, **der**, at the local coordinates $\xi = $ **samp**(**ig, 1**) and $\eta = $ **samp**(**ig, 2**)
    c. Evaluate the Jacobian **jac** = **der** ∗ **coord**
    d. Evaluate the determinant of the Jacobian as **d** = **det**(**jac**)
    e. Compute the inverse of the Jacobian as **jac1** = **inv**(**jac**)
    f. Compute the derivatives of the shape functions with respect to the global coordinates x and y as **deriv** = **jac1** ∗ **der**
    g. Use the function *formbee.m* to form the strain matrix **bee**
    h. Compute the strains as **eps** = **bee** ∗ **eld**
    i. Compute the stresses as **sigma** = **dee** ∗ **eps**
  3. Store the stresses in the matrix **SIGMA**(**nel, 3**)

The stresses computed at the centers of the elements are reorganized in a format suitable for plotting with the MATLAB graphic functions. In the present case, the stresses stored in the array **SIGMA**(**nel, 3**) are fed to the function *prepare_contour_data.m* listed in Appendix A.

For every node, the function locates all the elements surrounding it. Then the stresses are averaged and assigned to the node and stored in the matrices **ZX, ZY, ZT, Z1**, and **Z2** corresponding respectively to σ_{xx}, σ_{yy}, and τ_{xy} and the principal stresses σ_1 and σ_2. In this particular case, the matrix **ZX** and the vectors **XIG** and **YIG** are used in the MATLAB function **contourf** to produce a plot of the stresses σ_{xx}.

The results of the analysis are displayed in Figures 9.31 and 9.32. Figure 9.31 shows the deflection of the nodes situated along the center line (neutral axis). It can be clearly seen that the solution matches closely the analytical solution. Figure 9.32 displays a contour plot, the stresses in the x-direction. The stress gradient can be clearly seen. The stresses along the neutral axis are equal to zero.

9.5.5   ANALYSIS WITH ABAQUS USING THE LST

9.5.5.1   Interactive Edition

In this section, we will analyze the plate with a hole shown in Figure 9.33 using the linear strain triangle. The plate is made of aluminum with an elastic modulus of 70 GPa and a Poisson's ratio of 0.33. The plate is 5 mm thick and subject to a uniform pressure on both sides of 50 MPa. Since the plate presents two planes of symmetry in both geometry and loading, we will analyze a quarter only as shown in Figure 9.34. Indeed, whenever possible always take advantage of symmetry to simplify the model.

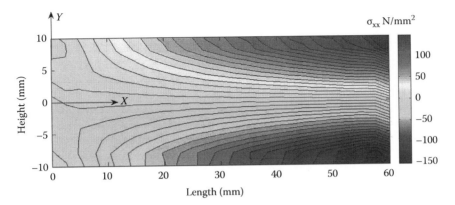

FIGURE 9.32   Stresses along the x-direction obtained with the LST element.

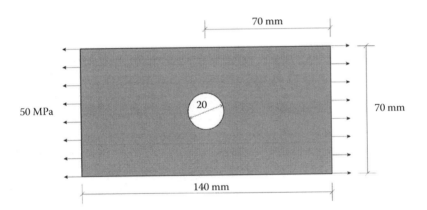

FIGURE 9.33   Aluminum plate with a hole.

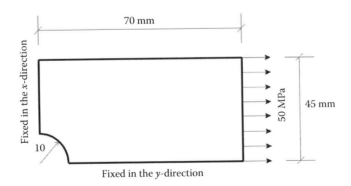

FIGURE 9.34   Making use of symmetry.

Start **Abaqus CAE**. Click on **Create Model Database**. On the main menu, click on **File** and set **Set Work Directory** to choose your working directory. Click on **Save As** and name the file **Plate_LST.cae**. On the left-hand-side menu, click on **Part** to begin creating the model. Name the part **Plate_LST**, check **2D Planar**, and check **Deformable** in the type. Choose **Shell** as the base feature. Enter an approximate size of 100 mm and sketch a quarter of the part as shown. In the sketcher menu, choose the **Create arc center and 2 end points** icon to create the arc, and **Create-Lines Rectangle** icon to create the edges. When finished, click on **Done** in the bottom-left corner of the viewport window (Figure 9.35).

FIGURE 9.35 Creating the Plate_LST Part.

Define a material named **Aluminum** with an elastic modulus of 70000 MPa and a Poisson's ratio of 0.32. Next, click on **Sections** to create a section named **Plate_section**. In the **Category** check **Solid**, and in the **Type**, check **Homogeneous**. Click on **Continue**. In the **Edit Section** dialog box, check **Plane stress/strain thickness** and enter 5 mm as the thickness. Click on **OK** (Figure 9.36).

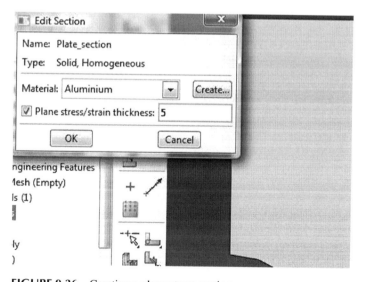

FIGURE 9.36 Creating a plane stress section.

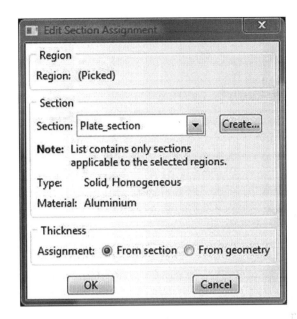

FIGURE 9.37 Editing section assignments.

Expand the menu under **Parts** and **Plate_LST** and double click on **Section Assignments**. With the mouse select the whole part. In the **Edit Section Assignments** dialog box, select **Plate_section** and click on **OK** (Figure 9.37).

In the model tree, double click on **Mesh** under the **Plate_LST**. In the main menu, under **Mesh**, click on **Mesh Controls**. In the dialog box, check **Tri** for **Element shape** and **Structured** for **Technique**. Click on **OK**. In the main menu, under **Mesh**, click on **Element Type**. In the dialog box, select **Standard** for element library, **Quadratic** for geometric order. The description of the element **CPS6M 6-node modified quadratic plane stress triangle** can be seen in the dialog box. Click on **OK** (Figure 9.38).

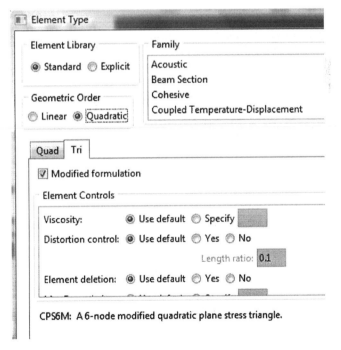

FIGURE 9.38 Mesh controls.

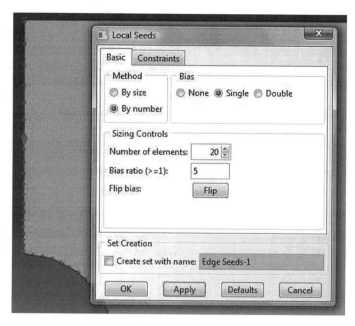

In the main menu, under **Seed**, click on **Edges**. Select the arc first. In the **Local seeds**, select by **number** and enter 15. Click on the vertical left edge, enter 20 and select simple for **bias**. The idea of this is to refine the mesh in the vicinity of the hole. Do the same for the other edges. When finished, click on **OK** and on **Done**. Under **Mesh**, click on **Part** and then **Yes** to mesh the part (Figure 9.39).

FIGURE 9.39   Seeding edge by size and simple bias.

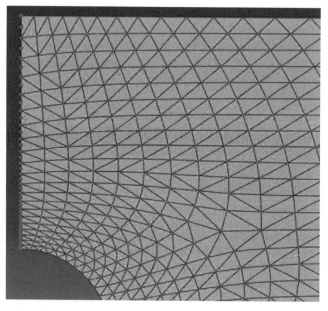

Expand the menu under **Plate_LST** and click on **Sets**. In the **Create set** dialog box, name the set **Left_Edge** and check **Node**. Click on **Continue**, and with the mouse select the nodes as shown in Figure 9.40. Repeat the procedure to create another node set that you will name **Bottom_Edge**.

FIGURE 9.40   Creating a node set.

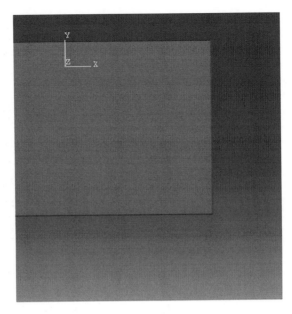

FIGURE 9.41   Creating a surface.

Expand the menu under **Plate_LST** and click on **Surfaces**. In the **Create Surface** dialog box, name the set **Loaded_Surface** and check **Geometry**. Click on **Continue**, and with the mouse select the left edge shown in Figure 9.41.

In the model tree, expand the **Assembly** and double click on **Instances**. Select **Plate_LST** for **Parts**, and click **OK**. In the model tree, expand **Steps** and **Initial**, and double click on **BC**. Name the boundary condition **Left_side**, select **Displacement/Rotation** for the type, and click on **Continue**. In the bottom-right corner of the viewport, click on sets and in the dialog box select **Plate_LST-1.Left_Edge**. Click on **Continue**. In the **Edit Boundary Condition** check **U1**. Click **OK**. Double click again on **BC**. Name the boundary condition **Bottom_side**, select **Displacement/Rotation** for the type, and click on **Continue**. In the bottom-right corner of the viewport, click on sets and in the dialog box select **Plate_LST-1.Bottom_Edge**. Click on **Continue**. In the **Edit Boundary Condition** check **U2**. Click **OK** (Figure 9.42).

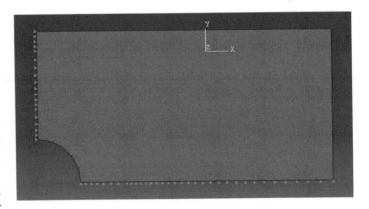

FIGURE 9.42   Imposing BC using node sets.

In the model tree, double click on **Steps**. Name the step **Apply_loads**. Set the procedure to **General** and select **Static, General**. Click on **Continue**. Give the step a description and click **OK**. In the model tree, under **steps**, and under **Apply_loads**, click on **Loads**. Name the load **Pressure** and select **Pressure** as the type. Click on **Continue**. In the right-bottom corner of the viewport, click on **Surfaces**. In the dialog box, select **loaded_Surface** and click on **Continue**. In the new dialog box, enter −50 MPa (Figure 9.43).

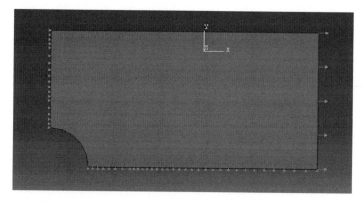

FIGURE 9.43   Imposing a pressure load on a surface.

In the model tree, expand the **Field Output Requests** and then double click on **F-Output-1**. **F-Output-1** is the default and is automatically generated when creating the step. Uncheck the variables **Contact** and select any other variable you wish to add to the field output. Click on **OK**. Under **Analysis**, right click on **Jobs** and then click on **Create**.

In the **Create Job** dialog box, name the job **Plate_LST** and click on **Continue**. In the **Edit Job** dialog box, enter a description for the job. Check **Full analysis**, select to run the job in **Background**, and check to start it **immediately**. Click **OK**. Expand the tree under **Jobs**, right click on **Plate_LST**. Then, click on **Submit**. If you get the following message **Plate_LST completed successfully** in the bottom window, then your job is free of errors and was executed properly.

Under the top menu, in the **Module** scroll to **Visualization**, and click to load **Abaqus Viewer**. On the main menu, under **File**, click **Open**, navigate to your working directory, and open the file **Plate_LST.odb**. Click on the **Common options** icon to display the **Common Plot options** dialog box. Under **labels**, check **Show Element labels** and **Show Node labels** if you wish to display elements and nodes' numbering. Click on the icon **Plot Contours on deformed shape** to display the deformed shape of the beam. Under the main menu, select **S** and **Max.In-Plane Principal** to plot the first principal stress as shown in Figure 9.44.

9.5.5.2 Keyword Edition

Except for simple geometries, it is very difficult to generate a mesh using keywords as we did previously. Hence, in this example, instead of writing an input file, we will simply open the one generated previously by Abaqus. Navigate into the working directory and locate the file **Plate_LST.inp** and open it with your preferred text editor. It is a very long file as it lists all the nodes, their coordinates, and all the elements with their connectivity. Note that the two node sets created are present as well as the surface. Scroll to the end of the file and locate:

```
**
** Name: Pressure    Type: Pressure
*Dsload
Loaded_Surface, P, -50.
**
```

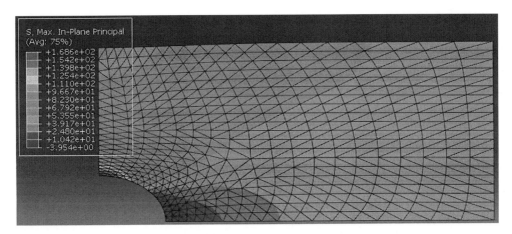

FIGURE 9.44 Plotting the maximum in-plane principal stress (under tension).

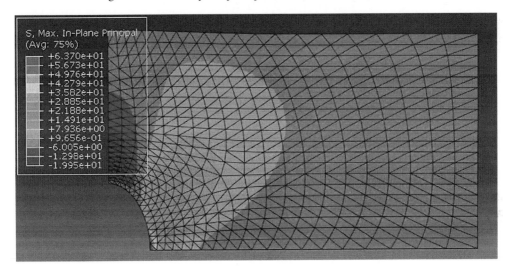

FIGURE 9.45 Plotting the maximum in-plane principal stress (under compression).

Change the value of -50 to 50 to apply a compressive pressure. Rename the file **Plate_LST_Keyword.inp**. Submit the job through the command line:

C:\WorkingDirectory>**Abaqus job=Plate_LST_Keyword inter**

When the job is successfully completed, start **Abaqus viewer** and open the file **Plate_LST_Keyword.odb**. Click on the **Common options** icon to display the **Common Plot options** dialog box. Under **labels**, check **Show Element labels** and **Show Node labels** if you wish to display elements and nodes' numbering. Click on the icon **Plot Contours on deformed shape** to display the deformed shape of the beam. Under the main menu, select **S** and **Max.In-Plane Principal** to plot the first principal stress, Figure 9.45. Now, compare with Figure 9.44.

9.6   THE BILINEAR QUADRILATERAL

The linear strain quadrilateral has four nodes and straight edges, as shown in Figure 9.46. Its shape functions have already been obtained in Chapter 7, and they are also given here:

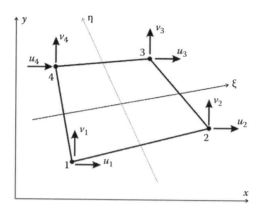

FIGURE 9.46   Linear quadrilateral element.

$$N_1(\xi, \eta) = 0.25(1 - \xi - \eta + \xi\eta)$$
$$N_2(\xi, \eta) = 0.25(1 + \xi - \eta - \xi\eta)$$
$$N_3(\xi, \eta) = 0.25(1 + \xi + \eta + \xi\eta) \tag{9.59}$$
$$N_4(\xi, \eta) = 0.25(1 - \xi + \eta - \xi\eta)$$

9.6.1 Displacement Field

The displacement field over the element is approximated as

$$u = N_1 u_1 + N_2 u_2 + N_3 u_3 + N_4 u_4 \tag{9.60}$$
$$v = N_1 v_1 + N_2 v_2 + N_3 v_3 + N_4 v_4 \tag{9.61}$$

or in a matrix form as

$$
\left\{ \begin{array}{c} u \\ v \end{array} \right\} =
\begin{bmatrix}
N_1 & 0 & | & N_2 & 0 & | & N_3 & 0 & | & N_4 & 0 \\
0 & N_1 & | & 0 & N_2 & | & 0 & N_3 & | & 0 & N_4
\end{bmatrix}
\left\{ \begin{array}{c} u_1 \\ v_1 \\ u_2 \\ v_2 \\ u_3 \\ v_3 \\ u_4 \\ v_4 \end{array} \right\} \tag{9.62}
$$

or more compactly as

$$\{U\} = [N]\{a\} \tag{9.63}$$

The element is isoparametric, therefore the shape functions $N_i(\xi, \eta)$ also define the geometrical transformation between the reference and the parent element. The coordinates x and y of any point of the parent element are given as

$$x = N_1 x_1 + N_2 x_2 + N_3 x_3 + N_4 x_4 \tag{9.64}$$

$$y = N_1 y_1 + N_2 y_2 + N_3 y_3 + N_4 y_4 \tag{9.65}$$

The Jacobian of the transformation is given as

$$[J] = \begin{bmatrix} \dfrac{\partial x}{\partial \xi} & \dfrac{\partial y}{\partial \xi} \\ \dfrac{\partial x}{\partial \eta} & \dfrac{\partial y}{\partial \eta} \end{bmatrix} = \begin{bmatrix} \displaystyle\sum_{i=1}^{4} \dfrac{\partial N_i}{\partial \xi} x_i & \displaystyle\sum_{i=1}^{4} \dfrac{\partial N_i}{\partial \xi} y_i \\ \displaystyle\sum_{i=1}^{4} \dfrac{\partial N_i}{\partial \eta} x_i & \displaystyle\sum_{i=1}^{4} \dfrac{\partial N_i}{\partial \eta} y_i \end{bmatrix}$$

After deriving and rearranging, the Jacobian is written in the form of a product of two matrices:

$$[J] = \frac{1}{4} \begin{bmatrix} -(1-\eta) & (1-\eta) & (1+\eta) & -(1+\eta) \\ -(1-\xi) & -(1+\xi) & (1+\xi) & (1-\xi) \end{bmatrix} \begin{bmatrix} x_1 & y_1 \\ x_2 & y_2 \\ x_3 & y_3 \\ x_4 & y_4 \end{bmatrix} \tag{9.66}$$

9.6.2 STRAIN MATRIX

Substituting for the displacements u and v in Equation (9.6) using Equation (9.64), the strain vector is obtained as

$$\{\epsilon\} = [B]\{a\} \tag{9.67}$$

with

$$[B] = \begin{bmatrix} \dfrac{\partial N_1}{\partial x} & 0 & \bigg| & \dfrac{\partial N_2}{\partial x} & 0 & \bigg| & \dfrac{\partial N_3}{\partial x} & 0 & \bigg| & \dfrac{\partial N_4}{\partial x} & 0 \\[2ex] 0 & \dfrac{\partial N_1}{\partial y} & \bigg| & 0 & \dfrac{\partial N_2}{\partial y} & \bigg| & 0 & \dfrac{\partial N_3}{\partial y} & \bigg| & 0 & \dfrac{\partial N_4}{\partial y} \\[2ex] \dfrac{\partial N_1}{\partial y} & \dfrac{\partial N_1}{\partial x} & \bigg| & \dfrac{\partial N_2}{\partial y} & \dfrac{\partial N_2}{\partial x} & \bigg| & \dfrac{\partial N_3}{\partial y} & \dfrac{\partial N_3}{\partial x} & \bigg| & \dfrac{\partial N_4}{\partial y} & \dfrac{\partial N_4}{\partial x} \end{bmatrix} \tag{9.68}$$

To evaluate the matrix $[B]$, it is necessary to relate the partial derivatives in the (x, y) coordinates to the local coordinates (ξ, η). The derivative of the shape functions can be written as follows using the chain rule:

$$\frac{\partial N_i}{\partial \xi} = \frac{\partial N_i}{\partial x} \frac{\partial x}{\partial \xi} + \frac{\partial N_i}{\partial y} \frac{\partial y}{\partial \xi} \tag{9.69}$$

$$\frac{\partial N_i}{\partial \eta} = \frac{\partial N_i}{\partial x} \frac{\partial x}{\partial \eta} + \frac{\partial N_i}{\partial y} \frac{\partial y}{\partial \eta} \tag{9.70}$$

which can be rewritten in matrix form as

$$\begin{Bmatrix} \dfrac{\partial N_i}{\partial \xi} \\ \dfrac{\partial N_i}{\partial \eta} \end{Bmatrix} = \begin{bmatrix} \dfrac{\partial x}{\partial \xi} & \dfrac{\partial y}{\partial \xi} \\ \dfrac{\partial x}{\partial \eta} & \dfrac{\partial y}{\partial \eta} \end{bmatrix} \begin{Bmatrix} \dfrac{\partial N_i}{\partial x} \\ \dfrac{\partial N_i}{\partial y} \end{Bmatrix} \tag{9.71}$$

The derivatives of the shape functions in the (x, y) system are obtained by inversing the previous equation:

$$\left\{ \begin{array}{c} \dfrac{\partial N_i}{\partial x} \\[2ex] \dfrac{\partial N_i}{\partial y} \end{array} \right\} = [J]^{-1} \begin{bmatrix} \dfrac{\partial x}{\partial \xi} & \dfrac{\partial y}{\partial \xi} \\[2ex] \dfrac{\partial x}{\partial \eta} & \dfrac{\partial y}{\partial \eta} \end{bmatrix} \tag{9.72}$$

In practice, as it was shown with the linear strain triangle, the matrix $[B]$ is not calculated but assembled from the values of $\frac{\partial N_i}{\partial x}$ and $\frac{\partial N_i}{\partial y}$ obtained with Equation (9.72).

9.6.3 STIFFNESS MATRIX

The stiffness matrix of the element is given by

$$[K_e] = \left[\int_{A_e} [B]^T [D][B] t \, dA \right] \tag{9.73}$$

The integration over the volume is evaluated using Gauss quadrature as

$$[K_e] = t \int_{-1}^{+1} \int_{-1}^{+1} [B(\xi, \eta)]^T [D][B(\xi, \eta)] det[J(\xi, \eta)] d\eta \, d\xi$$

$$= t \sum_{i=1}^{ngp} \sum_{j=1}^{ngp} W_i W_j [B(\xi_i, \eta_j)]^T [D][B(\xi_i, \eta_j)] det[J(\xi_i, \eta_j)] \tag{9.74}$$

where
  t represents the thickness of the element
  ngp the number of Gauss points

To integrate exactly the element, two Gauss points are required in each direction.

9.6.4 ELEMENT FORCE VECTOR

The element force vector is given by

$$\{f_e\} = \int_{A_e} [N]^T \{b\} t \, dA + \int_{L_e} [N]^T \{t\} t \, dl + \sum_i [N_{(\{x\} = \{\bar{x}\})}]^T \{P\}_i \tag{9.75}$$

Considering that the body forces b are due to gravity, the first term of Equation (9.75) is evaluated using Gauss quadrature:

$$\int_{A_e} [N]^T \{b\} t \, dA = t \sum_{i=1}^{ngp} \sum_{j=1}^{ngp} W_i W_j [N(\xi_i, \eta_j)]^T \left\{ \begin{array}{c} 0 \\ -\rho g \end{array} \right\} det[J(\xi_i, \eta_j)] \tag{9.76}$$

To evaluate the second and third terms of Equation (9.76), it is better to proceed with an example such as the one represented in Figure 9.47. The element is subject on side 3-4 to a surface traction q that has a normal component q_n and a tangential component q_t as well as two concentrated forces of magnitude P and $2P$ acting respectively on nodes 1 and 2.

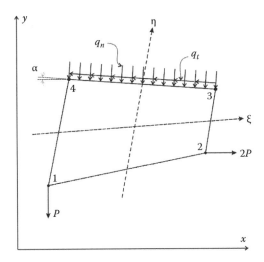

FIGURE 9.47   Element loading.

To be able to evaluate the second term on the right-hand side of Equation (9.76) that deals with the surface traction, it is necessary to define a sign convention.

When the nodes of an element are numbered anticlockwise, as shown in Figure 9.47, a tangential force, such as q_t, is positive if it acts anticlockwise. A normal force, such as q_n, is positive if it acts toward the interior of the element.

The components q_x and q_y of the loads q_n and q_t are given by

$$q_x = q_t dL \cos \alpha - q_n dL \sin \alpha = q_t dx - q_n dy$$
$$q_y = q_n dL \cos \alpha + q_t dL \sin \alpha = q_n dx + q_t dy$$

(9.77)

Since in this case the integration will be carried out along the side $(\xi, +1)$, then the following variable changes, $dx = \frac{\partial x}{\partial \xi} d\xi$ and $dy = \frac{\partial y}{\partial \eta} d\eta$ are appropriate. Substituting in Equation (9.77) yields

$$q_x = \left(q_t \frac{\partial x}{\partial \xi} - q_n \frac{\partial y}{\partial \xi} \right) d\xi$$
$$q_y = \left(q_n \frac{\partial x}{\partial \xi} + q_t \frac{\partial y}{\partial \xi} \right) d\xi$$

(9.78)

Then the second term on the right-hand side of Equation (9.76) is therefore obtained as

$$\int_{A_e} [N]^T \begin{Bmatrix} q_x \\ q_y \end{Bmatrix} dA = t \int_{L_{3-4}} [N(\xi, +1)]^T \begin{Bmatrix} q_x \\ q_y \end{Bmatrix} dl$$

$$= t \sum_{i=1}^{ngp} W_i [N(\xi_i, +1)]^T \begin{Bmatrix} \left(q_t \dfrac{\partial x(\xi_i, +1)}{\partial \xi} - q_n \dfrac{\partial y(\xi_i, +1)}{\partial \xi} \right) \\ \left(q_n \dfrac{\partial x(\xi_i, +1)}{\partial \xi} + q_t \dfrac{\partial y(\xi_i, +1)}{\partial \xi} \right) \end{Bmatrix}$$

(9.79)

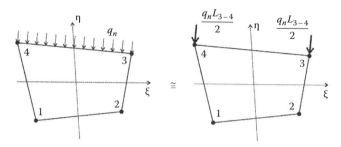

FIGURE 9.48   Equivalent nodal loading.

Remark: In practice, when the loads are uniformly distributed they are replaced by equivalent nodal loads as represented in Figure 9.48. The preceding development is to be used only if the shape of the loading is complicated.

The third term on the right-hand side of Equation (9.76) relates to concentrated loads applied at the nodes. At node 1, we have $N_1 = 1$, $N_2 = 0$, $N_3 = 0$, $N_4 = 0$, and at node 2, we have $N_1 = 0$, $N_2 = 1$, $N_3 = 0$, $N_4 = 0$. It follows therefore

$$\sum_{k=1}[N]_{x=x_k}\{P_k\} = \begin{bmatrix} 1 & 0 \\ 0 & 1 \\ 0 & 0 \\ 0 & 0 \\ 0 & 0 \\ 0 & 0 \\ 0 & 0 \\ 0 & 0 \end{bmatrix} \begin{Bmatrix} 0 \\ -P \end{Bmatrix} = \begin{bmatrix} 0 & 0 \\ 0 & 0 \\ 1 & 0 \\ 0 & 1 \\ 0 & 0 \\ 0 & 0 \\ 0 & 0 \\ 0 & 0 \end{bmatrix} \begin{Bmatrix} 2P \\ 0 \end{Bmatrix} = \begin{Bmatrix} 0 \\ -P \\ 2P \\ 0 \\ 0 \\ 0 \\ 0 \\ 0 \end{Bmatrix} \tag{9.80}$$

9.6.5   COMPUTER CODE: Q4_PLANE_STRESS.m

The program is virtually identical to its predecessor **CST_PLANE_STRESS.m,** except that the stiffness matrix is computed using numerical integration with Gauss quadrature. The size of some of the arrays has increased to account for extra degrees of freedom. In order to assess the performance of the element, we will analyze once again the cantilever beam shown in Figure 9.7. We will use 12 elements to discretize the domain, as shown in Figure 9.49. The nodes numbered 19, 20, and 21 represent the fixed end. The program is listed next.

9.6.5.1   Data Preparation

To read the data, we will use the M-file **Q4_COARSE_MESH_DATA.m** listed next.

FILE: Q4_COARSE_MESH_DATA.m

```
%  File:     Q4_COARSE_MESH_DATA
%
global nnd nel nne nodof eldof n ngp
global geom connec dee nf Nodal_loads
%
```

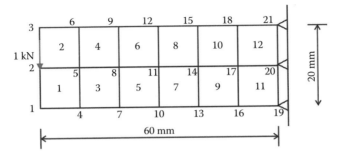

FIGURE 9.49 Finite element discretization with 4-nodded quadrilateral elements.

```
% To change the size of the mesh, alter the next statements
%
nnd = 21 ;                    % Number of nodes:
nel = 12;                      % Number of elements:
nne = 4 ;                     % Number of nodes per element:
nodof =2;                     % Number of degrees of freedom per node
ngp = 2                       % number of Gauss points
eldof = nne*nodof;            % Number of degrees of freedom per element
%
%
% Nodes coordinates x and y
geom = [0,     -10.0; ...        %    x and y coordinates of node 1
        0.0     0.0; ...          %    x and y coordinates of node 2
        0.0    10.0; ...        %   x and y coordinates of node 3
        10.0   -10.0; ...        %   x and y coordinates of node 4
        10.0    0.0; ...        %   x and y coordinates of node 5
        10.0   10.0; ...        %   x and y coordinates of node 6
        20.0   -10.0; ...        %   x and y coordinates of node 7
        20.0    0.0; ...        %   x and y coordinates of node 8
        20.0   10.0; ...        %   x and y coordinates of node 9
        30.0   -10.0; ...        %   x and y coordinates of node 10
        30.0    0.0; ...        %   x and y coordinates of node 11
        30.0   10.0; ...       %   x and y coordinates of node 12
        40.0   -10.0; ...        %   x and y coordinates of node 13
        40.0    0.0; ...        %   x and y coordinates of node 14
        40.0   10.0; ...        %   x and y coordinates of node 15
        50.0   -10.0; ...        %   x and y coordinates of node 16
        50.0    0.0; ...        %   x and y coordinates of node 17
        50.0   10.0; ...        %   x and y coordinates of node 18
        60.0   -10.0; ...        %   x and y coordinates of node 19
        60.0    0.0; ...        %   x and y coordinates of node 20
        60.0   10.0];            %   x and y coordinates of node 21
%
%
%
disp ('Nodes X-Y coordinates')
geom
%
% Element connectivity
connec= [ 1    4    5    2 ;...   % Element 1
          2    5    6    3 ;...   % Element 2
          4    7    8    5 ;...   % Element 3
          5    8    9    6 ;...   % Element 4
          7   10   11    8 ;...   % Element 5
          8   11   12    9 ;...   % Element 6
         10   13   14   11 ;...   % Element 7
         11   14   15   12 ;...   % Element 8
```

```
                    13    16    17    14 ;...    % Element 9
                    14    17    18    15 ;...    % Element 10
                    16    19    20    17 ;...    % Element 11
                    17    20    21    18];       % Element 12
%
%
disp ('Elements connectivity')
connec
%
E = 200000.;      % Elastic modulus in MPa
vu = 0.3;         % Poisson's ratio
thick = 5.;       % Beam thickness in mm
%
% Form the elastic matrix for plane stress
%
dee = formdsig(E,vu);
%
%
% Boundary conditions
%
nf = ones(nnd, nodof);     % Initialize the matrix nf to 1
nf(19,:) = [0    0];       % Node 19 is restrained in the x and y directions
nf(20,:) = [0    0];       % Node 20 is restrained in the x and y directions
nf(21,:) = [0    0];       % Node 21 is restrained in the x and y directions
%
% Counting of the free degrees of freedom
%
n=0;
for i=1:nnd
    for j=1:nodof
        if nf(i,j) ~= 0
            n=n+1;
            nf(i,j)=n;
        end
    end
end
%
% loading
%
Nodal_loads= zeros(nnd, 2);    % Initialize the matrix of nodal loads to 0
%
% Apply a concentrated at the node having x = 0, and y = 0.
%
Force = 1000.;   % N
%
Nodal_loads(1,:) = [0.   -Force];
```

The input data for this beam consist of

- **nnd = 21**; number of nodes
- **nel = 12**; number of elements
- **nne = 4**; number of nodes per element
- **nodof = 2**; number of degrees of freedom per node

The coordinates x and y of the nodes are given in the form of a matrix **geom(nnd, 2)**. The element connectivity is given in the matrix **connec(nel, 4)**. Note that the internal numbering of the nodes is anticlockwise.

As shown in Figure 9.49, nodes 19, 20, and 21 represent the fixed end of the cantilever which is fully fixed. The prescribed degrees of freedom of these nodes are assigned the digit 0. All the degrees of freedom of all the other nodes, which are free, are assigned the digit 1. The information on the boundary conditions is given in the matrix **nf**(**nnd**, **nodof**). The concentrated force of 1000 N is applied at node 2. The force will be assembled into the global force vector **fg** in the main program.

9.6.5.2 Main Program

The main program **Q4_PLANE_STRESS.m** is listed next.

```
% THIS PROGRAM USES AN 4-NODDED QUADRILATERAL ELEMENT FOR THE LINEAR ELASTIC
% STATIC ANALYSIS OF A TWO DIMENSIONAL PROBLEM
%
% Make these variables global so they can be shared by other functions
%
clc
clear all
global nnd nel nne nodof eldof n ngp
global geom connec dee nf Nodal_loads
%
 format long g
%
% To change the size of the problem or change the elastic properties
% supply another input file
%
Q4_COARSE_MESH_DATA
%
%%%%%%%%%%%%%%%%%%%%%%%%%%%% End of input%%%%%%%%%%%%%%%%%%%%%%%%%%%%%%%%%%%
%
% Assemble the global force vector
%
fg=zeros(n,1);
for i=1: nnd
    if nf(i,1) ~= 0
        fg(nf(i,1))= Nodal_loads(i,1);
    end
    if nf(i,2) ~= 0
        fg(nf(i,2))= Nodal_loads(i,2);
    end
end
%
%  Form the matrix containing the abscissas and the weights of Gauss points
%
ngp = 2;
samp=gauss(ngp);
%
% Numerical integration and assembly of the global stiffness matrix
%
%  initialize the global stiffness matrix to zero
kk = zeros(n, n);
%
for i=1:nel
    [coord,g] = elem_q4(i) ;    % coordinates of the nodes of element i,
                                % and its steering vector
    ke=zeros(eldof,eldof) ;     % Initialize the element stiffness matrix
                                % to zero
    for ig=1: ngp
        wi = samp(ig,2);
    for jg=1: ngp
        wj=samp(jg,2);
        [der,fun] = fmlin(samp, ig,jg);  % Derivative of shape functions
                                         %in local coordinates
        jac=der*coord;                   % Compute Jacobian matrix
        d=det(jac);                      % Compute determinant of Jacobian
                                         % matrix
        jac1=inv(jac);                   % Compute inverse of the Jacobian
        deriv=jac1*der;                  % Derivative of shape functions
```

```
                                    % in global coordinates
        bee=formbee(deriv,nne,eldof);      % Form matrix [B]
        ke=ke + d*thick*wi*wj*bee'*dee*bee; % Integrate stiffness matrix
    end
    end
    kk=form_kk(kk,ke, g);               % assemble global stiffness matrix
end
%
%
%%%%%%%%%%%%%%%%%%%%%%%%  End of assembly %%%%%%%%%%%%%%%%%%%%%%%%%%%%%%%%%
%
%
delta = kk\fg ;                         % solve for unknown displacements
%
disp('node        x_disp        y_disp ')     %
for i=1: nnd                                   %
    if nf(i,1) == 0                            %
        x_disp =0.;                            %
    else
        x_disp = delta(nf(i,1));               %
    end
%
    if nf(i,2) == 0                            %
        y_disp = 0.;                           %
    else
        y_disp = delta(nf(i,2));               %
    end
disp([i  x_disp  y_disp])               % Display displacements of each node
DISP(i,:) = [x_disp  y_disp]
end
%
%
ngp=1;                                  % Calculate stresses and strains at
                                        %the center of each element
samp=gauss(ngp);
%
for i=1:nel
    [coord,g] = elem_q4(i);     % coordinates of the nodes of element i,
                                % and its steering vector
    eld=zeros(eldof,1);         % Initialize element displacement to zero
    for m=1:eldof               %
        if g(m)==0              %
            eld(m)=0.;          %
        else                    %
            eld(m)=delta(g(m)); % Retrieve element displacement from the
                                % global displacement vector
        end
    end
%
    for ig=1: ngp
        wi = samp(ig,2);
    for jg=1: ngp
        wj=samp(jg,2);
        [der,fun] = fmlin(samp, ig,jg);     % Derivative of shape functions
                                            % in local coordinates
        jac=der*coord;                      % Compute Jacobian matrix
        jac1=inv(jac);                      % Compute inverse of the Jacobian
        deriv=jac1*der;                     % Derivative of shape functions
                                            % in global coordinates
        bee=formbee(deriv,nne,eldof);       % Form matrix [B]
        eps=bee*eld                         % Compute strains
        sigma=dee*eps                       % Compute stresses
    end
    end
    SIGMA(i,:)=sigma ;              % Store stresses for all elements
end
```

```
%
% Average stresses at nodes
%
[ZX, ZY, ZT, Z1, Z2]=stresses_at_nodes_Q4(SIGMA);
%
%
% Plot stresses in the x_direction
%
U2 = DISP(:,2);
cmin = min(U2);
cmax = max(U2);
caxis([cmin cmax]);
patch('Faces', connec, 'Vertices', geom, 'FaceVertexCData',U2,...
                       'Facecolor','interp','Marker','.');

colorbar;
```

9.6.5.3   Integration of the Stiffness Matrix

The stiffness matrix of the element is given by Equation (9.74). For each element, it is evaluated as follows:

1. For every element $i = 1$ to *nel*
2. Retrieve the coordinates of its nodes **coord(nne, 2)** and its steering vector **g(eldof)** using the function *elem_Q4.m*
3. Initialize the stiffness matrix to zero
   a. Loop over the Gauss points $ig = 1$ to *ngp*
   b. Retrieve the weight **wi** as **samp(ig, 2)**
      i. Loop over the Gauss points $jg = 1$ to *ngp*
      ii. Retrieve the weight **wj** as **samp(jg, 2)**
      iii. Use the function *fmlin.m* to compute the shape functions, vector **fun**, and their derivatives, matrix **der**, in local coordinates, $\xi = $ **samp(ig, 1)** and $\eta = $ **samp(jg, 1)**.
      iv. Evaluate the Jacobian **jac** = **der** ∗ **coord**
      v. Evaluate the determinant of the Jacobian as **d** = **det(jac)**
      vi. Compute the inverse of the Jacobian as **jac1** = **inv(jac)**
      vii. Compute the derivatives of the shape functions with respect to the global coordinates x and y as **deriv** = **jac1** ∗ **der**
      viii. Use the function *formbee.m* to form the strain matrix **bee**
      ix. Compute the stiffness matrix as **ke** = **ke** + **d** ∗ **thick** ∗ **wi** ∗ **wj** ∗ **bee′** ∗ **dee** ∗ **bee**
4. Assemble the stiffness matrix **ke** into the global matrix **kk**

The evaluation of the stiffness matrix requires the use of Gauss quadrature. To do so, the abscissas and the weight of the corresponding Gauss points need to be made available to the program. These are arranged in the array **samp(ngp, 2)** organized as follows:

$$\xi_i = samp(i, 1) \quad and \quad W_i = samp(i, 2) \tag{9.81}$$

The MATLAB function **gauss.m** is listed in Appendix A and can be used for up to **ngp = 4**. The function **elem_q4.m** is also listed in Appendix A. It returns the coordinates of the nodes of each element as well as its steering vector textbfg. The function **fmlin.m** also listed in Appendix A returns the shape functions, vector **fun**, and their derivatives, matrix **der**, in local coordinates.

9.6.5.4 Computation of the Stresses and Strains

Once the global system of equations is solved, we will compute the stresses at the centroid of the elements. For this we set $ngp = 1$.

1. For each element
2. Retrieve the coordinates of its nodes **coord(nne, 2)** and its steering vector **g(eldof)** using the function *elem_Q4.m*
3. Retrieve its nodal displacements **eld(eldof)** from the global vector of displacements **delta(n)**
   a. Loop over the Gauss points $ig = 1$ to ngp
   b. Loop over the Gauss points $jg = 1$ to ngp
   c. Use the function *fmlin.m* to compute the shape functions, vector **fun**, and their local derivatives, **der**, at the local coordinates $\xi = $ **samp(ig, 1)** and $\eta = $ **samp(jg, 1)**
   d. Evaluate the Jacobian **jac = der * coord**
   e. Evaluate the determinant of the Jacobian as **d = det(jac)**
   f. Compute the inverse of the Jacobian as **jac1 = inv(jac)**
   g. Compute the derivatives of the shape functions with respect to the global coordinates x and y as **deriv = jac1 * der**
   h. Use the function *formbee.m* to form the strain matrix **bee**
   i. Compute the strains as **eps = bee * eld**
   j. Compute the stresses as **sigma = dee * eps**
4. Store the stresses in the matrix **SIGMA(nel, 3)**

The stresses computed at the centers of the elements are averaged at the nodes using the function **Stresses_at_nodes_Q4.m**, listed in Appendix A, which returns σ_x, σ_x, τ_x, σ_1, and σ_2. In the present case, we can either feed any of the stresses or the displacements of the nodes to the MATLAB function **patch**, with the argument '*interp*' to interpolate between the values at the nodes and get contour plots.

Figures 9.50 and 9.51 show respectively the contours of the vertical displacement v_2 and of the stress σ_{xx}. It can be clearly seen that the displacement of the tip, equal to 0.104 mm, is very close to the exact displacement, equal to 1.108 mm obtained with Equation (9.41). On the other hand, the stresses are not correct. This is not a problem with the element but rather with the calculations of the stresses in the program. Indeed, in the program the stresses are calculated at the center of the elements then averaged at the nodes. The maximum stress of about 75 MPa represents the value at the center of the element.

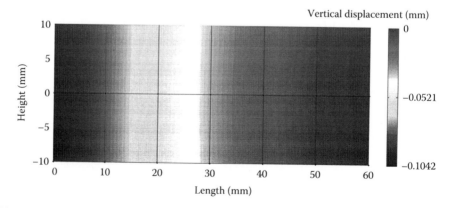

FIGURE 9.50   Contour of the vertical displacement v_2.

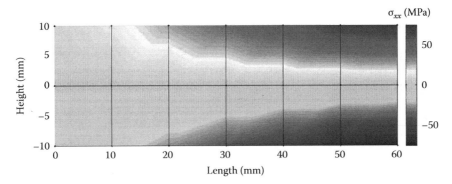

FIGURE 9.51   Contour of the stress σ_{xx}.

9.6.5.5   Program with Automatic Mesh Generation

To better model the stress gradient, we need to refine the mesh. In the new program named **Q4_PLANE_STRESS_MESH.m**, listed next, the mesh is automatically created by calling the function **Q4_mesh.m**. This function prepares the elements' connectivity and nodal geometry matrices, and is listed after the main program.

```
% THIS PROGRAM USES AN 4-NODDED QUADRILATERAL ELEMENT FOR THE LINEAR ELASTIC
% STATIC ANALYSIS OF A TWO DIMENSIONAL PROBLEM
%
% Make these variables global so they can be shared by other functions
%
clc
clear all
global nnd nel nne nodof eldof n ngp
global geom connec dee nf Nodal_loads
global Length Width NXE NYE X_origin Y_origin dhx dhy
%
 format long g
%
% To change the size of the mesh, alter the next statements
%
Length = 60.; % Length of the model
Width =20.;    % Width
NXE = 24;      % Number of rows in the x direction
NYE = 8;       % Number of rows in the y direction
dhx = Length/NXE; % Element size in the x direction
dhy = Width/NYE; % Element size in the x direction
X_origin = 0. ; % X origin of the global coordinate system
Y_origin = Width/2. ;   % Y origin of the global coordinate system
%
nne = 4;
nodof = 2;
eldof = nne*nodof;
%
Q4_mesh      % Generate the mesh
%
E = 200000.;     % Elastic modulus in MPa
vu = 0.3;        % Poisson's ratio
thick = 5.;       % Beam thickness in mm
%
% Form the elastic matrix for plane stress
%
dee = formdsig(E,vu);
%
%
% Boundary conditions
```

```
%
nf = ones(nnd, nodof);     % Initialize the matrix nf to 1
%
% Restrain in all directions the nodes situated @
% (x = Length)
%
for i=1:nnd
    if geom(i,1) == Length;
        nf(i,:) = [0 0];
    end
end
%
% Counting of the free degrees of freedom
%
n=0; for i=1:nnd
    for j=1:nodof
        if nf(i,j) ~= 0
            n=n+1;
            nf(i,j)=n;
        end
    end
end
%
% loading
%
Nodal_loads= zeros(nnd, 2);    % Initialize the matrix of nodal loads to 0
%
% Apply a concentrated at the node having x = 0, and y = 0.
%
Force = 1000.;  % N
%
for i=1:nnd
    if geom(i,1) == 0. && geom(i,2) == 0.
        Nodal_loads(i,:) = [0.  -Force];
    end
end
%
%%%%%%%%%%%%%%%%%%%%%%%%%%%% End of input%%%%%%%%%%%%%%%%%%%%%%%%%%%%%%%%%%%
%
% Assemble the global force vector
%
fg=zeros(n,1);
for i=1: nnd
    if nf(i,1) ~= 0
        fg(nf(i,1))= Nodal_loads(i,1);
    end
    if nf(i,2) ~= 0
        fg(nf(i,2))= Nodal_loads(i,2);
    end
end
%
%  Form the matrix containing the abscissas and the weights of Gauss points
%
ngp = 2;
samp=gauss(ngp);
%
% Numerical integration and assembly of the global stiffness matrix
%
%  initialize the global stiffness matrix to zero
kk = zeros(n, n);
%
for i=1:nel
    [coord,g] = elem_q4(i) ;       % coordinates of the nodes of element i,
                                   %  and its steering vector
    ke=zeros(eldof,eldof) ;        % Initialize the element stiffness
                                   %  matrix to zero
    for ig=1: ngp
```

```
            wi = samp(ig,2);
      for jg=1: ngp
            wj=samp(jg,2);
            [der,fun] = fmlin(samp, ig,jg);   % Derivative of shape functions
                                              % in local coordinates
            jac=der*coord;                    % Compute Jacobian matrix
            d=det(jac);                       % Compute determinant of Jacobian
                                              %  matrix
            jac1=inv(jac);                    % Compute inverse of the Jacobian
            deriv=jac1*der;                   % Derivative of shape functions in
                                              % global coordinates
            bee=formbee(deriv,nne,eldof);     %  Form matrix [B]
            ke=ke + d*thick*wi*wj*bee'*dee*bee; % Integrate stiffness matrix
      end
      end
      kk=form_kk(kk,ke, g);                   % assemble global stiffness matrix
end
%
%
%%%%%%%%%%%%%%%%%%%%%%%%%%  End of assembly %%%%%%%%%%%%%%%%%%%%%%%%%%%%%%%%%%
%
%
delta = kk\fg ;                              % solve for unknown displacements
%
disp('node       x_disp       y_disp ')      %
for i=1: nnd                                  %
    if nf(i,1) == 0                           %
        x_disp =0.;                           %
    else                                      %
        x_disp = delta(nf(i,1));              %
    end                                       %
%
    if nf(i,2) == 0                           %
        y_disp = 0.;                          %
    else                                      %
        y_disp = delta(nf(i,2));              %
    end                                       %
disp([i  x_disp  y_disp])                     % Display displacements of each node
DISP(i,:) = [x_disp  y_disp]
end

%
%
ngp=1;                                       % Calculate stresses and strains at
                                             %the center of each element
samp=gauss(ngp);
%
for i=1:nel
    [coord,g] = elem_q4(i);                  % coordinates of the nodes of element i,
                                             % and its steering vector
    eld=zeros(eldof,1);                      % Initialize element displacement to zero
    for m=1:eldof                            %
        if g(m)==0                           %
            eld(m)=0.;                       %
        else                                 %
            eld(m)=delta(g(m));              % Retrieve element displacement from the
                                             % global displacement vector
        end
    end
%
    for ig=1: ngp
        wi = samp(ig,2);
    for jg=1: ngp
        wj=samp(jg,2);
        [der,fun] = fmlin(samp, ig,jg); % Derivative of shape functions in
                                        % local coordinates
        jac=der*coord;                  % Compute Jacobian matrix
```

```
            jac1=inv(jac);                 % Compute inverse of the Jacobian
            deriv=jac1*der;                % Derivative of shape functions in
                                           % global coordinates
            bee=formbee(deriv,nne,eldof);  % Form matrix [B]
            eps=bee*eld                    % Compute strains
            sigma=dee*eps                  % Compute stresses
        end
        end
    SIGMA(i,:)=sigma ;                     % Store stresses for all elements
end
%
% Average stresses at nodes
%
[ZX, ZY, ZT, Z1, Z2]=stresses_at_nodes_Q4(SIGMA);
%
%
% Plot stresses in the x_direction
%
U2 = DISP(:,2);
cmin = min(U2);
cmax = max(U2);
caxis([cmin cmax]);
patch('Faces', connec, 'Vertices', geom, 'FaceVertexCData',U2, ...
      'Facecolor','interp','Marker','.');
colorbar;
```

Q4_mesh.m

```
% This module generates a mesh of linear quadrilateral elements
%
global nnd nel nne  nodof eldof  n
global geom connec dee nf Nodal_loads
global Length Width NXE NYE X_origin Y_origin dhx dhy
%
%
nnd = 0;
k = 0;
for i = 1:NXE
    for j=1:NYE
        k = k + 1;
        n1 = j + (i-1)*(NYE + 1);
        geom(n1,:) = [(i-1)*dhx - X_origin    (j-1)*dhy - Y_origin ];
        n2 = j + i*(NYE+1);
        geom(n2,:) = [i*dhx - X_origin      (j-1)*dhy - Y_origin  ];
        n3 = n1 + 1;
        geom(n3,:) = [(i-1)*dhx - X_origin     j*dhy - Y_origin  ];
        n4 = n2 + 1;
        geom(n4,:) = [i*dhx- X_origin      j*dhy - Y_origin     ];
        nel = k;
        connec(nel,:) = [n1  n2  n4  n3];
        nnd = n4;
        end
end
%
```

The variables **NXE** and **NYE** represent respectively the number of intervals along the x and y directions, as shown in Figure 9.52. For each interval i and j, four nodes n_1, n_2, n_3, and n_4 and one element are created. The element has nodes n_1, n_2, n_4, n_3. In total the number of elements and nodes created are respectively equal to $nel = NXE \times NYE$, and $nnd = (NXE+1) \times (NYE+1)$. The module also returns the matrices $geom(nnd, 2)$ and $connec(nel, nne)$. The results obtained with the fine mesh are displayed in Figures 9.53 and 9.54 respectively as contour plots of the vertical displacement v_2 and the stress σ_{xx}. The stress values are more accurate. They are very similar to those obtained with the linear strain triangular element shown in Figure 9.13.

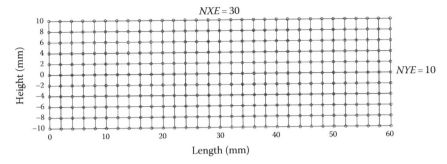

FIGURE 9.52   Automatic mesh generation with the Q4 element.

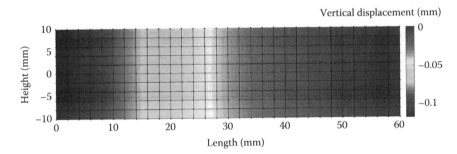

FIGURE 9.53   Contour of the vertical displacement v_2.

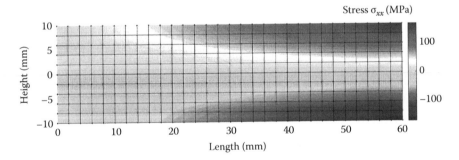

FIGURE 9.54   Contour of the stresses along the x-axis σ_{xx}.

9.6.6   ANALYSIS WITH ABAQUS USING THE Q4 QUADRILATERAL

9.6.6.1   Interactive Edition

In this section, we will analyze the cantilever beam shown in Figure 9.7 with the Abaqus interactive edition. We keep the same geometrical properties, $C = 10$ mm, $L = 60$ mm, $t = 5$ mm, the same mechanical properties, a Young's modulus of 200000 MPa and a Poisson's ratio of 0.3 and the same loading; a concentrated force P of 1000 N.

Start **Abaqus CAE**. Click on **Create Model Database**. On the main menu, click on **File** and set **Set Work Directory** to choose your working directory. Click on **Save As** and name the file **BEAM_Q4.cae**. On the left-hand-side menu, click on **Part** to begin creating the model. Name the part **Beam_Q4**, check **2D Planar**, and check **Deformable** in the type. Choose **Shell** as the base feature. Enter an approximate size of 100 mm and click on **Continue**. In the sketcher menu, choose the **Create-Lines Rectangle** icon to begin drawing the geometry of the beam. Click on **Done** in the bottom-left corner of the viewport window (Figure 9.55).

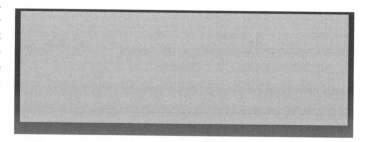

FIGURE 9.55   Creating the Beam_Q4 Part.

If we want to make sure that we will have nodes lying on the neutral axis of the beam, it is advisable to partition the beam along the neutral axis. On the main menu, click on **Tools** then on **Partition**. In the dialog box, check **Face** in **Type**, and **Use shortest path between 2 points** in **Method**. Select the two end points as shown in Figure 9.56, and in the prompt area, click on **Create partition**.

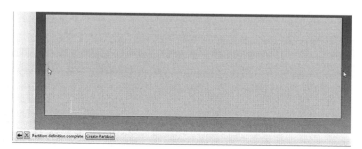

FIGURE 9.56   Creating a partition.

Define a material named steel with an elastic modulus of 200000 MPa and a Poisson's ratio of 0.3. Next, click on **Sections** to create a section named **Beam_section_Q4**. In the **Category** check **Solid**, and in the **Type**, check **Homogeneous**. Click on **Continue**. In the **Edit Section** dialog box, check **Plane stress/strain thickness** and enter 5 mm as the thickness. Click on **OK** (Figure 9.57).

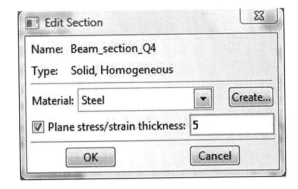

FIGURE 9.57   Creating a plane stress section.

Expand the menu under **Parts** and **BEAM_Q4**, and double click on **Section Assignments**. With the mouse select the whole part. In the **Edit Section Assignments** dialog box, select **Beam_section_Q4**, and click on **OK** (Figure 9.58).

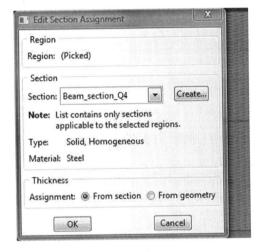

FIGURE 9.58   Editing section assignments.

In the model tree, double click on **Mesh** under the **BEAM_Q4**. In the main menu, under **Mesh**, click on **Mesh Controls**. In the dialog box, check **Quad** for **Element shape** and **Structured** for **Technique**. Click on **OK** (Figure 9.59).

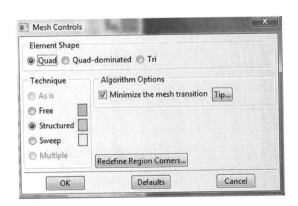

FIGURE 9.59   Mesh controls.

In the main menu, under **Mesh**, click on **Element Type**. With the mouse select all the part in the viewport. In the dialog box, select **Standard** for element library, **Linear** for geometric order. In **Quad**, check **Reduced integration**. The description of the element **CPS4R: A 4-node bilinear plane stress quadrilateral, reduced integration, hourglass control** can be seen in the dialog box. Click on **OK** (Figure 9.60).

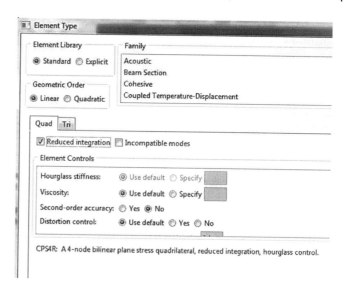

FIGURE 9.60   Selecting element type.

In the main menu, under **Seed**, click on **Part**. In the dialog box, enter 5 for **Approximate global size**. Click on **OK** and on **Done** (Figure 9.61).

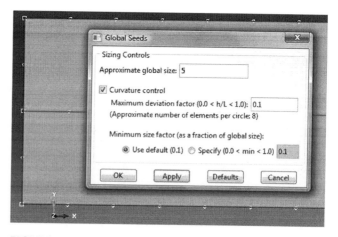

FIGURE 9.61   Seeding part by size.

In the main menu, under **Mesh**, click on **Part**. In the prompt area, click on **Yes**. In the main menu select **View**, then **Part Display Options**. In the **Part Display Options**, under **Mesh**, check **Show node labels** and **Show element labels**. Click **Apply**. The element and node labels will appear in the viewport (Figure 9.62).

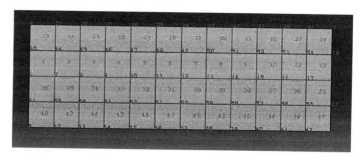

FIGURE 9.62   Mesh.

In the model tree, expand the **Assembly** and double click on **Instances**. Select **BEAM_Q4** for **Parts** and click **OK**. In the model tree, expand **Steps** and **Initial** and double click on **BC**. Name the boundary condition **FIXED**, select **Symmetry/ Antisymmetry/Encastre** for the type, and click on **Continue**. Keep the shift key down, and with the mouse select the right edge and click on **Done** in the prompt area. In the **Edit Boundary Condition** check **ENCASTRE**. Click **OK** (Figure 9.63).

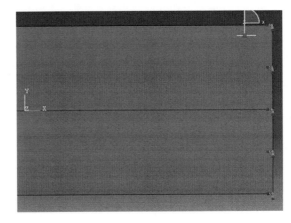

FIGURE 9.63   Imposing BC using geometry.

In the model tree, double click on **Steps**. Name the step **Apply_loads**. Set the procedure to **General**, and select **Static, General**. Click on **Continue**. Give the step a description and click **OK**. In the model tree, under **steps**, and under **Apply_loads**, click on **Loads**. Name the load **Point_Load** and select **Concentrated Force** as the type. Click on **Continue**. Using the mouse click on the middle of the left edge and click on **Done** in the prompt area. In the **Edit Load** dialog box, enter −1000 for **CF2**. Click **OK** (Figure 9.64).

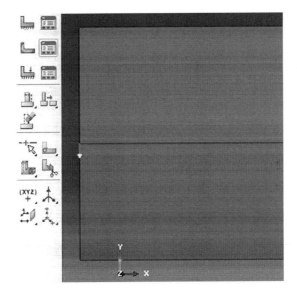

FIGURE 9.64   Imposing a concentrated force using geometry.

In the model tree, expand the **Field Output Requests** and then double click on **F-Output-1**. **F-Output-1** is the default and is automatically generated when creating the step. Uncheck the variables **Contact** and select any other variable you wish to add to the field output. Click on **OK**. Under **Analysis**, right click on **Jobs** and then click on **Create**.

In the **Create Job** dialog box, name the job **BEAM_Q4** and click on **Continue**. In the **Edit Job** dialog box, enter a description for the job. Check **Full analysis**, select to run the job in **Background**, and check to start it **immediately**. Click **OK**. Expand the tree under **Jobs**, right click on **BEAM_Q4**. Then, click on **Submit**. If you get the following message **BEAM_Q4 completed successfully** in the bottom window, then your job is free of errors and was executed properly. Under the top menu, in the **Module** scroll to **Visualization**, and click to load **Abaqus Viewer**. On the main menu, under **File**, click **Open**, navigate to your working directory, and open the file **BEAM_Q4.odb**. It should

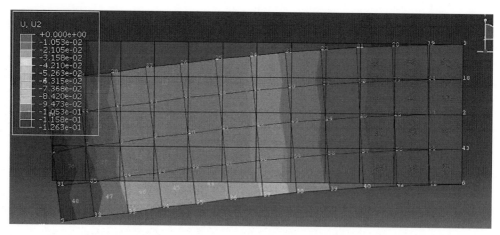

FIGURE 9.65   Plotting displacements on deformed and undeformed shapes.

have the same name as the job you submitted. Click on the **Common options** icon to display the **Common Plot options** dialog box. Under **labels**, check **Show Element labels** and **Show Node labels** to display elements and nodes' numbering. Click on the icon **Plot Contours on both shapes** to display the deformed shape of the beam. Under the main menu, select **U** and **U2** to plot the vertical displacement. It can be seen that the displacement of the left edge is equal to −**0.1263** mm, which is almost similar to the analytical solution and the results obtained with the MATLAB code (Figure 9.65).

In the menu bar, click on **Report** and **Field Output**. In the **Report Field Output** dialog box, for **Position** select **Unique nodal**, check **U1**, and **U2** under **U: Spatial displacement**. Then, click on **Set up**. Click on **Select** to navigate to your working directory. Name the file **BEAM_Q4.rpt**. Uncheck **Append to file** and click **OK**. Use your favorite text editor and open the file **BEAM_Q4.rpt**, which should be the same as the one listed next:

```
********************************************************
Field Output Report, written Tue Jun 07 14:16:55 2011

Source 1
---------

   ODB: C:/ABAQUS_FILES/BEAM_Q4.odb
   Step: Apply_loads
   Frame: Increment        1: Step Time =     1.000

Loc 1 : Nodal values from source 1

Output sorted by column "Node Label".

Field Output reported at nodes for part: BEAM_Q4-1

            Node          U.U1           U.U2
            Label         @Loc 1         @Loc 1
------------------------------------------------------
              1     -329.597E-18    -126.304E-03
              2           -0.       -106.912E-36
              3     -1.90702E-33    -399.459E-36
              4     -28.2845E-03    -124.280E-03
              5      28.2845E-03    -124.280E-03
              6      1.90702E-33    -399.459E-36
              7      208.167E-18    -105.542E-03
              8     -176.942E-18    -95.1550E-03
```

9	60.2816E-18	-77.1076E-03
10	-63.7511E-18	-66.2657E-03
11	-18.8276E-18	-50.6013E-03
12	2.05998E-18	-40.7127E-03
13	-37.2966E-18	-28.0818E-03
14	14.2030E-18	-19.9475E-03
15	-35.9684E-18	-11.1797E-03
16	9.86624E-18	-5.59480E-03
17	-11.6891E-18	-1.24520E-03
18	-2.18595E-33	-47.0851E-36
19	-4.78542E-03	-2.79632E-03
20	-8.88433E-03	-6.86775E-03
21	-12.5353E-03	-12.2002E-03
22	-16.1799E-03	-20.9244E-03
23	-18.9209E-03	-28.9212E-03
24	-21.7734E-03	-41.4322E-03
25	-23.7241E-03	-51.1911E-03
26	-25.7807E-03	-66.7828E-03
27	-26.8659E-03	-77.4101E-03
28	-28.3401E-03	-95.3045E-03
29	-28.3412E-03	-106.447E-03
30	-14.5331E-03	-120.582E-03
31	14.5331E-03	-120.582E-03
32	28.3412E-03	-106.447E-03
33	28.3401E-03	-95.3045E-03
34	26.8659E-03	-77.4101E-03
35	25.7807E-03	-66.7828E-03
36	23.7241E-03	-51.1911E-03
37	21.7734E-03	-41.4322E-03
38	18.9209E-03	-28.9212E-03
39	16.1799E-03	-20.9244E-03
40	12.5353E-03	-12.2002E-03
41	8.88433E-03	-6.86775E-03
42	4.78542E-03	-2.79632E-03
43	2.18595E-33	-47.0851E-36
44	-14.3650E-03	-110.266E-03
45	-13.5899E-03	-91.3903E-03
46	-13.3946E-03	-80.4582E-03
47	-12.5300E-03	-63.5695E-03
48	-11.7534E-03	-53.1450E-03
49	-10.5325E-03	-38.9201E-03
50	-9.35962E-03	-29.8439E-03
51	-7.72590E-03	-19.0591E-03
52	-6.15508E-03	-12.1474E-03
53	-4.20098E-03	-5.59630E-03
54	-1.92738E-03	-1.67195E-03
55	1.92738E-03	-1.67195E-03
56	4.20098E-03	-5.59630E-03
57	6.15508E-03	-12.1474E-03
58	7.72590E-03	-19.0591E-03
59	9.35962E-03	-29.8439E-03
60	10.5325E-03	-38.9201E-03
61	11.7534E-03	-53.1450E-03
62	12.5300E-03	-63.5695E-03
63	13.3946E-03	-80.4582E-03
64	13.5899E-03	-91.3903E-03
65	14.3650E-03	-110.266E-03

Minimum	-28.3412E-03	-126.304E-03
At Node	29	1
Maximum	28.3412E-03	-47.0851E-36
At Node	32	43
Total	-388.578E-18	-3.15015

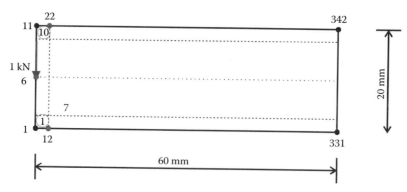

FIGURE 9.66   Generating a mesh manually in Abaqus.

9.6.6.2   Keyword Edition

In this section, we will use a text editor to prepare an input file for the cantilever beam. We will refine the mesh by using 10 elements along the y-axis and 30 elements along the longitudinal direction. In total, there will be 300 elements and 342 nodes. The corner nodes are shown in Figure 9.66.

The file is named **BEAM_Q4_Keyword.inp** and is listed next:

```
*Heading
 Analysis of cantilever beam as a plane stress problem using
 the 4-node bilinear quadrilateral
*Preprint, echo=YES
**
**
** Node generation
**
**
*NODE
 1,    0.,      0.
 11,    0.,     20.
 331,   60.,     0.
 342,   60.,    20.
*NGEN,NSET=Left_Edge
 1,11
*NGEN,NSET=Right_Edge
 331,342
*NFILL
Left_Edge,Right_Edge,30,11
*NSET, NSET = Loaded_node
6
**
** Element generation
**
*ELEMENT,TYPE=CPS4R
1, 1, 12, 13, 2
*ELGEN, ELSET = All_Elements
1, 10, 1, 1, 30, 11, 10
**
*MATERIAL, NAME =STEEL
*ELASTIC
200000., 0.3
*SOLID SECTION, ELSET = All_Elements, MATERIAL = STEEL
5.
**
** BOUNDARY CONDITIONS
**
**
*Boundary
```

```
Right_Edge, encastre
**
** STEP: Apply_Loads
**
*Step, name=Apply_Loads
*Static
1., 1., 1e-05, 1.
**
** LOADS
**
*Cload
Loaded_node, 2, -1000.
**
**
** OUTPUT REQUESTS
**
**
*Output, field, variable=PRESELECT
**
*Output, history, variable=PRESELECT
*End Step
```

1. The input file always starts with the keyword ***HEADING**, which in this case is entered as **Analysis of cantilever beam as a plane stress problem using the 4-node bilinear quadrilateral**.
2. Using ***Preprint, echo=YES** will allow to print an echo of the input file to the file with an extension ***.dat**
3. Using the keyword ***Node**, we define the four corner nodes 1, 11, 331, and 342, as shown in Figure 9.66.
4. Using the keyword ***NGEN**, we generate the nodes located on the left edge. In the data line, we enter the number of the first end node 1, which has been previously defined, then the number of the second end node 11, which also must have been previously defined, followed by the increment in the numbers between each node along the line, which in this case is the default 1. We then group the nodes in a set named **Left_Edge**.
5. Using the keyword ***NGEN** again, we generate the nodes located on the right edge and group them in a set named **Right_Edge**.
6. Using the keyword ***NFILL**, we generate all the remaining nodes by filling in nodes between two bounds. In the data line, we enter first the node sets **Left_Edge** and **Right_Edge** followed by the number of intervals along each line between bounding nodes, in this case 30, and the increment in node numbers from the node number at the first bound set end, which in this case is 11.
7. Using the keyword ***NSET, NSET = Loaded_node**, we create a node set containing node 6. This will be used to apply the concentrated load of 1000 N.
8. Using the keyword ***ELEMENT** and **Type = CPS4R**, which stands for a continuum plane stress four node quadrilateral, we define element 1 as well as its connectivity.
9. Using the keyword ***ELGEN** we generate all the elements that we group in the set **All_elements**. The keyword ***ELGEN** requires in its data line:
   a. Master element number.
   b. Number of elements to be defined in the first row generated, including the master element.
   c. Increment in node numbers of corresponding nodes from element to element in the row. The default is 1.
   d. Increment in element numbers in the row. The default is 1.
   e. If necessary, copy this newly created master row to define a layer of elements.
   f. Number of rows to be defined, including the master row. The default is 1.
   g. Increment in node numbers of corresponding nodes from row to row.

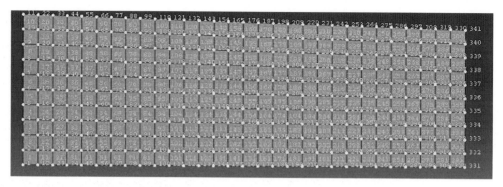

FIGURE 9.67 Mesh generated with the keyword edition.

h. Increment in element numbers of corresponding elements from row to row.
i. If necessary, copy this newly created master layer to define a block of elements (only necessary for a 3D mesh).
j. Number of layers to be defined, including the master layer. The default is 1.
k. Increment in node numbers of corresponding nodes from layer to layer.
l. Increment in element numbers of corresponding elements from layer to layer.
10. Using the keywords *Material and *elastic, we define a material named **steel** having an elastic modulus of 200,000 MPa and a Poisson's ratio of 0.3.
11. Using the keyword *solid section, we assign the material **steel** to all the elements, and in the data line we enter the thickness of the domain, which in this case is 5 mm.
12. Using the created node sets, we impose the boundary conditions with the keyword *Boundary. We fully fix the node set **Right_Edge** by using **encastre**.
13. Next using the keyword *step, we create a step named **Apply_Loads**. The keyword *static indicates that it will be a general static analysis.
14. Using the keyword *cload, we apply a concentrated load of −1000 N in the direction 2 to the node in node set **Loaded_node**.
15. Using the keywords *Output, field, variable=PRESELECT, and *Output, history, variable=PRESELECT we request the default variables for both field and history outputs.
16. Finally, we end the step and the file with *End Step.

At the command line type **Abaqus job=BEAM_Q4_Keyword inter** is followed by **Return**. If you get an error, open the file with extension *.dat to see what type of error. To load the visualization model, type **Abaqus Viewer** at the command line (Figure 9.67).

On the main menu, under **File**, click **Open**, navigate to your working directory, and open the file **BEAM_Q4_Keyword.odb**. Click on the **Common options** icon to display the **Common Plot options** dialog box. Under **labels**, check **Show Element labels** and **Show Node labels** to display the mesh generated. Uncheck **Show Element labels** and **Show Node labels**, then click on the icon **Plot Deformed Shape** to display the deformed shape of the beam. On the main menu, click on **Results** then on **Field Output** to open the **Field Output** dialog box. Choose **U Spatial displacements at nodes**. For component, choose $U2$ to plot the vertical displacement (Figure 9.68).

9.7 THE 8-NODE QUADRILATERAL

9.7.1 FORMULATION

The 8-nodded quadrilateral element has curved sides, which makes it very useful in modeling structures with curved edges (Figure 9.69). The element shape functions are given as

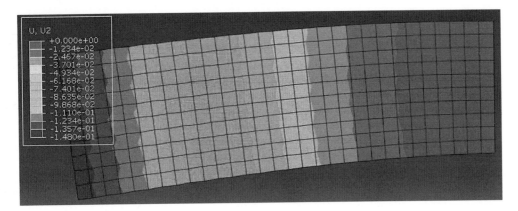

FIGURE 9.68 Displacement contour.

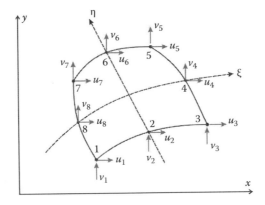

FIGURE 9.69 Eight-nodded isoparametric element.

$$
\begin{Bmatrix}
N_1(\xi,\eta) \\
N_2(\xi,\eta) \\
N_3(\xi,\eta) \\
N_4(\xi,\eta) \\
N_5(\xi,\eta) \\
N_6(\xi,\eta) \\
N_7(\xi,\eta) \\
N_8(\xi,\eta)
\end{Bmatrix}
=
\begin{Bmatrix}
-0.25(1-\xi)(1-\eta)(1+\xi+\eta) \\
0.50(1-\xi^2)(1-\eta) \\
-0.25(1+\xi)(1-\eta)(1-\xi+\eta) \\
0.50(1+\xi)(1-\eta^2) \\
-0.25(1+\xi)(1+\eta)(1-\xi-\eta) \\
0.50(1-\xi^2)(1+\eta) \\
-0.25(1-\xi)(1+\eta)(1+\xi-\eta) \\
0.50(1-\xi)(1-\eta^2)
\end{Bmatrix}
\tag{9.82}
$$

The displacement field over the element is approximated as

$$
u = N_1 u_1 + N_2 u_2 + N_3 u_3 + N_4 u_4 + N_5 u_5 + N_6 u_6 + N_7 u_7 + N_8 u_8 \tag{9.83}
$$

$$
v = N_1 v_1 + N_2 v_2 + N_3 v_3 + N_4 v_4 + N_5 v_5 + N_6 v_6 + N_7 v_7 + N_8 v_8 \tag{9.84}
$$

or in a matrix form as

$$
\begin{Bmatrix} u \\ v \end{Bmatrix} = \begin{bmatrix} N_1 & 0 & | & N_2 & 0 & | & \cdots & \cdots & | & N_8 & 0 \\ 0 & N_1 & | & 0 & N_2 & | & \cdots & \cdots & | & 0 & N_8 \end{bmatrix} \begin{Bmatrix} u_1 \\ v_1 \\ u_2 \\ v_2 \\ \vdots \\ \vdots \\ u_8 \\ v_8 \end{Bmatrix}
\tag{9.85}
$$

or more compactly as

$$
\{U\} = [N]\{a\}
\tag{9.86}
$$

The element is isoparametric, therefore the shape functions $N_i(\xi, \eta)$ also define the geometrical transformation between the reference and the parent element. The coordinates x and y of any point of the parent element are given as

$$
x = N_1 x_1 + N_2 x_2 + \cdots + N_8 x_8
\tag{9.87}
$$
$$
y = N_1 y_1 + N_2 y_2 + \cdots + N_8 y_8
\tag{9.88}
$$

The Jacobian of the transformation is given as

$$
[J] = \begin{bmatrix} \dfrac{\partial x}{\partial \xi} & \dfrac{\partial y}{\partial \xi} \\ \dfrac{\partial x}{\partial \eta} & \dfrac{\partial y}{\partial \eta} \end{bmatrix} = \begin{bmatrix} \sum\limits_{i=1}^{8} \dfrac{\partial N_i}{\partial \xi} x_i & \sum\limits_{i=1}^{8} \dfrac{\partial N_i}{\partial \xi} y_i \\ \sum\limits_{i=1}^{8} \dfrac{\partial N_i}{\partial \eta} x_i & \sum\limits_{i=1}^{8} \dfrac{\partial N_i}{\partial \eta} y_i \end{bmatrix}
$$

After deriving and rearranging, the Jacobian is written in the form of a product of two matrices:

$$
[J] = \begin{bmatrix} \dfrac{\partial N_1}{\partial \xi} & \dfrac{\partial N_2}{\partial \xi} & \cdots & \dfrac{\partial N_8}{\partial \xi} \\ \dfrac{\partial N_1}{\partial \eta} & \dfrac{\partial N_2}{\partial \eta} & \cdots & \dfrac{\partial N_8}{\partial \eta} \end{bmatrix} \begin{bmatrix} x_1 & y_1 \\ x_2 & y_2 \\ \vdots & \vdots \\ x_8 & y_8 \end{bmatrix}
\tag{9.89}
$$

The strain matrix $[B]$ is obtained as

$$
[B] = \begin{bmatrix} \dfrac{\partial N_1}{\partial x} & 0 & | & \dfrac{\partial N_2}{\partial x} & 0 & | & \cdots & \cdots & | & \dfrac{\partial N_8}{\partial x} & 0 \\ 0 & \dfrac{\partial N_1}{\partial y} & | & 0 & \dfrac{\partial N_2}{\partial y} & | & \cdots & \cdots & | & 0 & \dfrac{\partial N_4}{\partial y} \\ \dfrac{\partial N_1}{\partial y} & \dfrac{\partial N_1}{\partial x} & | & \dfrac{\partial N_2}{\partial y} & \dfrac{\partial N_2}{\partial x} & | & \cdots & \cdots & | & \dfrac{\partial N_4}{\partial y} & \dfrac{\partial N_4}{\partial x} \end{bmatrix}
\tag{9.90}
$$

The stiffness matrix is obtained in the same way as for the bilinear quadrilateral element except that it has got a dimension of 16×16 as there are 16 degrees of freedom per element.

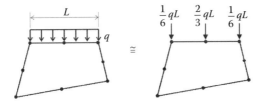

FIGURE 9.70   Equivalent nodal loads.

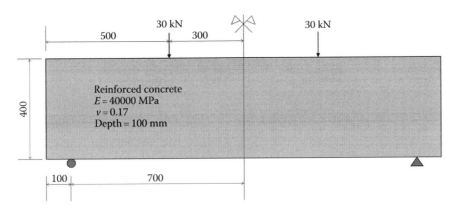

FIGURE 9.71   Geometry and loading.

9.7.2   EQUIVALENT NODAL FORCES

When the shape of the loading on an element edge is complicated, the integration process detailed in Section 9.6.4 should be used. However, if the loads are uniformly distributed then the equivalent nodal loads shown in Figure 9.70 can be used.

9.7.3   PROGRAM Q8_PLANE_STRESS.m

The program is virtually identical to its predecessor **Q4_PLANE_STRESS.m** except that some of the arrays have slightly bigger dimensions because of the increased number of degrees of freedom per element. In order to assess the performance of the element, we will analyze the simply supported deep beam subject to four-point bending shown in Figure 9.71. Taking advantage of symmetry, only half the model is analyzed. We will use 32 elements to discretize the domain as shown in Figure 9.72. The nodes numbered 113–121 represent the mid-span. These nodes are allowed to displace vertically but not horizontally. The program is listed next.

9.7.3.1   Data Preparation

To read the data, we will use the M-file **Q8_COARSE_MESH_DATA.m** listed next.

FILE:Q8_COARSE_MESH_DATA.m

```
%
%%%%%%%%%%%%%%%%%%%%%%%%%%% Beginning of data input %%%%%%%%%%%%%%%%%%%%%%%%%%%%%
%
global nnd nel nne nodof eldof  n ngp
global geom connec dee nf  Nodal_loads

%
nnd = 121 ;              % Number of nodes:
```

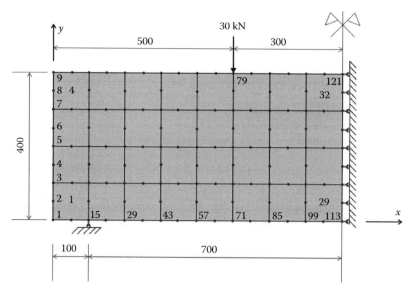

FIGURE 9.72 Coarse mesh.

```
nel = 32;                      % Number of elements:
nne = 8 ;                      % Number of nodes per element:
nodof =2;                      % Number of degrees of freedom per node
ngp = 2                        % number of Gauss points
eldof = nne*nodof;             % Number of degrees of freedom per element
%
% Thickness of the domain
thick = 100.
%
%   Nodes coordinates x and y
%
geom = [ 0        0   ; ...          %    x and y coordinates of node 1
         0        50  ; ...
         0        100 ; ...
         0        150 ; ...
         0        200 ; ...
         0        250 ; ...
         0        300 ; ...
         0        350 ; ...
         0        400 ; ...
         50       0   ; ...
         50       100 ; ...
         50       200 ; ...
         50       300 ; ...
         50       400 ; ...
         100      0   ; ...
         100      50  ; ...
         100      100 ; ...
         100      150 ; ...
         100      200 ; ...
         100      250 ; ...
         100      300 ; ...
         100      350 ; ...
         100      400 ; ...
         150      0   ; ...
         150      100 ; ...
         150      200 ; ...
         150      300 ; ...
         150      400 ; ...
         200      0   ; ...
         200      50  ; ...
```

```
200      100  ;  ...
200      150  ;  ...
200      200  ;  ...
200      250  ;  ...
200      300  ;  ...
200      350  ;  ...
200      400  ;  ...
250      0    ;  ...
250      100  ;  ...
250      200  ;  ...
250      300  ;  ...
250      400  ;  ...
300      0    ;  ...
300      50   ;  ...
300      100  ;  ...
300      150  ;  ...
300      200  ;  ...
300      250  ;  ...
300      300  ;  ...
300      350  ;  ...
300      400  ;  ...
350      0    ;  ...
350      100  ;  ...
350      200  ;  ...
350      300  ;  ...
350      400  ;  ...
400      0    ;  ...
400      50   ;  ...
400      100  ;  ...
400      150  ;  ...
400      200  ;  ...
400      250  ;  ...
400      300  ;  ...
400      350  ;  ...
400      400  ;  ...
450      0    ;  ...
450      100  ;  ...
450      200  ;  ...
450      300  ;  ...
450      400  ;  ...
500      0    ;  ...
500      50   ;  ...
500      100  ;  ...
500      150  ;  ...
500      200  ;  ...
500      250  ;  ...
500      300  ;  ...
500      350  ;  ...
500      400  ;  ...
550      0    ;  ...
550      100  ;  ...
550      200  ;  ...
550      300  ;  ...
550      400  ;  ...
600      0    ;  ...
600      50   ;  ...
600      100  ;  ...
600      150  ;  ...
600      200  ;  ...
600      250  ;  ...
600      300  ;  ...
600      350  ;  ...
600      400  ;  ...
650      0    ;  ...
650      100  ;  ...
650      200  ;  ...
650      300  ;  ...
```

```
             650       400 ; ...
             700       0   ; ...
             700       50  ; ...
             700       100 ; ...
             700       150 ; ...
             700       200 ; ...
             700       250 ; ...
             700       300 ; ...
             700       350 ; ...
             700       400 ; ...
             750       0   ; ...
             750       100 ; ...
             750       200 ; ...
             750       300 ; ...
             750       400 ; ...
             800       0   ; ...
             800       50  ; ...
             800       100 ; ...
             800       150 ; ...
             800       200 ; ...
             800       250 ; ...
             800       300 ; ...
             800       350 ; ...
             800       400]   ;  %   x and y coordinates of node 121
%
% Element connectivity
%

  connec = [1    10  15  16  17  11  3   2   ; ...  % Element 1
            3    11  17  18  19  12  5   4   ; ...
            5    12  19  20  21  13  7   6   ; ...
            7    13  21  22  23  14  9   8   ; ...
            15   24  29  30  31  25  17  16  ; ...
            17   25  31  32  33  26  19  18  ; ...
            19   26  33  34  35  27  21  20  ; ...
            21   27  35  36  37  28  23  22  ; ...
            29   38  43  44  45  39  31  30  ; ...
            31   39  45  46  47  40  33  32  ; ...
            33   40  47  48  49  41  35  34  ; ...
            35   41  49  50  51  42  37  36  ; ...
            43   52  57  58  59  53  45  44  ; ...
            45   53  59  60  61  54  47  46  ; ...
            47   54  61  62  63  55  49  48  ; ...
            49   55  63  64  65  56  51  50  ; ...
            57   66  71  72  73  67  59  58  ; ...
            59   67  73  74  75  68  61  60  ; ...
            61   68  75  76  77  69  63  62  ; ...
            63   69  77  78  79  70  65  64  ; ...
            71   80  85  86  87  81  73  72  ; ...
            73   81  87  88  89  82  75  74  ; ...
            75   82  89  90  91  83  77  76  ; ...
            77   83  91  92  93  84  79  78  ; ...
            85   94  99  100 101 95  87  86  ; ...
            87   95  101 102 103 96  89  88  ; ...
            89   96  103 104 105 97  91  90  ; ...
            91   97  105 106 107 98  93  92  ; ...
            99   108 113 114 115 109 101 100 ; ...
            101  109 115 116 117 110 103 102 ; ...
            103  110 117 118 119 111 105 104 ; ...
            105  111 119 120 121 112 107 106 ];     % Element 32
%
% Material properties
%
E=40000; vu=0.17;           % Young's modulus and Poisson's ratio
%
% Form the matrix of elastic properties
%
```

```
dee=formdsig(E,vu);          % Matrix of elastic properties for plane stress
%
% Boundary conditions
%
nf = ones(nnd, nodof);                       % Initialize the matrix nf to 1
%
for i=1:nnd
    if geom(i,1) == 800.;
        nf(i,:) = [0 1];
    end
     if geom(i,1) == 100. && geom(i,2) == 0.;
        nf(i,:) = [1 0];
    end

end
%
% Counting of the free degrees of freedom
%
n=0;
for i=1:nnd
    for j=1:nodof
        if nf(i,j) ~= 0
            n=n+1;
            nf(i,j)=n;
        end
    end
end
disp ('Nodal freedom')
nf
disp ('Total number of active degrees of freedom')
n
%
% loading
%
Nodal_loads = zeros(nnd, 2);
Nodal_loads(79,2)=-30000.;          % Vertical load on node 79
%
% End input
```

The input data for this beam consist of

- **nnd = 121**; number of nodes
- **nel = 32**; number of elements
- **nne = 8**; number of nodes per element
- **nodof = 2**; number of degrees of freedom per node

The coordinates x and y of the nodes are given in the form of a matrix **geom(nnd, 2)**. The element connectivity is given in the matrix **connec(nel, 8)**. Note that the internal numbering of the nodes is anticlockwise.

As shown in Figure 9.72, nodes 113–121 are fixed in the x-direction only. Node 15, which represents the simple support, is fixed in the y-direction only. The information on the boundary conditions is given in the matrix **nf(nnd, nodof)**. The concentrated force of 30000 N is applied at node 79. Notice the negative sign to indicate that the force acts in the negative y-direction. The force will be assembled into the global force vector **fg** in the main program.

9.7.3.2 Main Program

The main program **Q8_PLANE_STRESS.m** is listed next.

```
% THIS PROGRAM USES AN 8-NODDED QUADRILATERAL ELEMENT FOR THE LINEAR ELASTIC
% STATIC ANALYSIS OF A TWO DIMENSIONAL PROBLEM
```

```
%
% Make these variables global so they can be shared by other functions
%
clc
clear all
global nnd nel nne  nodof eldof  n ngp
global geom connec dee nf Nodal_loads
%
   format long g
%
% This is where the to input the data in the form of a file with
% an extension .m
%
Q8_coarse_mesh_data
%
%%%%%%%%%%%%%%%%%%%%%%%%%%% End of input%%%%%%%%%%%%%%%%%%%%%%%%%%%%%%%%%
%
% Assemble the global force vector
%
fg=zeros(n,1);
for i=1: nnd
    if nf(i,1) ~= 0
        fg(nf(i,1))= Nodal_loads(i,1);
    end
    if nf(i,2) ~= 0
        fg(nf(i,2))= Nodal_loads(i,2);
    end
end
%
%  Form the matrix containing the abscissas and the weights of Gauss points
%
samp=gauss(ngp);
%
% Numerical integration and assembly of the global stiffness matrix
%
%  initialize the global stiffness matrix to zero
kk = zeros(n, n);
%
for i=1:nel
    [coord,g] = elem_q8(i) ;       % coordinates of the nodes of element i,
                                   % and its steering vector
    ke=zeros(eldof,eldof) ;        % Initialize the element stiffness
                                   % matrix to zero
    for ig=1: ngp
        wi = samp(ig,2);
    for jg=1: ngp
        wj=samp(jg,2);
        [der,fun] = fmquad(samp, ig,jg); % Derivative of shape functions
                                         % in local coordinates
        jac=der*coord;             % Compute Jacobian matrix
        d=det(jac);                % Compute determinant of Jacobian matrix
        jac1=inv(jac);             % Compute inverse of the Jacobian
        deriv=jac1*der;            % Derivative of shape functions in
                                   % global coordinates
        bee=formbee(deriv,nne,eldof);    % Form matrix [B]
        ke=ke + d*thick*wi*wj*bee'*dee*bee; % Integrate stiffness matrix
    end
    end
    kk=form_kk(kk,ke, g);          % assemble global stiffness matrix
end
%
%
%%%%%%%%%%%%%%%%%%%%%%%%%  End of assembly %%%%%%%%%%%%%%%%%%%%%%%%%%%%%%%%
%
%
```

```
delta = kk\fg ;                          % solve for unknown displacements
disp('node        x_disp        y_disp ') %
for i=1: nnd                             %
    if nf(i,1) == 0                      %
        x_disp =0.;                      %
    else
        x_disp = delta(nf(i,1));         %
    end
%
    if nf(i,2) == 0                      %
        y_disp = 0.;                     %
    else
        y_disp = delta(nf(i,2));         %
    end
disp([i   x_disp  y_disp]) ;             % Display displacements of each node
DISP(i,:) = [x_disp  y_disp];
end
%
%
ngp=1;                                   % Calculate stresses and strains at
                                         % the center of each element

samp=gauss(ngp);
%
for i=1:nel
    [coord,g] = elem_q8(i);              % coordinates of the nodes of element i, and its steering
                                           vector
    eld=zeros(eldof,1);                  % Initialize element displacement to zero
    for m=1:eldof                        %
        if g(m)==0                       %
            eld(m)=0.;                   %
        else                             %
            eld(m)=delta(g(m));          % Retrieve element displacement from the global displacement
                                           vector
        end
    end
%
    for ig=1: ngp
        wi = samp(ig,2);
    for jg=1: ngp
        wj=samp(jg,2);
        [der,fun] = fmquad(samp, ig,jg); % Derivative of shape functions in local coordinates
        jac=der*coord;                   % Compute Jacobian matrix
        jac1=inv(jac);                   % Compute inverse of the Jacobian
        deriv=jac1*der;                  % Derivative of shape functions in global coordinates
        bee=formbee(deriv,nne,eldof);    %  Form matrix [B]
        eps=bee*eld                      % Compute strains
        sigma=dee*eps                    % Compute stresses
    end
    end
    SIGMA(i,:)=sigma ;                   % Store stresses for all elements
end
%
%
[ZX, ZY, ZT, Z1, Z2]=stresses_at_nodes_Q8(SIGMA);
 U2 = DISP(:,2);
%
%
% Choose one the quantities ( U2, ZX, ZY, ZT, Z1, Z2) to plot
%

 cmin = min(ZT);
 cmax = max(ZT);
 caxis([cmin cmax]);
 patch('Faces', connec, 'Vertices', geom, 'FaceVertexCData',ZT,...
                          'Facecolor','interp','Marker','.');

 colorbar;
```

9.7.3.3   Integration of the Stiffness Matrix

The computation of the stiffness matrix is carried out in the same fashion as for the linear quadrilateral element except that the function *elem_Q4.m* is replaced by *elem_Q8.m* and *fmlin.m* by *fmquad.m*. The exact integration of the stiffness matrix requires 3 Gauss points in each direction.

9.7.3.4   Results with the Coarse Mesh

Figures 9.73 through 9.75 show respectively the contours of the vertical displacement v_2, the stress σ_{xx}, and the shear stress τ_{xy}. The stresses are calculated at the centers of the elements and averaged at the nodes. The program predicts a displacement at mid-span equal to 0.15 mm. To check whether this result is accurate, consider the present deep beam as a slender beam and use the engineering beam theory to calculate the mid-span deflection. For the slender beam with a stiffness EI shown in Figure 9.76, the mid-span deflection is obtained as

$$\delta_{max} = \frac{Pa(3L^2 - 4a^2)}{24EI} \tag{9.91}$$

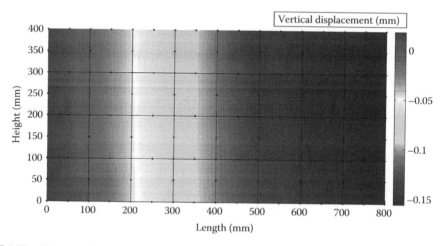

FIGURE 9.73   Contour of the vertical displacement v_2.

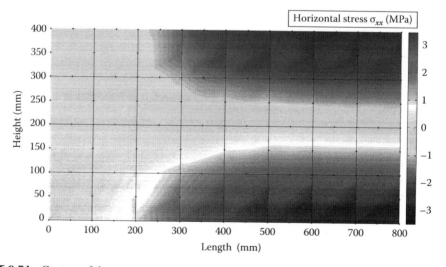

FIGURE 9.74   Contour of the stress σ_{xx}.

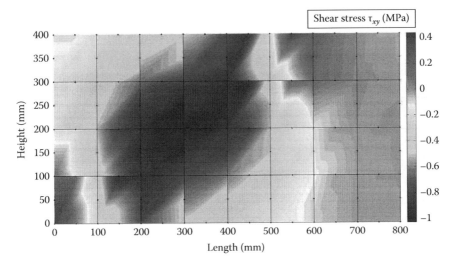

FIGURE 9.75   Contour of the stress τ_{xy}.

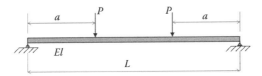

FIGURE 9.76   Slender beam under 4-point bending.

According to Equation (9.91), the mid-span displacement is equal to 0.12 mm, which is less than the value of 0.15 mm obtained with the program. This is somewhat logical since the engineering beam theory, on which the analytical formula is based, does not take into account the extra shear deflections that develop in deep beams. We can, therefore, confidently affirm that the displacement obtained with the program is as good as can be obtained with a coarse mesh. The contour of the horizontal stress σ_{xx} in Figure 9.74 looks acceptable: compression at the top and tension at the bottom with the neutral axis is free of any stress. The contour of the shear stress τ_{xy} is also acceptable. A shear band can be seen between the support and the load application point. Elsewhere, the shear stresses are quite negligible.

9.7.3.5   Program with Automatic Mesh Generation

In the program **Plane_Q8_MESH.m**, the mesh is automatically generated with the module **Q8_mesh.m**. This module prepares the elements' connectivity and nodal geometry matrices and is listed next after the main program.

Plane_Q8_mesh.m

```
% THIS PROGRAM USES AN 8-NODDED QUADRILATERAL ELEMENT FOR THE LINEAR ELASTIC
% STATIC ANALYSIS OF A TWO DIMENSIONAL PROBLEM. IT CONTAINS AN AUTOMATIC
% MESH GENERATION MODULE Q8_mesh.m
%
% Make these variables global so they can be shared by other functions
%
clc
clear all
global nnd nel nne  nodof eldof  n ngp
global geom connec dee nf Nodal_loads
```

```
global Length Width NXE NYE X_origin Y_origin dhx dhy
%
 format long g
%
% To change the size of the problem alter the next lines
%
%
Length = 800.; % Length of the model
Width = 400.;    % Width
NXE = 32;        % Number of rows in the x direction
NYE = 16;        % Number of rows in the y direction
dhx = Length/NXE; % Element size in the x direction
dhy = Width/NYE;  % Element size in the x direction
X_origin = 0. ;  % X origin of the global coordinate system
Y_origin = 0. ;   % Y origin of the global coordinate system
%
nne = 8;
nodof = 2;
eldof = nne*nodof;
ngp = 2;
%
Q8_mesh       % Generate the mesh
%
E = 40000.;      % Elastic modulus in MPa
vu = 0.17;        % Poisson's ratio
thick = 100.;      % Beam thickness in mm
%
% Form the elastic matrix for plane stress
%
dee = formdsig(E,vu);
%
%
% Boundary conditions
%
nf = ones(nnd, nodof);    % Initialize the matrix nf to 1
%
% Restrain in all directions the nodes situated @
% (x = Length)
%
for i=1:nnd
    if geom(i,1) == Length;
        nf(i,:) = [0 1];
    end
     if geom(i,1) == 100. && geom(i,2) == 0.;
        nf(i,:) = [1 0];
    end

end
%
% Counting of the free degrees of freedom
%
n=0;
for i=1:nnd
    for j=1:nodof
        if nf(i,j) ~= 0
            n=n+1;
            nf(i,j)=n;
        end
    end
end
%
% loading
%
Nodal_loads= zeros(nnd, 2);   % Initialize the matrix of nodal loads to 0
%
% Apply a concentrated at the node having x = 0, and y = 0.
%
```

```
Force = 30000.;   % N
%
for i=1:nnd
    if geom(i,1) == 500. && geom(i,2) == 400.
        Nodal_loads(i,:) = [0.   -Force]; % Force acting in negative
                                           % direction
    end
end
%
%%%%%%%%%%%%%%%%%%%%%%%%%%% End of input%%%%%%%%%%%%%%%%%%%%%%%%%%%%%%%%%%%
%
% Assemble the global force vector
%
fg=zeros(n,1);
for i=1: nnd
    if nf(i,1) ~= 0
        fg(nf(i,1))= Nodal_loads(i,1);
    end
    if nf(i,2) ~= 0
        fg(nf(i,2))= Nodal_loads(i,2);
    end
end
%
%  Form the matrix containing the abscissas and the weights of Gauss points
%
samp=gauss(ngp);
%
% Numerical integration and assembly of the global stiffness matrix
%
%  initialize the global stiffness matrix to zero
kk = zeros(n, n);
%
for i=1:nel
    [coord,g] = elem_q8(i) ;          % coordinates of the nodes of element i,
                                      % and its steering vector
    ke=zeros(eldof,eldof) ;           % Initialize the element stiffness
                                      % matrix to zero

    for ig=1: ngp
        wi = samp(ig,2);
        for jg=1: ngp
            wj=samp(jg,2);
            [der,fun] = fmquad(samp, ig,jg);  % Derivative of shape functions
                                              % in local coordinates
            jac=der*coord;           % Compute Jacobian matrix
            d=det(jac);              % Compute determinant of Jacobian matrix
            jac1=inv(jac);           % Compute inverse of the Jacobian
            deriv=jac1*der;          % Derivative of shape functions in
                                     % global coordinates
            bee=formbee(deriv,nne,eldof);     %  Form matrix [B]
            ke=ke + d*thick*wi*wj*bee'*dee*bee; % Integrate stiffness matrix
        end
    end
    kk=form_kk(kk,ke, g);            % assemble global stiffness matrix
end
%
%
%%%%%%%%%%%%%%%%%%%%%%%%%%%%  End of assembly %%%%%%%%%%%%%%%%%%%%%%%%%%%%%%%%%
%
%
delta = kk\fg ;                              % solve for unknown displacements
disp('node      x_disp       y_disp ')  %
for i=1: nnd                                 %
    if nf(i,1) == 0                          %
        x_disp =0.;                          %
    else                                     %
        x_disp = delta(nf(i,1));             %
    end
```

```
%
    if nf(i,2) == 0                          %
        y_disp = 0.;                         %
    else
        y_disp = delta(nf(i,2));             %
    end
disp([i  x_disp  y_disp]) ;             % Display displacements of each node
DISP(i,:) = [x_disp  y_disp];
end
%
%
ngp=1;                                       % Calculate stresses and strains at
                                             % the center of each element
samp=gauss(ngp);
%
for i=1:nel
    [coord,g] = elem_q8(i);             % coordinates of the nodes of element i,
                                        % and its steering vector
    eld=zeros(eldof,1);                 % Initialize element displacement to zero
    for m=1:eldof                       %
        if g(m)==0                      %
            eld(m)=0.;                  %
        else                            %
            eld(m)=delta(g(m));         % Retrieve element displacement from the
                                        % global displacement vector
        end
    end
%
    for ig=1: ngp
        wi = samp(ig,2);
    for jg=1: ngp
        wj=samp(jg,2);
        [der,fun] = fmquad(samp, ig,jg); % Derivative of shape functions in
                                         % local coordinates
        jac=der*coord;                   % Compute Jacobian matrix
        jac1=inv(jac);                   % Compute inverse of the Jacobian
        deriv=jac1*der;                  % Derivative of shape functions
                                         % in global coordinates
        bee=formbee(deriv,nne,eldof);    % Form matrix [B]
        eps=bee*eld                      % Compute strains
        sigma=dee*eps                    % Compute stresses
    end
    end
    SIGMA(i,:)=sigma ;            % Store stresses for all elements
end
%
%
[ZX, ZY, ZT, Z1, Z2]=stresses_at_nodes_Q8(SIGMA);
%
%
% Plot stresses in the x_direction
%
U2 = DISP(:,2);
cmin = min(U2);
cmax = max(U2);
caxis([cmin cmax]);
patch('Faces', connec, 'Vertices', geom, 'FaceVertexCData',U2,...
                          'Facecolor','interp','Marker','.');
colorbar;
```

Q8_mesh.m

```
% This module generates a mesh of 8-nodded quadrilateral elements
%
global nnd nel nne  nodof eldof  n
```

```
global geom connec dee nf Nodal_loads
global Length Width NXE NYE X_origin Y_origin dhx dhy
%
%
nnd = 0;
k = 0;
for i = 1:NXE
   for j=1:NYE
     k = k + 1;
%
     n1 = (i-1)*(3*NYE+2)+2*j - 1;
     n2 = i*(3*NYE+2)+j - NYE - 1;
     n3 = i*(3*NYE+2)+2*j-1;
     n4 = n3 + 1;
     n5 = n3 + 2;
     n6 = n2 + 1;
     n7 = n1 + 2;
     n8 = n1 + 1;
%
     geom(n1,:) = [(i-1)*dhx - X_origin     (j-1)*dhy - Y_origin ];
     geom(n3,:) = [i*dhx - X_origin       (j-1)*dhy - Y_origin  ];
     geom(n2,:) = [(geom(n1,1)+geom(n3,1))/2   (geom(n1,2)+geom(n3,2))/2];
     geom(n5,:) = [i*dhx- X_origin        j*dhy - Y_origin ];
     geom(n4,:) = [(geom(n3,1)+ geom(n5,1))/2  (geom(n3,2)+ geom(n5,2))/2];
     geom(n7,:) = [(i-1)*dhx - X_origin       j*dhy - Y_origin  ];
     geom(n6,:) = [(geom(n5,1)+ geom(n7,1))/2  (geom(n5,2)+ geom(n7,2))/2];
     geom(n8,:) = [(geom(n1,1)+ geom(n7,1))/2  (geom(n1,2)+ geom(n7,2))/2];
%
     nel = k;
     nnd = n5;
     connec(k,:) = [n1 n2 n3 n4 n5 n6 n7 n8];
   end
end
```

The variables **NXE** and **NYE** represent respectively the number of intervals along the x and y directions as shown in Figure 9.77. For each interval i and j, one element with nodes $n_1, \ldots, n_8$ is created. The module returns the matrices $geom(nnd, 2)$ and $connec(nel, nne)$ as well as the number of elements nel and the number of nodes nnd. The results obtained with the fine mesh ($NXE = 32$

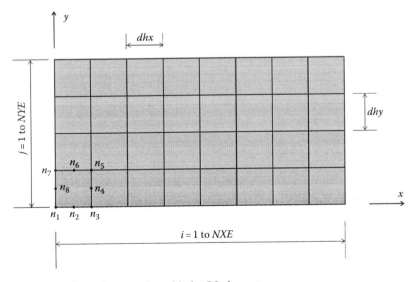

FIGURE 9.77   Automatic mesh generation with the Q8 element.

and *NYE* = 16) are displayed in Figures 9.78 through 9.80 respectively as contour plots of the vertical displacement v_2, the stress σ_{xx}, and the shear stress τ_{xy}. The stresses are calculated at the centers of the elements and averaged at the nodes. More details can be obtained with a finer mesh; for example, notice the stress concentration at the load application point and the shape of the shear band.

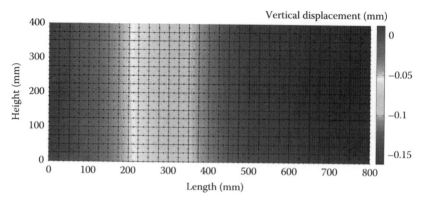

FIGURE 9.78   Contour of the vertical displacement v_2.

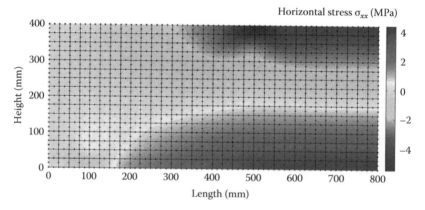

FIGURE 9.79   Contour of the stress σ_{xx}.

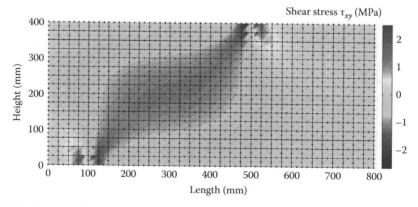

FIGURE 9.80   Contour of the stress τ_{xy}.

9.7.4 ANALYSIS WITH ABAQUS USING THE Q8 QUADRILATERAL

In this section, we will analyze the simply supported deep beam subject to four-point bending shown in Figure 9.71. Taking advantage of symmetry, only half the model is analyzed. We will use an element size of 25 mm so we could compare the results with those obtained previously.

Start **Abaqus CAE**. Click on **Create Model Database**. On the main menu, click on **File** and set **Set Work Directory** to choose your working directory. Click on **Save As** and name the file **Deep_Beam_Q8.cae**. On the left-hand-side menu, click on **Part** to begin creating the model. Name the part **DEEP_Beam_Q8**, check **2D Planar**, and check **Deformable** in the type. Choose **Shell** as the base feature. Enter an approximate size of 1000 mm and click on **Continue**. In the sketcher menu, choose the **Create-Lines Rectangle** icon to begin drawing the geometry of the beam. Click on **Done** in the bottom-left corner of the viewport window (Figure 9.81).

FIGURE 9.81   Creating the Deep_Beam_Q8 Part.

Define a material named RConcrete with an elastic modulus of 40000 MPa and a Poisson's ratio of 0.17. Next, click on **Sections** to create a section named **Beam_section_Q8**. In the **Category** check **Solid**, and in the **Type**, check **Homogeneous**. Click on **Continue**. In the **Edit Section** dialog box, check **Plane stress/strain thickness** and enter 100 mm as the thickness. Click on **OK** (Figure 9.82).

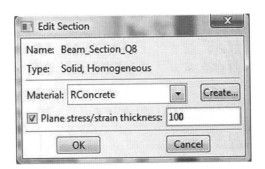

FIGURE 9.82   Creating a plane stress section.

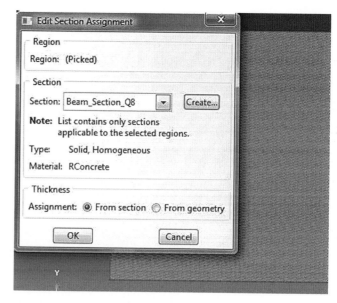

Expand the menu under **Parts** and **Deep_Beam_Q8**, and double click on **Section Assignments**. With the mouse select the whole part. In the **Edit Section Assignments** dialog box, select **Beam_section_Q8** and click on **OK** (Figure 9.83).

FIGURE 9.83   Editing section assignments.

In the model tree, double click on **Mesh** under the **Deep_Beam_Q8**. In the main menu, under **Mesh**, click on **Mesh Controls**. In the dialog box, check **Quad** for **Element shape** and **Structured** for **Technique**. Click on **OK**. Under **Mesh**, click on **Element Type**. In the dialog box, select **Standard** for element library, **Quadratic** for geometric order. In **Quad**, check **Reduced integration**. The description of the element **CPS8R: A 8-node biquadratic plane stress quadrilateral, reduced integration** can be seen in the dialog box. Click on **OK** (Figure 9.84).

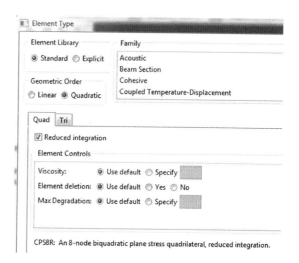

FIGURE 9.84   Mesh controls and element type.

In the main menu, under **Seed**, click on **Part**. In the dialog box, enter 25 for **Approximate global size**. Click on **OK** and on **Done**. In the main menu, under **Mesh**, click on **Part**. In the prompt area, click on **Yes** (Figure 9.85).

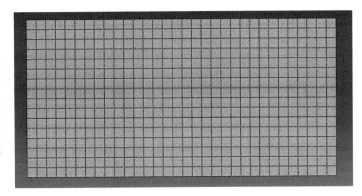

FIGURE 9.85   Mesh.

Under **Part**, in the left-hand-side menu, click on **Sets**. In the dialog box, name the set **Loaded_node** and check **Node** for **Type**. Click on **Continue**. In the viewport, locate the node situated at 300 mm from the right edge, which is the centerline of the beam. Click on **Done** (Figure 9.86).

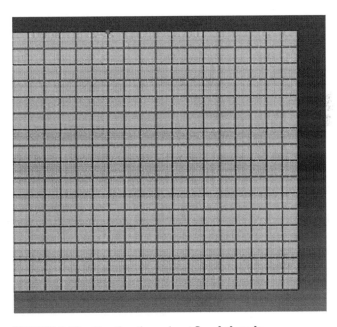

FIGURE 9.86   Creating the node set **Loaded_node**.

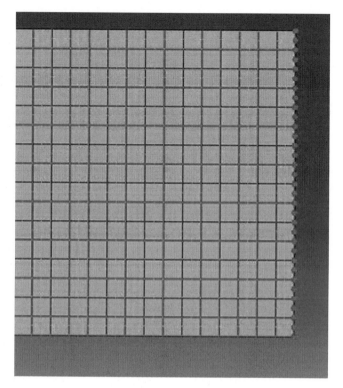

Repeat the procedure, and this time name the node set **Centerline**. In the viewport, locate all the nodes situated on the right edge. Click on **Done** (Figure 9.87).

FIGURE 9.87   Creating the node set **Centerline**.

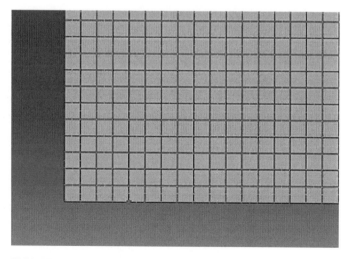

Repeat the procedure and this time name the node set **Support**. In the viewport, locate the node situated at 100 mm from the left bottom corner. Click on **Done** (Figure 9.88).

FIGURE 9.88   Creating the node set **Support**.

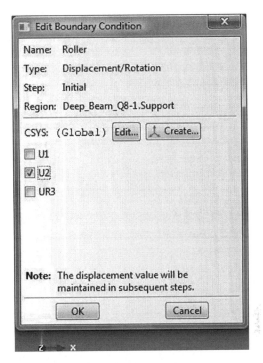

In the model tree, expand the **Assembly** and double click on **Instances**. Select **Deep_Beam_Q8** for **Parts**, and click **OK**. In the model tree, expand **Steps** and **Initial**, and double click on **BC**. Name the boundary condition **Roller**, select **Displacement/Rotation** for the type, and click on **Continue**. In the bottom-right corner of the viewport, click on **Sets**, and select **Deep_Beam_Q8-1.Support** and click on **Continue**. In the **Edit Boundary Condition** check **U2**. Click **OK**. Repeat the procedure again, this time select the set **Deep_Beam_Q8-1.Centerline** and click on **Continue**. In the **Edit Boundary Condition** check **U1**. Click **OK** (Figure 9.89).

FIGURE 9.89   Imposing BC using a node set.

In the model tree, double click on **Steps**. Name the step **Apply_loads**. Set the procedure to **General** and select **Static, General**. Click on **Continue**. Click on **OK**. In the model tree, under **steps**, and under **Apply_loads**, click on **Loads**. Name the load **Point_Load** and select **Concentrated force** as the type. Click on **Continue**. In the bottom-right corner of the viewport, click on sets and select **Deep_Beam_Q8-1.Loaded_ node**. In the **Edit Load** dialog box, enter −30000 for **CF2**. Click **OK** (Figure 9.90).

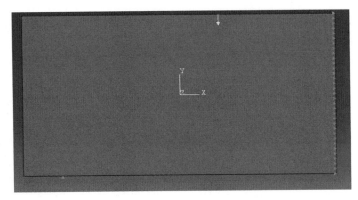

FIGURE 9.90   BC and loads.

Under **Analysis**, right click on **Jobs** and then click on **Create**.

In the **Create Job** dialog box, name the job **Deep_Beam_Q8** and click on **Continue**. In the **Edit Job** dialog box, enter a description for the job. Check **Full analysis**, select to run the job in **Background** and check to start it **immediately**. Click **OK**. Expand the tree under **Jobs**, right click on **Deep_Beam_Q8**. Then, click on **Submit**. If you get the following message **Deep_Beam_Q8**

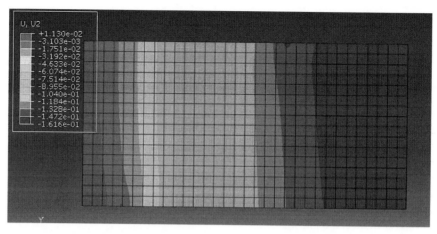

FIGURE 9.91   Contour of the vertical displacement.

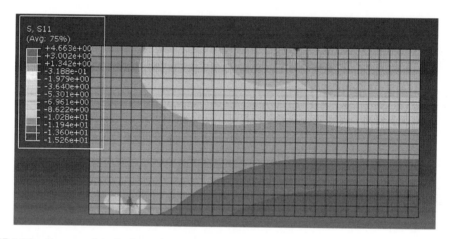

FIGURE 9.92   Contour of the horizontal stress σ_{xx}.

completed successfully in the bottom window, then your job is free of errors and was executed properly.

Under the top menu, in the **Module** scroll to **Visualization**, and click to load **Abaqus Viewer**. On the main menu, under **File**, click **Open**, navigate to your working directory, and open the file **Deep_Beam_Q8.odb**. It should have the same name as the job you submitted. Click on the icon **Plot** on **Undeformed shape**. Under the main menu, select **U** and **U2** to plot the vertical displacement (Figure 9.91). It can be seen that the displacement contour is exactly the same as that obtained with the MATLAB code (Figure 9.78).

Under the main menu, select **S** and **S11** to plot σ_{xx} (Figure 9.92). Again, the contour is very similar to that shown in Figure 9.79.

9.8   SOLVED PROBLEM WITH MATLAB®

9.8.1   STRIP FOOTING WITH THE CST ELEMENT

Figure 9.93 represents a strip footing on a sandy soil with an elastic modulus $E = 10^5 \text{ kN/m}^2$ and a Poisson's ratio $\mu = 0.3$. The footing is 2 m wide and supports a uniformly distributed load of 5 kN/m^2. Five meters beneath the footing the soil is made up of a solid rock formation that can be

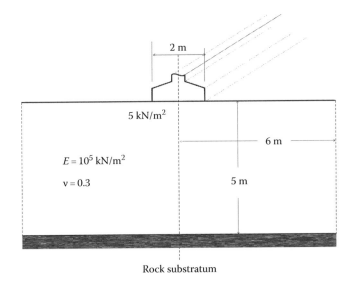

FIGURE 9.93 Strip footing.

considered very stiff. In addition, assume that 6 m away from the center of the footing the horizontal displacement of the soil is negligible.

Consider an element length of 0.5 m, analyze the footing using both the CST and LST elements:

- Plot the vertical deflection of the center line as a function of depth
- Produce a contour of the second principal stress σ_2

The finite strip footing is a three-dimensional solid. However, the longitudinal direction is very important, which therefore prevents thickness change. The ends of the strip foundations are prevented from moving in the z-direction, then the displacement w is equal to zero. At the mid-span of the footing, by symmetry, w must be also equal to zero. Therefore, we assume that w is zero everywhere and the displacements u and v are functions of x and y only. Such a state is characterized by $\epsilon_{zz} = \epsilon_{xz} = \epsilon_{yz} = 0$ and it is a state of plane strain. The function **formdeps.m** is used to generate the matrix of the elastic properties. In addition, the geometry of the footing is symmetrical, therefore only the right half is discretized as shown in Figure 9.94.

The domain is discretized using 12 intervals along the x-direction, $NXE = 12$, and 10 along the y-direction, $NYE = 10$. These give an element size of 0.5 m in both directions as shown in Figure 9.95.

The boundary conditions of restrained nodes are generated using their coordinates as follows:

- The nodes directly beneath the center of the footing, $x = 0$, and the nodes situated on the right boundary, $x = Length$, are restrained in the x-direction

```
if geom(i,1) == 0.  | geom(i,1) == Length;
    nf(i,:) = [0 1];
end
```

- The nodes situated on the rocky substratum, $y = 0$, are restrained in all directions

```
if geom(i,2) == 0. ;
    nf(i,:) = [0  0];
end
```

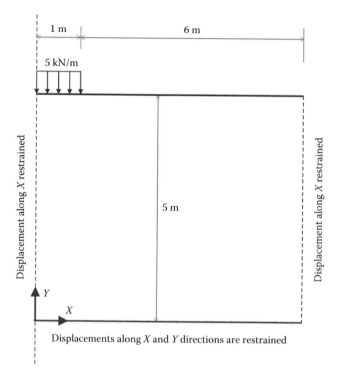

FIGURE 9.94   Strip footing model.

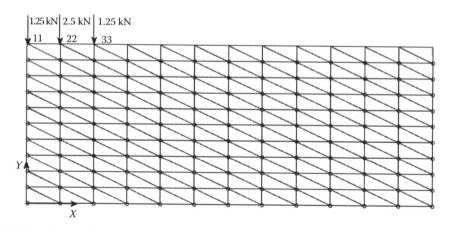

FIGURE 9.95   Mesh with the CST element.

The mesh generating function **T3_mesh.m** does not actually generate the loading. This was added manually to the figure. Indeed, since **T3_mesh.m** numbers the nodes in the y-direction, it is not difficult to see in Figure 9.95 that nodes 11, 22, and 33 are the loaded nodes. Equivalent statically concentrated loads are applied to these nodes as follows:

```
Nodal_loads(11,:) = [0.  -1.25];
Nodal_loads(22,:) = [0.  -2.50];
Nodal_loads(33,:) = [0.  -1.25];
```

CST_STRIP_FOOTING.m

```
% THIS PROGRAM USES AN 3-NODE LINEAR TRIANGULAR ELEMENT FOR THE
% LINEAR ELASTIC STATIC ANALYSIS OF A TWO DIMENSIONAL PROBLEM
% IT INCLUDES AN AUTOMATIC MESH GENERATION
%
% Make these variables global so they can be shared by other functions
%
clear all
clc
global nnd nel nne  nodof eldof  n
global geom  dee nf Nodal_loads
global Length Width NXE NYE X_origin Y_origin
%
 format long g
%
%
% To change the size of the problem or change elastic properties
%     supply another input file
%
Length = 6.; % Length of the model
Width =5.;     % Width
NXE = 12;        % Number of rows in the x direction
NYE = 10;        % Number of rows in the y direction
dhx = Length/NXE; % Element size in the x direction
dhy = Width/NYE;  % Element size in the x direction
X_origin = 0. ;  % X origin of the global coordinate system
Y_origin = 0. ;   % Y origin of the global coordinate system
%
nne = 3;
nodof = 2;
eldof = nne*nodof;
%
T3_mesh ;            % Generate the mesh
%
% Material
%
E = 100000.;      % Elastic modulus in MPa
vu = 0.3;         % Poisson's ratio
thick = 1.;        % Beam thickness in mm
%
% Form the elastic matrix for plane strain
%
dee = formdeps(E,vu);
%
%
% Boundary conditions
%
nf = ones(nnd, nodof);     % Initialize the matrix nf to 1
%
% Restrain in the x-direction the nodes situated @
% (x = 0  or x = Length)
%
for i=1:nnd
    if geom(i,1) == 0.  | geom(i,1) == Length;
        nf(i,:) = [0 1];
    end
end
%
% Restrain in all directions the nodes situated @
% (y = 0)
%
for i=1:nnd
    if geom(i,2) == 0. ;
        nf(i,:) = [0 0];
    end
end
```

```
%
% Counting of the free degrees of freedom
%
n=0;
for i=1:nnd
    for j=1:nodof
        if nf(i,j) ~= 0
            n=n+1;
            nf(i,j)=n;
        end
    end
end
%
% loading
%
Nodal_loads= zeros(nnd, 2); % Initialize the matrix of nodal loads to 0
%
% Apply equivalent concentrated loads on nodes 11, 22, and 33 in the
% y-direction.
%
   Nodal_loads(11,:) = [0.  -1.25];
   Nodal_loads(22,:) = [0.  -2.50];
   Nodal_loads(33,:) = [0.  -1.25];
%
%%%%%%%%%%%%%%%%%%%%%%%%%%% End of input%%%%%%%%%%%%%%%%%%%%%%%%%%%%%%%%%%
%
% Assemble the global force vector
%
fg=zeros(n,1);
for i=1: nnd
    if nf(i,1) ~= 0
        fg(nf(i,1))= Nodal_loads(i,1);
    end
    if nf(i,2) ~= 0
        fg(nf(i,2))= Nodal_loads(i,2);
    end
end
%
% Assembly of the global stiffness matrix
%
%  initialize the global stiffness matrix to zero
%
kk = zeros(n, n);
%
for i=1:nel
    [bee,g,A] = elem_T3(i);      % Form strain matrix, and stering vector
    ke=thick*A*bee'*dee*bee;     % Compute stiffness matrix
    kk=form_kk(kk,ke, g);        % assemble global stiffness matrix
end
%
%
%%%%%%%%%%%%%%%%%%%%%%%%%%%%  End of assembly %%%%%%%%%%%%%%%%%%%%%%%%%%%%%%%%
%
%
delta = kk\fg ;               % solve for unknown displacements
%
for i=1: nnd                                %
    if nf(i,1) == 0                         %
        x_disp =0.;                         %
    else
        x_disp = delta(nf(i,1));            %
    end
%
    if nf(i,2) == 0                         %
        y_disp = 0.;                        %
    else
        y_disp = delta(nf(i,2));            %
```

```
        end
        node_disp(i,:) =[x_disp  y_disp];
end
%
%
% Retrieve the y_disp of the nodes located on center line beneath
% the footing
%
k = 0;
vertical_disp=zeros(1,NYE+1);
for i=1:nnd;

    if geom(i,1)== 0.
        k=k+1;
        y_coord(k) = geom(i,2);
        vertical_disp(k)=node_disp(i,2);
    end
end
%
for i=1:nel
    [bee,g,A] = elem_T3(i);        % Form strain matrix, and stering vector
    eld=zeros(eldof,1);            % Initialize element displacement to zero
    for m=1:eldof
        if g(m)==0
         eld(m)=0.;
        else                           %
         eld(m)=delta(g(m));       % Retrieve element displacement
        end
    end
%
    eps=bee*eld;                   % Compute strains
    EPS(i,:)=eps ;                 % Store strains for all elements
    sigma=dee*eps;                 % Compute stresses
    SIGMA(i,:)=sigma ;             % Store stresses for all elements
end
%
% Calculate the principal stresses
%
SIG1=zeros(nel,1); SIG2=zeros(nel,1);
for i = 1:nel
    SIG1(i)=(SIGMA(i,1)+SIGMA(i,2))/2 + ...
            sqrt(((SIGMA(i,1)+SIGMA(i,2))/2)^2 +SIGMA(i,3)^2);
    SIG2(i)=(SIGMA(i,1)+SIGMA(i,2))/2 - ...
            sqrt(((SIGMA(i,1)+SIGMA(i,2))/2)^2 +SIGMA(i,3)^2);
end
cmin = min(SIG2);
cmax = max(SIG2);
caxis([cmin cmax]);
patch('Faces', connec, 'Vertices', geom, 'FaceVertexCData',SIG2, ...
      'Facecolor','flat','Marker','o');
colorbar;
%
plottools;
```

The computed results are shown in Figure 9.96. A patch plot of the principal stress σ_2 as well as the vertical displacements of the nodes situated just below the center of the footing are shown. Both the maximum displacement and maximum stress, respectively equal to 0.12 mm and 8 kN/m^2, occur just below the footing.

9.8.2 STRIP FOOTING WITH THE LST ELEMENT

The domain is also discretized using 12 intervals along the x-direction, $NXE = 12$, and 10 along the y-direction, $NYE = 10$. These give an element size of 0.5 m in both directions as shown in

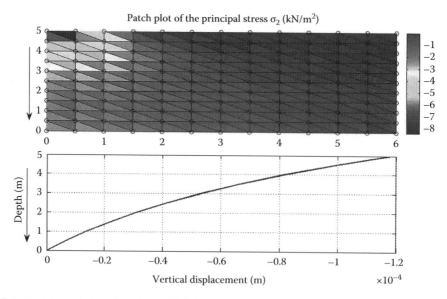

FIGURE 9.96 Computed result with the CST element.

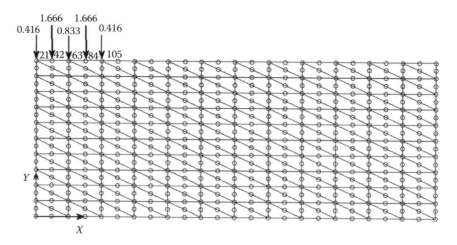

FIGURE 9.97 Mesh with the LST element.

Figure 9.97. The boundary conditions of restrained nodes are entered in the same way as done previously using the nodal coordinates.

The mesh generating function **T6_mesh.m** does not generate the loading. This was added manually as shown in Figure 9.97. The function **T6_mesh.m** numbers the nodes in the y-direction, therefore it is not difficult to see in Figure 9.97 that nodes 21, 42, 63, 84, and 105 are the loaded nodes. The equivalent statically concentrated loads are calculated as shown in Figure 9.98, and they are entered as follows:

```
Nodal_loads(21,:)  = [0.  -0.416];
Nodal_loads(42,:)  = [0.  -1.666];
Nodal_loads(63,:)  = [0.  -0.833];
Nodal_loads(84,:)  = [0.  -1.666];
Nodal_loads(105,:) = [0.  -0.416];
```

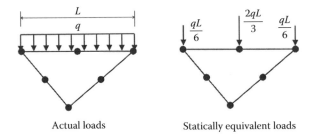

Actual loads   Statically equivalent loads

FIGURE 9.98 Statically equivalent loads for the LST element.

LST_STRIP_FOOTING.m

```
% THIS PROGRAM USES A 6-NODE LINEAR TRIANGULAR ELEMENT FOR THE
% LINEAR ELASTIC STATIC ANALYSIS OF A TWO DIMENSIONAL PROBLEM
% IT INCLUDES AN AUTOMATIC MESH GENERATION
%
% Make these variables global so they can be shared by other functions
%
clear all
clc
global nnd nel nne  nodof eldof  n
global connec geom dee nf Nodal_loads XIG YIG
global Length Width NXE NYE X_origin Y_origin
%
 format long g
%
%
% To change the size of the problem or change elastic properties
%     supply another input file
%
Length = 6.; % Length of the model
Width =5.;     % Width
NXE = 12;       % Number of rows in the x direction
NYE = 10;       % Number of rows in the y direction
XIG = zeros(2*NXE+1,1); YIG=zeros(2*NYE+1,1); % Vectors holding grid coordinates
dhx = Length/NXE; % Element size in the x direction
dhy = Width/NYE;  % Element size in the x direction
X_origin = 0. ;  % X origin of the global coordinate system
Y_origin = 0. ;   % Y origin of the global coordinate system
%
nne = 6;
nodof = 2;
eldof = nne*nodof;
%
T6_mesh ;          % Generate the mesh
%
% Material
%
E = 100000.;       % Elastic modulus in MPa
vu = 0.3;          % Poisson's ratio
thick = 1.;        % Beam thickness in mm
nhp = 3;           % Number of sampling points
%
% Form the elastic matrix for plane stress
%
dee = formdeps(E,vu);
%
%
% Boundary conditions
%
nf = ones(nnd, nodof);     % Initialize the matrix nf to 1
```

```
%
% Restrain in the x-direction the nodes situated @
% (x = 0  or x = Length)
%
for i=1:nnd
    if geom(i,1) == 0.  | geom(i,1) == Length;
        nf(i,:) = [0 1];
    end
end
%
% Restrain in all directions the nodes situated @
% (y = 0)
%
for i=1:nnd
    if geom(i,2) == 0. ;
        nf(i,:) = [0 0];
    end
end
%
%
% Counting of the free degrees of freedom
%
n=0;
for i=1:nnd
    for j=1:nodof
        if nf(i,j) ~= 0
            n=n+1;
            nf(i,j)=n;
        end
    end
end
%
% loading
%
Nodal_loads= zeros(nnd, 2);   % Initialize the matrix of nodal loads to 0
%
%
% Apply equivalent concentrated loads on nodes 21, 42, 63, 84 and 105 in the
% y-direction.
%
   Nodal_loads(21,:) = [0.  -0.416];
   Nodal_loads(42,:) = [0.  -1.666];
   Nodal_loads(63,:) = [0.  -0.833];
   Nodal_loads(84,:) = [0.  -1.666];
   Nodal_loads(105,:) = [0.  -0.416];
%
%
%%%%%%%%%%%%%%%%%%%%%%%%%%% End of input%%%%%%%%%%%%%%%%%%%%%%%%%%%%%%%%%%%%
%
% Assemble the global force vector
%
fg=zeros(n,1);
for i=1: nnd
    if nf(i,1) ~= 0
        fg(nf(i,1))= Nodal_loads(i,1);
    end
    if nf(i,2) ~= 0
        fg(nf(i,2))= Nodal_loads(i,2);
    end
end
%
% Assembly of the global stiffness matrix
%
%
%  Form the matrix containing the abscissas and the weights of Hammer points
%
samp=hammer(nhp);
```

```
%
%   initialize the global stiffness matrix to zero
%
kk = zeros(n, n);
%
for i=1:nel
    [coord,g] = elem_T6(i);   % Form strain matrix, and stering vector
    ke=zeros(eldof,eldof) ;   % Initialize the element stiffness matrix to zero
    for ig = 1:nhp
        wi = samp(ig,3);
        [der,fun] = fmT6_quad(samp, ig);
        jac = der*coord;
        d = det(jac);
        jac1=inv(jac);         % Compute inverse of the Jacobian
        deriv=jac1*der;        % Derivative of shape functions in global coordinates
        bee=formbee(deriv,nne,eldof);   % Form matrix [B]
        ke=ke + d*thick*wi*bee'*dee*bee; % Integrate stiffness matrix
    end
    kk=form_kk(kk,ke, g);                  % assemble global stiffness matrix
end
%
%
%%%%%%%%%%%%%%%%%%%%%%%%%  End of assembly %%%%%%%%%%%%%%%%%%%%%%%%%%%%%%%%%%%%
%
%
delta = kk\fg ;              % solve for unknown displacements
%
for i=1: nnd                                  %
    if nf(i,1) == 0                           %
        x_disp =0.;                           %
    else
        x_disp = delta(nf(i,1));              %
    end
%
    if nf(i,2) == 0                           %
        y_disp = 0.;                          %
    else
        y_disp = delta(nf(i,2));              %
    end
    node_disp(i,:) =[x_disp  y_disp];
end
%
%
% Retrieve the x_coord and y_disp of the nodes located on the neutral axis
%
k = 0;
for i=1:nnd;

    if geom(i,1)== 0.
        k=k+1;
        y_coord(k) = geom(i,2);
        vertical_disp(k)=node_disp(i,2);
    end
end
%
nhp = 1;   % Calculate stresses at the centroid of the element
samp=hammer(nhp);
%
for i=1:nel
    [coord,g] = elem_T6(i);             % Retrieve coordinates and stering vector
    eld=zeros(eldof,1);                 % Initialize element displacement to zero
    for m=1:eldof                       %
        if g(m)==0                      %
            eld(m)=0.;                  %
        else                            %
            eld(m)=delta(g(m));         % Retrieve element displacement from
                                        % the global displacement vector
```

```
            end
        end
%
    for ig=1: nhp
        [der,fun] = fmT6_quad(samp,ig); % Derivative of shape functions in local coordinates
            jac=der*coord;                  % Compute Jacobian matrix
            jac1=inv(jac);                  % Compute inverse of the Jacobian
            deriv=jac1*der;                 % Derivative of shape functions in global coordinates
            bee=formbee(deriv,nne,eldof);   %  Form matrix [B]
            eps=bee*eld;                    % Compute strains
            sigma=dee*eps ;                 % Compute stresses
        end
    SIGMA(i,:)=sigma ;                  % Store stresses for all elements
end
%
% Prepare stresses for plotting
%
[ZX, ZY, ZT, Z1, Z2]=prepare_contour_data(SIGMA);
%
%  Plot mesh using patches
%
% patch('Faces',connec,'Vertices',geom,'FaceVertexCData',hsv(nel), ...
%         'Facecolor','none','Marker','o');
%
% Plot stresses in the x_direction
%
[C,h]= contourf(XIG,YIG,Z2,40);
%clabel(C,h);
colorbar plottools;
```

    The computed results with the LST element are shown in Figure 9.99. A contour plot of the principal stress σ_2 as well as the vertical displacements of the nodes situated just below the center of the footing are shown. Like with the CST element, both the maximum displacement and maximum stress, respectively equal to 0.12 mm and 8 kN/m^2, occur just below the footing.

9.8.3   BRIDGE PIER WITH THE Q8 ELEMENT

Using the code **Q8_PLANE_STRESS.m**, analyze the bridge pier shown in Figure 9.100. It is subject to six concentrated loads of 170 kN each. The material is reinforced concrete with an elastic

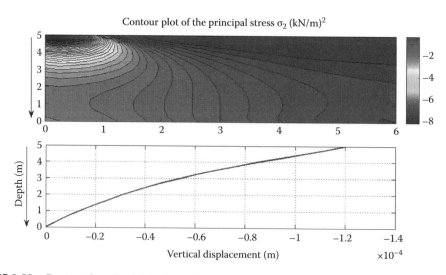

FIGURE 9.99   Computed result with the LST element.

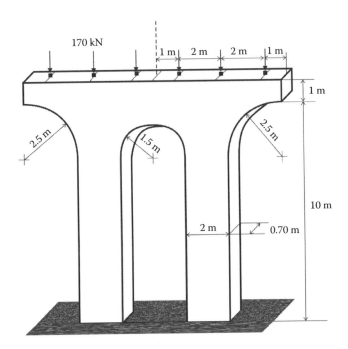

FIGURE 9.100   Bridge pier.

modulus of 50000 MPa and a Poisson's ratio of 0.17. Assume the base support as extremely rigid. The first step consists of finding ways of simplifying the model. The thickness of the pier is equal to 0.70 m, which is relatively small compared to the horizontal and vertical dimensions. The pier can therefore be analyzed as a plane stress problem. In addition, both the loading and geometry are symmetrical, therefore only half the pier can be analyzed. The second step consists of identifying the boundary conditions on the model as shown in Figure 9.101. The third step consists of constructing an appropriate mesh of the domain. Indeed, the quality of the numerical results depends very much on the quality of the mesh. However, this is probably the most difficult and time-consuming task in any finite element analysis specially when complex geometries are considered. Like in the present case, the domain is not regular, therefore the mesh generation routine **Q8_mesh.m** presented previously cannot be used since it was written for regular rectangular domains. To mesh the present domain, the Abaqus interactive edition was used. As seen previously, Abaqus generates an input file. The nodal coordinates and elements' connectivity are imported into MATLAB. However, this is not a straightforward procedure. Indeed, as shown in Figure 9.102, within the 8-node quadrilateral element, Abaqus numbers differently the nodes.

The following in an excerpt of the Abaqus input file **pier.inp**, which lists the connectivity of elements 1 to 10:

```
*Element, type=CPS8R
  1,   65,   67,  117,   64,  173,  174,  175,  176
  2,   67,   65,   66,  116,  173,  177,  178,  179
  3,   62,   20,  114,   68,  180,  181,  182,  183
  4,  118,  113,   69,  120,  184,  185,  186,  187
  5,  145,  158,  161,  156,  188,  189,  190,  191
  6,   70,   69,    2,    1,  192,  193,  194,  195
  7,  141,   78,   79,  152,  196,  197,  198,  199
  8,  120,   69,   70,  119,  186,  192,  200,  201
  9,  112,   73,   60,   61,  202,  203,  204,  205
 10,   81,  137,  121,  167,  206,  207,  208,  209
 ......
```

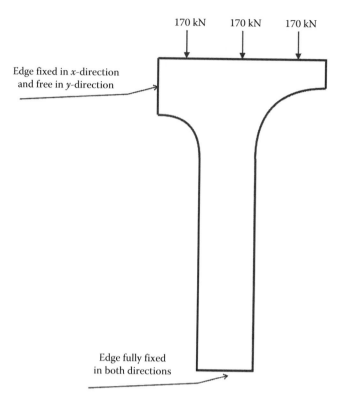

FIGURE 9.101   Bridge pier model.

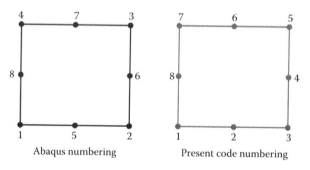

FIGURE 9.102   Element internal node numbering.

For these data to be used in the MATLAB code **Q8_PLANE_STRESS.m**, they are rearranged as follows:

```
connec =   [65   173 67   174 117 175 64   176 ;...     % Element 1
            67   173 65   177 66   178 116 179 ;...
            62   180 20   181 114 182 68    183 ;...
            118 184 113 185 69    186 120 187 ;...
            145 188 158 189 161 190 156 191 ;...
            70   192 69   193 2    194 1    195 ;...
            141 196 78   197 79   198 152 199 ;...
            120 186 69   192 70   200 119 201 ;...
            112 202 73   203 60   204 61    205 ;...
            81   206 137 207 121 208 167 209 ;...
. . . . .
```

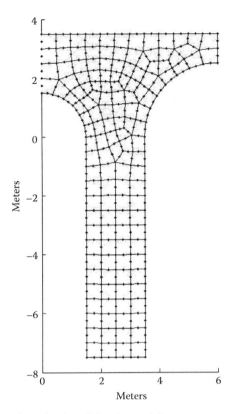

FIGURE 9.103   Finite element discretization of the pier model.

Note how the columns are swapped to comply with the MATLAB code numbering scheme. This can be achieved by importing the input file to Microsoft Excel and rearranging the columns manually or by a writing a MATLAB code that reads the input file and rearranges the columns.

The data consist of 138 elements and 481 nodes. The details are given in the file **PIER_Q8_data.m**, and the actual mesh is shown in Figure 9.103. Note that a consistent set of units is used: dimensions in meters, forces in kN, and Young's modulus in kN/m^2. All the nodes situated at $x = 0$ are fixed in the x-directions, and all the nodes situated at $y = -7.5$ forming the base are fixed in both directions. Nodes 18, 19, and 20, situated respectively at $(x = 5.\ \text{m}, y = 3.5\ \text{m})$, $(x = 3.\ \text{m}, y = 3.5\ \text{m})$, and $(x = 1.\ \text{m}, y = 3.5\ \text{m})$, are each subject to a vertical force of -170 kN.

PIER_Q8_data.m

```
%%%%%%%%%%%%%%%%%%%%%%%%%%% Beginning of data input %%%%%%%%%%%%%%%%%%%%%%%%%%%%%
%
% Data file for the bridge pier analysis using 8-nodded quadrilaterals
%
global nnd nel nne nodof eldof  n ngp
global geom connec dee nf  Nodal_loads

%
nnd = 481 ;                 % Number of nodes:
nel = 138;                   % Number of elements:
nne = 8 ;                   % Number of nodes per element:
nodof =2;                   % Number of degrees of freedom per node
ngp = 2                     % number of Gauss points
eldof = nne*nodof;          % Number of degrees of freedom per element
%
% Thickness of the domain
```

```
thick = 0.7   ;  % Thickness in meters
% Material properties
%
E=50.e6;  vu=0.17;       % Young's modulus (kN/m^2)and Poisson's ratio
%
% Form the matrix of elastic properties
%
dee=formdsig(E,vu);   % Matrix of elastic properties plane strain
%
%  Nodes coordinates x and y
%
geom = [1.4489        0.3882       ;...    x and y coordinates of node
481
          1.2990        0.7500       ;...
          1.0607        1.0607       ;...
          0.7500        1.2990       ;...
          0.3882        1.4489       ;...
          1.5000        0.0000       ;...
          0.0000        1.5000       ;...
          1.5000       -7.5000       ;...
          3.5000       -7.5000       ;...
          3.5000        0.0000       ;...
          3.5480        0.4877       ;...
          3.9213        1.3889       ;...
          4.2322        1.7678       ;...
          4.6111        2.0787       ;...
          5.5123        2.4520       ;...
          6.0000        2.5000       ;...
          6.0000        3.5000       ;...
          5.0000        3.5000       ;...
          3.0000        3.5000       ;...
          1.0000        3.5000       ;...
          0.0000        3.5000       ;...
          1.5000       -0.5000       ;...
          1.5000       -1.0000       ;...
          1.5000       -1.5000       ;...
          1.5000       -2.0000       ;...
          1.5000       -2.5000       ;...
          1.5000       -3.0000       ;...
          1.5000       -3.5000       ;...
          1.5000       -4.0000       ;...
          1.5000       -4.5000       ;...
          1.5000       -5.0000       ;...
          1.5000       -5.5000       ;...
          1.5000       -6.0000       ;...
          1.5000       -6.5000       ;...
          1.5000       -7.0000       ;...
          2.0000       -7.5000       ;...
          2.5000       -7.5000       ;...
          3.0000       -7.5000       ;...
          3.5000       -7.0000       ;...
          3.5000       -6.5000       ;...
          3.5000       -6.0000       ;...
          3.5000       -5.5000       ;...
          3.5000       -5.0000       ;...
          3.5000       -4.5000       ;...
          3.5000       -4.0000       ;...
          3.5000       -3.5000       ;...
          3.5000       -3.0000       ;...
          3.5000       -2.5000       ;...
          3.5000       -2.0000       ;...
          3.5000       -1.5000       ;...
          3.5000       -1.0000       ;...
          3.5000       -0.5000       ;...
          3.6903        0.9567       ;...
          5.0433        2.3097       ;...
          6.0000        3.0000       ;...
```

```
5.5000        3.5000        ;...
4.5000        3.5000        ;...
4.0000        3.5000        ;...
3.5000        3.5000        ;...
2.5000        3.5000        ;...
2.0000        3.5000        ;...
1.5000        3.5000        ;...
0.5000        3.5000        ;...
0.0000        3.0000        ;...
0.0000        2.5000        ;...
0.0000        2.0000        ;...
0.5279        2.5104        ;...
1.4959        3.0626        ;...
1.5745        0.8916        ;...
1.7401        0.4784        ;...
3.0678        0.0527        ;...
3.2060        0.5869        ;...
2.4709        3.0574        ;...
3.6579        1.6005        ;...
4.4579        2.3344        ;...
4.5316        2.9797        ;...
4.9797        3.1580        ;...
1.8298        0.0390        ;...
1.9024       -0.4486        ;...
1.9821       -0.9532        ;...
3.9824        2.0741        ;...
2.9942       -1.9923        ;...
2.0045       -2.5012        ;...
3.0014       -3.5050        ;...
2.0045       -4.0063        ;...
3.0035       -5.0060        ;...
2.0042       -5.5061        ;...
3.0033       -6.0056        ;...
2.0026       -6.0047        ;...
3.0027       -6.5043        ;...
2.0013       -6.5036        ;...
3.0016       -7.0020        ;...
2.0002       -7.0015        ;...
3.0035       -5.5056        ;...
2.0042       -5.0056        ;...
2.0038       -4.5049        ;...
3.0033       -4.5053        ;...
3.0027       -4.0051        ;...
2.0039       -3.5054        ;...
2.0034       -3.0033        ;...
2.9993       -3.0030        ;...
2.9970       -2.4998        ;...
2.0057       -1.9916        ;...
2.0044       -1.4742        ;...
2.9993       -1.4785        ;...
3.0327       -0.9731        ;...
3.0998       -0.4957        ;...
3.3175        1.1463        ;...
3.9957        3.0076        ;...
3.4702        2.9078        ;...
2.9566        3.0118        ;...
1.9863        3.0733        ;...
1.3008        1.2396        ;...
1.0014        3.0384        ;...
0.9828        1.5734        ;...
0.5929        1.9562        ;...
0.5039        3.0118        ;...
1.5590        1.4193        ;...
2.0705        0.5842        ;...
1.9288        1.0906        ;...
3.6925        2.5511        ;...
4.8666        2.6297        ;...
```

```
2.5004      -2.4996     ;...
2.5046      -4.0070     ;...
2.5049      -5.5071     ;...
2.5027      -6.5046     ;...
2.5035      -6.0055     ;...
2.5047      -5.0066     ;...
2.5043      -4.5058     ;...
2.5031      -3.5058     ;...
2.5017      -3.0033     ;...
2.4994      -1.9860     ;...
2.5006      -1.4524     ;...
2.5053      -0.8546     ;...
2.7641      -0.4931     ;...
2.8417       1.1161     ;...
3.9934       2.5661     ;...
1.3049       1.7687     ;...
2.8367       0.6671     ;...
3.1567       1.5168     ;...
2.1675       0.1239     ;...
2.5011      -7.0019     ;...
2.6413      -0.1504     ;...
2.4210       2.6375     ;...
2.8474       2.5634     ;...
1.9773       2.6633     ;...
1.5193       2.6440     ;...
1.0320       2.5819     ;...
2.0271       1.5224     ;...
2.4001       1.0934     ;...
2.4406       0.6514     ;...
2.2846      -0.3356     ;...
2.8123       1.5885     ;...
1.9850       2.2736     ;...
1.1200       2.1537     ;...
3.2207       2.4235     ;...
1.5744       2.2513     ;...
2.7330       2.1697     ;...
2.3705       2.2499     ;...
1.6578       1.8873     ;...
3.0443       2.0121     ;...
2.3874       1.5056     ;...
2.0083       1.8949     ;...
2.3454       1.8741     ;...
5.3776       2.9409     ;...
4.3237       2.6067     ;...
3.5630       2.2212     ;...
2.5059       0.2226     ;...
3.3638       1.8329     ;...
1.7419       1.5990     ;...
2.7962       0.3080     ;...
2.6233       1.8785     ;...
0.2640       2.5052     ;...
0.5159       2.7611     ;...
0.2520       3.0059     ;...
0.0000       2.7500     ;...
0.0000       2.2500     ;...
0.2964       1.9781     ;...
0.5604       2.2333     ;...
1.2500       3.5000     ;...
1.0007       3.2692     ;...
1.2487       3.0505     ;...
1.4980       3.2813     ;...
1.4299       1.3294     ;...
1.4377       1.0656     ;...
1.7516       0.9911     ;...
1.7439       1.2549     ;...
2.7902       2.3665     ;...
2.8887       2.0909     ;...
```

```
3.1325      2.2178      ;...
3.0341      2.4934      ;...
1.6573      0.6850      ;...
1.4368      0.8208      ;...
1.3858      0.5740      ;...
1.5945      0.4333      ;...
1.9987      0.0815      ;...
1.8661     -0.2048      ;...
2.0935     -0.3921      ;...
2.2260     -0.1058      ;...
1.9053      0.5313      ;...
1.9996      0.8374      ;...
2.2286      3.0653      ;...
2.4854      3.2787      ;...
2.2500      3.5000      ;...
1.9932      3.2866      ;...
3.9879      2.3201      ;...
3.8430      2.5586      ;...
3.6277      2.3861      ;...
3.7727      2.1477      ;...
5.1221      2.7853      ;...
4.9550      2.4697      ;...
5.2743      2.3924      ;...
5.4449      2.6964      ;...
4.6623      2.4821      ;...
4.5345      2.2065      ;...
4.8215      2.2048      ;...
5.4388      3.2205      ;...
5.2500      3.5000      ;...
4.9898      3.3290      ;...
5.1786      3.0494      ;...
6.0000      2.7500      ;...
5.6888      2.9705      ;...
5.7550      2.4880      ;...
1.6649      0.0195      ;...
1.5000     -0.2500      ;...
1.7012     -0.4743      ;...
3.2999     -0.4979      ;...
3.0663     -0.7344      ;...
3.2664     -0.9866      ;...
3.5000     -0.7500      ;...
1.9423     -0.7009      ;...
1.5000     -0.7500      ;...
1.7411     -0.9766      ;...
4.2202      2.2043      ;...
4.1073      1.9210      ;...
4.4140      1.9325      ;...
3.2496     -1.4892      ;...
3.5000     -1.2500      ;...
3.0160     -1.2258      ;...
2.9956     -2.2460      ;...
2.7468     -1.9891      ;...
2.4999     -2.2428      ;...
2.7487     -2.4997      ;...
3.2497     -3.0015      ;...
3.5000     -2.7500      ;...
3.2485     -2.4999      ;...
2.9981     -2.7514      ;...
3.0020     -3.7550      ;...
2.7522     -3.5054      ;...
2.5038     -3.7564      ;...
2.7536     -4.0061      ;...
3.2517     -4.5026      ;...
3.5000     -4.2500      ;...
3.2513     -4.0026      ;...
3.0030     -4.2552      ;...
3.0035     -5.2558      ;...
```

```
2.7541        -5.0063       ;...
2.5048        -5.2569       ;...
2.7542        -5.5063       ;...
2.2531        -6.0051       ;...
2.0019        -6.2541       ;...
2.2520        -6.5041       ;...
2.5031        -6.2551       ;...
2.2507        -7.0017       ;...
2.5019        -6.7533       ;...
2.0007        -6.7525       ;...
3.2517        -6.0028       ;...
3.0030        -6.2549       ;...
3.2514        -6.5021       ;...
3.5000        -6.2500       ;...
2.7527        -6.5045       ;...
2.7534        -6.0056       ;...
2.5042        -5.7563       ;...
3.0034        -5.7556       ;...
3.0022        -6.7531       ;...
3.2508        -7.0010       ;...
3.5000        -6.7500       ;...
2.0001        -7.2507       ;...
1.7501        -7.0007       ;...
1.5000        -7.2500       ;...
1.7500        -7.5000       ;...
1.7506        -6.5018       ;...
1.5000        -6.7500       ;...
3.2500        -7.5000       ;...
3.5000        -7.2500       ;...
3.0008        -7.2510       ;...
3.2517        -5.0030       ;...
3.2517        -5.5028       ;...
3.5000        -5.2500       ;...
2.2545        -5.5066       ;...
2.2545        -5.0061       ;...
2.0042        -5.2559       ;...
2.5045        -4.7562       ;...
3.0034        -4.7556       ;...
2.7538        -4.5055       ;...
2.2541        -4.5054       ;...
2.0040        -4.7553       ;...
3.2507        -3.5025       ;...
3.5000        -3.7500       ;...
2.2546        -4.0067       ;...
2.2535        -3.5056       ;...
2.0042        -3.7558       ;...
2.5024        -3.2546       ;...
3.0004        -3.2540       ;...
2.7505        -3.0031       ;...
2.2525        -3.0033       ;...
2.0036        -3.2544       ;...
3.2471        -1.9961       ;...
3.5000        -2.2500       ;...
2.2524        -2.5004       ;...
2.2525        -1.9888       ;...
2.0051        -2.2464       ;...
2.5000        -1.7192       ;...
2.2525        -1.4633       ;...
2.0051        -1.7329       ;...
2.5030        -1.1535       ;...
2.2437        -0.9039       ;...
1.9933        -1.2137       ;...
2.6347        -0.6739       ;...
2.7690        -0.9138       ;...
2.9320        -0.4944       ;...
3.2839         0.0264       ;...
3.0838        -0.2215       ;...
```

```
3.5000        -0.2500        ; ...
1.9780         1.3065        ; ...
2.1645         1.0920        ; ...
2.3938         1.2995        ; ...
2.2072         1.5140        ; ...
4.7557         3.0688        ; ...
4.7500         3.5000        ; ...
4.5158         3.2399        ; ...
6.0000         3.2500        ; ...
5.7500         3.5000        ; ...
3.2134         2.9598        ; ...
3.4851         3.2039        ; ...
3.2500         3.5000        ; ...
2.9783         3.2559        ; ...
2.7138         3.0346        ; ...
2.7500         3.5000        ; ...
1.7411         3.0679        ; ...
1.7500         3.5000        ; ...
1.1807         1.1501        ; ...
1.1900         0.9131        ; ...
0.5020         3.2559        ; ...
0.2500         3.5000        ; ...
0.0000         3.2500        ; ...
1.1418         1.4065        ; ...
0.8664         1.4362        ; ...
0.9131         1.1900        ; ...
0.8565         2.0550        ; ...
0.7878         1.7648        ; ...
1.1438         1.6711        ; ...
1.2125         1.9612        ; ...
0.1958         1.4872        ; ...
0.4905         1.7026        ; ...
0.0000         1.7500        ; ...
1.4319         1.5940        ; ...
0.7800         2.5461        ; ...
1.0167         2.8101        ; ...
0.7527         3.0251        ; ...
3.7896         1.4947        ; ...
3.4877         1.3734        ; ...
3.5039         1.0515        ; ...
3.7952         1.1785        ; ...
1.4872         0.1958        ; ...
1.7850         0.2587        ; ...
3.3770         0.5373        ; ...
3.1369         0.3198        ; ...
3.5120         0.2450        ; ...
3.6076         0.7257        ; ...
3.2618         0.8666        ; ...
2.8392         0.8916        ; ...
2.6209         1.1048        ; ...
2.4204         0.8724        ; ...
2.6386         0.6593        ; ...
3.8201         1.8373        ; ...
3.4634         2.0270        ; ...
3.5109         1.7167        ; ...
4.0675         1.5860        ; ...
4.2636         2.9936        ; ...
4.2500         3.5000        ; ...
3.9978         3.2538        ; ...
4.6991         2.8047        ; ...
2.7027        -0.3218        ; ...
2.8545        -0.0488        ; ...
1.5000        -1.2500        ; ...
1.7522        -1.4871        ; ...
3.7330         2.9577        ; ...
3.7500         3.5000        ; ...
1.7528        -1.9958        ; ...
```

```
1.5000     -2.2500     ;...
1.7522     -2.5006     ;...
3.5000     -1.7500     ;...
2.9968     -1.7354     ;...
1.5000     -2.7500     ;...
1.7517     -3.0017     ;...
2.0039     -2.7523     ;...
1.7519     -3.5027     ;...
1.5000     -3.7500     ;...
1.7523     -4.0032     ;...
3.5000     -3.2500     ;...
1.5000     -4.2500     ;...
1.7519     -4.5025     ;...
2.0041     -4.2556     ;...
1.7521     -5.0028     ;...
1.5000     -5.2500     ;...
1.7521     -5.5031     ;...
3.5000     -4.7500     ;...
1.5000     -5.7500     ;...
1.7513     -6.0023     ;...
2.0034     -5.7554     ;...
3.5000     -5.7500     ;...
1.5000     -6.2500     ;...
2.7514     -7.0020     ;...
2.5006     -7.2509     ;...
2.7500     -7.5000     ;...
2.2500     -7.5000     ;...
2.5045     -4.2564     ;...
1.5000     -4.7500     ;...
2.5010     -2.7515     ;...
1.5000     -3.2500     ;...
2.7500     -1.4655     ;...
1.5000     -1.7500     ;...
2.4732      0.4370     ;...
2.6511      0.2653     ;...
2.8164      0.4876     ;...
2.3950     -0.5951     ;...
2.4629     -0.2430     ;...
2.1768      1.8845     ;...
2.0177      1.7087     ;...
2.3664      1.6899     ;...
3.5814      2.7294     ;...
3.9945      2.7868     ;...
3.0796      1.1312     ;...
3.2371      1.3315     ;...
2.9845      1.5527     ;...
2.8270      1.3523     ;...
2.4459      2.8474     ;...
2.6342      2.6004     ;...
2.9020      2.7876     ;...
1.5076      2.8533     ;...
1.7483      2.6537     ;...
1.9818      2.8683     ;...
1.5468      2.4477     ;...
1.7797      2.2624     ;...
1.9812      2.4685     ;...
0.7500      3.5000     ;...
1.2757      2.6130     ;...
0.5740      1.3858     ;...
2.1992      2.6504     ;...
2.1190      0.3541     ;...
2.3367      0.1733     ;...
2.2555      0.6178     ;...
3.3455      2.6656     ;...
4.1586      2.5864     ;...
4.4277      2.7932     ;...
2.5736      0.0361     ;...
```

```
          3.0213        0.6270        ;...
          3.2603        1.6748        ;...
          4.3908        2.4706        ;...
          1.0760        2.3678        ;...
          1.3472        2.2025        ;...
          2.9320        0.1804        ;...
          2.9283        1.8003        ;...
          3.2041        1.9225        ;...
          2.3579        2.0620        ;...
          2.1778        2.2618        ;...
          1.9967        2.0842        ;...
          2.5517        2.2098        ;...
          2.4844        1.8763        ;...
          2.6782        2.0241        ;...
          1.4813        1.8280        ;...
          1.6161        2.0693        ;...
          1.8845        1.5607        ;...
          1.6504        1.5091        ;...
          2.5999        1.5471        ;...
          2.7178        1.7335        ;...
          2.3957        2.4437        ;...
          1.6998        1.7431        ;...
          3.3919        2.3224        ;...
          1.8330        1.8911        ];    %    x and y coordinates of node 481
%
% Element connectivity
%
  connec = [65   173 67   174 117 175 64    176 ;...    % Element 1
            67   173 65   177 66  178 116 179 ;...
            62   180 20   181 114 182 68  183 ;...
            118 184 113 185 69  186 120 187 ;...
            145 188 158 189 161 190 156 191 ;...
            70   192 69   193 2   194 1   195 ;...
            141 196 78   197 79  198 152 199 ;...
            120 186 69   192 70  200 119 201 ;...
            112 202 73   203 60  204 61  205 ;...
            81   206 137 207 121 208 167 209 ;...
            165 210 122 211 54  212 15  213 ;...
            54   211 122 214 75  215 14  216 ;...
            165 217 56   218 18  219 77  220 ;...
            16   221 55   222 165 213 15  223 ;...
            79   197 78   224 6   225 22  226 ;...
            52   227 107 228 106 229 51  230 ;...
            80   231 79   226 22  232 23  233 ;...
            14   215 75   234 81  235 13  236 ;...
            105 237 50   238 51  229 106 239 ;...
            102 240 82   241 132 242 123 243 ;...
            101 244 47   245 48  246 102 247 ;...
            98   248 84   249 130 250 124 251 ;...
            97   252 44   253 45  254 98  255 ;...
            94   256 86   257 128 258 125 259 ;...
            127 260 89   261 91  262 126 263 ;...
            93   264 142 265 126 262 91  266 ;...
            41   267 88   268 90  269 40  270 ;...
            127 263 126 271 90  268 88  272 ;...
            125 273 127 272 88  274 94  259 ;...
            40   269 90   275 92  276 39  277 ;...
            36   278 93   279 35  280 8   281 ;...
            93   266 91   282 34  283 35  279 ;...
            38   284 9    285 39  276 92  286 ;...
            43   287 86   256 94  288 42  289 ;...
            87   290 125 258 128 291 95  292 ;...
            129 293 128 257 86  294 97  295 ;...
            95   291 128 293 129 296 96  297 ;...
            46   298 84   248 98  254 45  299 ;...
            85   300 124 250 130 301 99  302 ;...
            131 303 130 249 84   304 101 305 ;...
```

```
99   301 130 303 131 306 100 307 ;...
49   308 82  240 102 246 48  309 ;...
83   310 123 242 132 311 103 312 ;...
103  311 132 313 133 314 104 315 ;...
133  316 134 317 80  318 104 314 ;...
135  319 134 320 106 228 107 321 ;...
10   322 71  323 107 227 52  324 ;...
149  325 120 326 150 327 162 328 ;...
76   329 77  219 18  330 57  331 ;...
55   332 17  333 56  217 165 222 ;...
111  334 110 335 59  336 19  337 ;...
73   338 111 337 19  339 60  203 ;...
68   340 112 205 61  341 62  183 ;...
69   185 113 342 3   343 2   193 ;...
64   175 117 344 63  345 21  346 ;...
113  347 115 348 4   349 3   342 ;...
155  350 116 351 115 352 138 353 ;...
7    354 5   355 116 178 66  356 ;...
113  184 118 357 138 352 115 347 ;...
117  174 67  358 148 359 114 360 ;...
12   361 74  362 108 363 53  364 ;...
1    365 6   224 78  366 70  195 ;...
11   367 72  368 71  322 10  369 ;...
72   367 11  370 53  363 108 371 ;...
139  372 136 373 150 374 151 375 ;...
74   376 81  209 167 377 169 378 ;...
74   361 12  379 13  235 81  376 ;...
109  380 76  331 57  381 58  382 ;...
77   329 76  383 122 210 165 220 ;...
143  384 135 321 107 323 71  385 ;...
23   386 24  387 104 318 80  233 ;...
110  388 109 382 58  389 59  335 ;...
83   312 103 390 25  391 26  392 ;...
82   308 49  393 50  237 105 394 ;...
26   395 27  396 100 397 83  392 ;...
85   302 99  398 28  399 29  400 ;...
84   298 46  401 47  244 101 304 ;...
29   402 30  403 96  404 85  400 ;...
87   292 95  405 31  406 32  407 ;...
86   287 43  408 44  252 97  294 ;...
32   409 33  410 89  411 87  407 ;...
88   267 41  412 42  288 94  274 ;...
33   413 34  282 91  261 89  410 ;...
38   286 92  414 142 415 37  416 ;...
93   278 36  417 37  415 142 264 ;...
124  418 129 295 97  255 98  251 ;...
30   419 31  405 95  297 96  403 ;...
123  420 131 305 101 247 102 243 ;...
27   421 28  398 99  307 100 396 ;...
133  313 132 241 82  394 105 422 ;...
24   423 25  390 103 315 104 387 ;...
133  422 105 239 106 320 134 316 ;...
151  424 168 425 171 426 139 375 ;...
152  427 134 319 135 384 143 428 ;...
164  429 163 430 149 328 162 431 ;...
110  432 121 207 137 433 109 388 ;...
136  434 108 435 140 436 153 437 ;...
73   438 144 439 145 440 111 338 ;...
68   441 147 442 146 443 112 340 ;...
147  444 157 445 154 446 146 442 ;...
117  360 114 181 20  447 63  344 ;...
114  359 148 448 147 441 68  182 ;...
5    449 4   348 115 351 116 355 ;...
112  443 146 450 144 438 73  202 ;...
70   366 78  196 141 451 119 200 ;...
141  452 168 424 151 453 119 451 ;...
119  453 151 374 150 326 120 201 ;...
```

```
                  156 454 110 334 111 440 145 191 ;...
                  137 455 166 456 76  380 109 433 ;...
                  123 310 83  397 100 306 131 420 ;...
                  124 300 85  404 96  296 129 418 ;...
                  125 290 87  411 89  260 127 273 ;...
                  92  275 90  271 126 265 142 414 ;...
                  141 199 152 428 143 457 168 452 ;...
                  108 434 136 372 139 458 72  371 ;...
                  140 435 108 362 74  378 169 459 ;...
                  166 460 75  214 122 383 76  456 ;...
                  148 461 155 462 157 444 147 448 ;...
                  72  458 139 426 171 463 71  368 ;...
                  161 464 153 436 140 459 169 465 ;...
                  163 429 164 466 159 467 154 468 ;...
                  158 469 159 466 164 470 172 471 ;...
                  138 472 160 473 157 462 155 353 ;...
                  155 461 148 358 67  179 116 350 ;...
                  120 325 149 474 170 475 118 187 ;...
                  136 437 153 476 162 327 150 373 ;...
                  152 198 79  231 80  317 134 427 ;...
                  161 189 158 471 172 477 153 464 ;...
                  146 446 154 467 159 478 144 450 ;...
                  118 475 170 479 160 472 138 357 ;...
                  121 432 110 454 156 480 167 208 ;...
                  158 188 145 439 144 478 159 469 ;...
                  154 445 157 473 160 481 163 468 ;...
                  161 465 169 377 167 480 156 190 ;...
                  162 476 153 477 172 470 164 431 ;...
                  75  460 166 455 137 206 81  234 ;...
                  171 425 168 457 143 385 71  463 ;...
                  170 474 149 430 163 481 160 479 ];    % Element 138
%
%
% Boundary conditions
%
nf = ones(nnd, nodof);                      % Initialize the matrix nf to 1
%
for i=1:nnd
    if geom(i,1) == 0.;
        nf(i,:) = [0 1];
    end
     if geom(i,2) == -7.5 ;
        nf(i,:) = [0 0];
      end
end
%
% Counting of the free degrees of freedom
%
n=0;
for i=1:nnd
    for j=1:nodof
        if nf(i,j) ~= 0
            n=n+1;
            nf(i,j)=n;
        end
      end
end
disp ('Nodal freedom')
nf
disp ('Total number of active degrees of freedom')
n
%
% loading
%
Nodal_loads = zeros(nnd, 2);
Nodal_loads(18,2)=-170.;            % Vertical load on node 18
Nodal_loads(19,2)=-170.;            % Vertical load on node 19
```

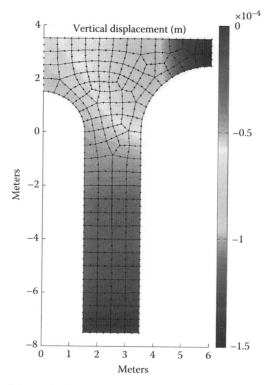

FIGURE 9.104   Contour of the vertical displacement.

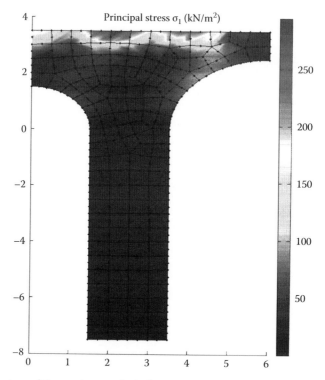

FIGURE 9.105   Contour of the maximum principal stress σ_1.

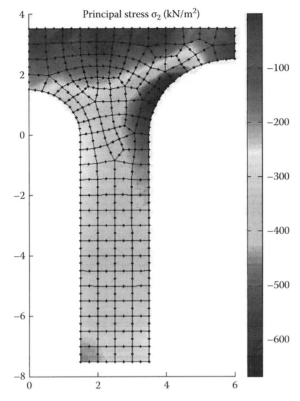

FIGURE 9.106 Contour of the minimum principal stress σ_2.

```
Nodal_loads(20,2)=-170.;        % Vertical load on node 20
%
% End input
```

To run this example, in the program **Q8_PLANE_STRESS.m**, replace **Q8_coarse_mesh_data** with **PIER_Q8_data.m**.

The obtained results are displayed in Figures 9.104 through 9.106 respectively as contour plots of the vertical displacement v_2, the first principal stress σ_1, and the second principal stress σ_2. The contours of the principal stresses may not be very accurate since they are calculated at the centers of the elements and averaged at the nodes. More details can be obtained with a finer mesh.

10 Axisymmetric Problems

10.1 DEFINITION

An axisymmetric problem is a three-dimensional problem that can be solved using a two-dimensional model provided that it posses a symmetry of revolution in both geometry, material properties and loading, and it can lend itself to a cylindrical coordinate. The circular footing on a semi-infinite soil mass shown in Figure 10.1 is a typical example of a three-dimensional problem that can be classified as axisymmetric. The only displacements required to define its behavior are the ones in the r and z directions, denoted by u and v, respectively. They are not a function of θ.

10.2 STRAIN–DISPLACEMENT RELATIONSHIP

Unlike plane stress/strain analysis, in axisymmetric problems a fourth component of the strain ϵ_θ (and hence σ_θ) must be considered in addition to the plane stress/strain components, ϵ_{rr}, ϵ_{zz}, and γ_{zr} (and stresses σ_{rr}, σ_{zz}, and τ_{zr}), as shown in Figure 10.2. The strains ϵ_{rr}, ϵ_{zz}, and γ_{zr} are related to the displacements u and v in the same way as for a plane stress/strain problem. It follows therefore:

$$\epsilon_{rr} = \frac{\partial u}{\partial r} \tag{10.1}$$

$$\epsilon_{zz} = \frac{\partial v}{\partial z} \tag{10.2}$$

$$\gamma_{rz} = \frac{\partial u}{\partial z} + \frac{\partial v}{\partial r} \tag{10.3}$$

The tangential or hoop strain depends only on the radial displacement u. The new length of the arc AB in Figure 10.3 is equal to $(r + u)\, d\theta$, the tangential strain is then given as

$$\epsilon_\theta = \frac{(r + u)\, d\theta - r\, d\theta}{r\, d\theta} = \frac{u}{r} \tag{10.4}$$

Rewriting Equations (10.1) and (10.4) in a matrix form yields

$$\begin{Bmatrix} \epsilon_{rr} \\ \epsilon_{zz} \\ \epsilon_\theta \\ \gamma_{rz} \end{Bmatrix} = \begin{bmatrix} \dfrac{\partial}{\partial x} & 0 \\ 0 & \dfrac{\partial}{\partial y} \\ 1/r & 0 \\ \dfrac{\partial}{\partial y} & \dfrac{\partial}{\partial x} \end{bmatrix} \begin{Bmatrix} u \\ v \end{Bmatrix} \tag{10.5}$$

or in a more compact form as

$$\{\epsilon\} = [L]U \tag{10.6}$$

where $[L]$ is a linear operator matrix.

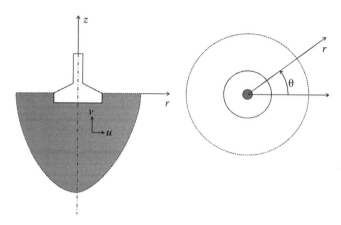

FIGURE 10.1 Typical axisymmetric problem.

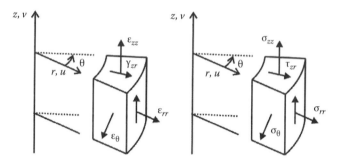

FIGURE 10.2 Strains and corresponding stresses in an axisymmetric solid.

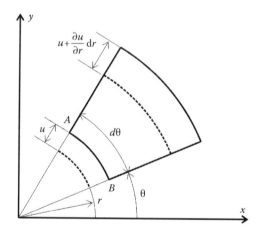

FIGURE 10.3 Tangential strain.

10.3 STRESS–STRAIN RELATIONS

In three dimensions, the generalized Hooke's law for an isotropic material with a modulus of elasticity E and a Poisson's ratio ν is given in terms of the elasticity matrix by Equation (5.136) and in terms of the compliance matrix by Equation (5.137). In an axisymmetric problem, the shear strains $\gamma_{r\theta}$ and $\gamma_{z\theta}$ and the shear stresses $\tau_{r\theta}$ and $\tau_{z\theta}$ all vanish because of the radial symmetry. Hence, Equation (5.136) is rewritten only in terms of the four stresses σ_{rr}, σ_{zz}, σ_{θ}, and τ_{zr}, and the four strains ϵ_{rr}, ϵ_{zz}, ϵ_θ, and γ_{zr}; that is,

$$\begin{Bmatrix} \sigma_{rr} \\ \sigma_{zz} \\ \sigma_\theta \\ \tau_{rz} \end{Bmatrix} = \frac{E}{(1+\nu)(1-2\nu)} \begin{bmatrix} 1-\nu & \nu & \nu & 0 \\ \nu & 1-\nu & \nu & 0 \\ \nu & \nu & 1-\nu & 0 \\ 0 & 0 & 0 & \dfrac{(1-2\nu)}{2} \end{bmatrix} \begin{Bmatrix} \epsilon_{rr} \\ \epsilon_{zz} \\ \epsilon_\theta \\ \gamma_{rz} \end{Bmatrix} \qquad (10.7)$$

10.4   FINITE ELEMENT FORMULATION

10.4.1   DISPLACEMENT FIELD

For an element having n nodes, the components of the displacement vector are interpolated using nodal approximations

$$u = N_1 u_1 + N_2 u_2 + \cdots + N_n u_n \qquad (10.8)$$

$$v = N_1 v_1 + N_2 v_2 + \cdots + N_n v_n \qquad (10.9)$$

which, when written in a matrix form, yields

$$\begin{Bmatrix} u \\ v \end{Bmatrix} = \begin{bmatrix} N_1 & 0 & | & N_2 & 0 & | & \ldots & | & N_n & 0 \\ 0 & N_1 & | & 0 & N_2 & | & \ldots & | & 0 & N_n \end{bmatrix} \begin{Bmatrix} u_1 \\ v_1 \\ u_2 \\ v_2 \\ \vdots \\ u_n \\ v_n \end{Bmatrix} \qquad (10.10)$$

or simply as

$$\{U\} = [N]a \qquad (10.11)$$

10.4.2   STRAIN MATRIX

Substituting for $\{U\}$ using Equation (10.10), the strain–displacement Equation (10.7) becomes

$$\{\epsilon\} = [B]\{a\} \qquad (10.12)$$

with

$$[B] = \begin{bmatrix} \dfrac{\partial N_1}{\partial x} & 0 & | & \dfrac{\partial N_2}{\partial x} & 0 & | & \ldots & | & \dfrac{\partial N_n}{\partial x} & 0 \\ 0 & \dfrac{\partial N_1}{\partial y} & | & 0 & \dfrac{\partial N_2}{\partial y} & | & \ldots & | & 0 & \dfrac{\partial N_n}{\partial y} \\ \dfrac{\partial N_1}{r} & 0 & | & \dfrac{\partial N_2}{r} & 0 & | & \ldots & | & \dfrac{\partial N_n}{r} & 0 \\ \dfrac{\partial N_1}{\partial y} & \dfrac{\partial N_1}{\partial x} & | & \dfrac{\partial N_2}{\partial y} & \dfrac{\partial N_2}{\partial x} & | & \ldots & | & \dfrac{\partial N_n}{\partial y} & \dfrac{\partial N_n}{\partial x} \end{bmatrix} \qquad (10.13)$$

10.4.3 Stiffness Matrix

The stiffness matrix is given as

$$[K_e] = \left[\int_{V_e} [B]^T [D][B] \, dv \right] = \left[\int \int \int_{V_e} [B]^T [D][B] r \, dr \, d\theta \, dz \right] \tag{10.14}$$

which, when integrated over one radian, becomes

$$[K_e] = \left[\int \int_{A_e} [B]^T [D][B] r \, dr \, dz \right] \tag{10.15}$$

10.4.4 Nodal Force Vectors

10.4.4.1 Body Forces

The nodal force vector for body forces such as gravity when integrated over one radian is given as

$$\{f_b\} = \int \int_{A_e} [N]^T \begin{Bmatrix} b_r \\ b_z \end{Bmatrix} r \, dr \, dz \tag{10.16}$$

10.4.4.2 Surface Forces Vector

Surface forces indicate traction forces around the external surface of the body. When integrated over one radian, the nodal vector is written as

$$\{f_s\} = \int_L [N]^T \begin{Bmatrix} t_r \\ t_z \end{Bmatrix} r \, dl \tag{10.17}$$

where dl represents the elemental length around the boundary of the element. When, for a unit pressure, Equation (10.17) is integrated over a linear or quadratic element, the equivalent nodal forces are shown in Figure 10.4 [3].

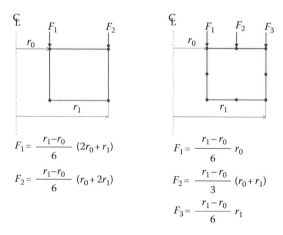

FIGURE 10.4   Axisymmetric equivalent nodal loads.

10.4.4.3 Concentrated Loads

For axisymmetric bodies, concentrated forces are actually line loads around the circumference of the body. When integrated over one radian, the equivalent nodal forces vector of the line loads are written as

$$\{f_c\} = \Sigma_i [N]^T r_i \begin{Bmatrix} P_r \\ P_z \end{Bmatrix}_i \tag{10.18}$$

where P_r and P_z are the radial and vertical components of the line force P_i.

10.4.4.4 Example

The thick walled annulus shown in Figure 10.5 has an internal diameter of 400 mm and an external diameter of 700 mm. It is subject on its top surface to a pressure of 0.5 N/mm² and to a line load at it base of 4 N/mm. Find the equivalent nodal loads on the element represented. The element has four nodes, each having two degrees of freedom. The vector of nodal loads has a dimension of 8. Nodes 4 and 3 are loaded by the 0.5 N/mm² pressure load. Using the equations shown in Figure 10.5, the vertical components F_1 and F_2 can be calculated as follows:

$$F_1 = -0.5 \times \frac{r_1 - r_0}{6}(2r_0 + r_1) = -35000\,\text{N} \tag{10.19}$$

$$F_2 = -0.5 \times \frac{r_1 - r_0}{6}(r_0 + 2r_1) = -45000\,\text{N} \tag{10.20}$$

Node 2 is loaded by a radial line load of 4 N/mm. Using Equation (10.18), the horizontal load acting at node 2 is obtained as

$$F_r = 700 \times 4 = 2800\,\text{N} \tag{10.21}$$

Hence, the vector of nodal forces for the element can be written as

$$\{F_e\}^T = \begin{bmatrix} 0. & 0. & 2800. & 0. & 0. & -45000 & 0. & -35000 \end{bmatrix} \tag{10.22}$$

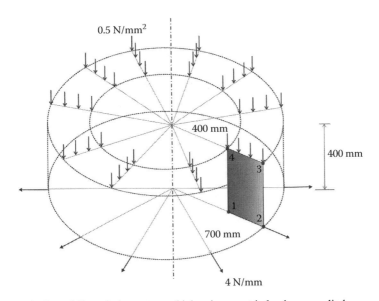

FIGURE 10.5 Typical quadrilateral element on which axisymmetric loads are applied.

10.5   PROGRAMMING

Figure 10.6 represents a circular footing on a sandy soil with an elastic modulus $E = 10^5 \, kN/m^2$ and a Poisson's ratio $\nu = 0.3$. The footing is 2 m in radius and supports a load of 200 kN. Nine meters beneath the footing, the soil is made up of a solid rock formation that can be considered very stiff. Assume that 7 m away from the centerline of the footing the horizontal displacement of the soil is negligible. Consider an element length of 0.5 m, analyze the footing using both the 6-node triangle and the 8-node quadrilateral elements.

Figure 10.7 shows the geometrical domain and the boundary conditions. Because of symmetry, only half the domain will be discretized. Nodes on the centerline will only displace in the vertical direction. Idem for the nodes placed at a 7 m radius because the horizontal movement of the soil at this distance is assumed negligible. The nodes placed at a depth of 9 m are fixed in all directions because the rock substratum is assumed in-deformable. The 200 kN is also transformed into an equivalent uniformly distributed load of 63.662 kN/m².

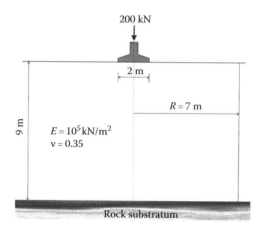

FIGURE 10.6   Circular footing on a sandy soil.

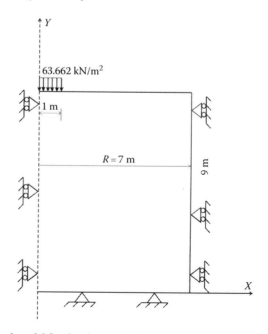

FIGURE 10.7   Geometrical model for the circular footing.

10.5.1   COMPUTER CODE: AXI_SYM_T6.m

The following program, *AXI_SYM_T6.m*, is an adaptation of the plane stress/strain program *LST_PLANE_STRESS_MESH.m* to axisymmetric conditions. The program is listed next and includes the automatic mesh generation function *T6_mesh.m*.

```
% THIS PROGRAM USES A 6-NODE TRIANGULAR ELEMENT FOR THE
% LINEAR ELASTIC STATIC ANALYSIS OF AN AXISYMMETRIC PROBLEM.
% IT INCLUDES AN AUTOMATIC MESH GENERATION
%
% Make these variables global so they can be shared by other functions
%
clear all
clc
global nnd nel nne  nodof eldof   n
global connec geom dee nf Nodal_loads
global Length Width NXE NYE X_origin Y_origin
%
 format long g
%
%
% To change the size of the problem or change elastic properties
%    supply another input file
%
Length = 7.; % Length of the model
Width =9.;    % Width
NXE = 14;        % Number of rows in the x direction
NYE = 18;        % Number of rows in the y direction
dhx = Length/NXE; % Element size in the r direction
dhy = Width/NYE;  % Element size in the z direction
X_origin = 0. ;  % r origin of the global coordinate system
Y_origin = 0. ;   % z origin of the global coordinate system
%
nne = 6;
nodof = 2;
eldof = nne*nodof;
%
T6_mesh ;          % Generate the mesh
%
% Material
%
E = 100000.;      % Elastic modulus in kN/m2
vu = 0.35;        % Poisson's ratio
nhp = 3;          % Number of sampling points
%
% Form the elastic matrix for plane stress
%
dee = formdax(E,vu);
%
%
% Boundary conditions
%
nf = ones(nnd, nodof);      % Initialize the matrix nf to 1
%
%
for i=1:nnd

    if geom(i,1) == 0 || geom(i,1) == Length
    nf(i,:) = [0  1]; % Restrain in direction r the nodes situated @
                    % (x = 0) and (x = Length)
    end
%
    if geom(i,2) == 0;
    nf(i,:) = [0 0]; % Restrain in all directions the nodes situated @
                    % (y = 0) Rock substratum
    end
```

```
end
%
% Counting of the free degrees of freedom
%
n=0;
for i=1:nnd
    for j=1:nodof
        if nf(i,j) ~= 0
            n=n+1;
            nf(i,j)=n;
        end
    end
end
%
% loading
%
Nodal_loads= zeros(nnd, 2);   % Initialize the matrix of nodal loads to 0
%
% Apply an equivalent nodal load to the nodes located at
% (r = 0, z = 9.), (r = 0.25, z = 9.), and (r = 0.5, z = 9.)
% (r = .75, z = 9.), (r = 1., z = 9.)
%
pressure = 63.662 ; % kN/m^2
%
for i=1:nnd
    if geom(i,1) == 0. && geom(i,2) == 9.
        Nodal_loads(i,:) = pressure*[0.   0.];
    elseif geom(i,1) == 0.25 && geom(i,2) == 9.
        Nodal_loads(i,:) = pressure*[0.   -0.0833];
    elseif geom(i,1) == 0.5 && geom(i,2) == 9.
        Nodal_loads(i,:) = pressure*[0.   (-0.0833-0.0833)];
    elseif geom(i,1) == 0.75 && geom(i,2) == 9.
        Nodal_loads(i,:) = pressure*[0.   -0.25];
    elseif geom(i,1) == 1. && geom(i,2) == 9.
        Nodal_loads(i,:) = pressure*[0.   -0.0833];
    end
 end
%
%%%%%%%%%%%%%%%%%%%%%%%%%%% End of input%%%%%%%%%%%%%%%%%%%%%%%%%%%%%%%%%%%%
%
% Assemble the global force vector
%
fg=zeros(n,1); for i=1: nnd
    if nf(i,1) ~= 0
        fg(nf(i,1))= Nodal_loads(i,1);
    end
    if nf(i,2) ~= 0
        fg(nf(i,2))= Nodal_loads(i,2);
    end
end
%
% Assembly of the global stiffness matrix
%
%
%  Form the matrix containing the abscissas and the weights of Hammer points
%
samp=hammer(nhp);
%
%  initialize the global stiffness matrix to zero
%
kk = zeros(n, n);
%
for i=1:nel
    [coord,g] = elem_T6(i);  % Form strain matrix, and steering vector
    ke=zeros(eldof,eldof) ;  % Initialize the element stiffness matrix to zero
     for ig = 1:nhp
        wi = samp(ig,3);
```

```
            [der,fun] = fmT6_quad(samp, ig);
            jac = der*coord;
            d = det(jac);
            jac1=inv(jac);          % Compute inverse of the Jacobian
            deriv=jac1*der;         % Derivative of shape functions in global coordinates
            [bee,radius]=formbee_axi(deriv,nne,fun, coord,eldof); %  Form matrix [B]
            ke=ke + d*wi*bee'*dee*bee*radius; % Integrate stiffness matrix
        end
     kk=form_kk(kk,ke, g);                        % assemble global stiffness matrix
end
%
%
%%%%%%%%%%%%%%%%%%%%%%%%  End of assembly %%%%%%%%%%%%%%%%%%%%%%%%%%%%%%%
%
%
delta = kk\fg ;               % solve for unknown displacements
%
for i=1: nnd                             %
    if nf(i,1) == 0                      %
        x_disp =0.;                      %
    else
        x_disp = delta(nf(i,1));         %
    end
%
    if nf(i,2) == 0                      %
        y_disp = 0.;                     %
    else
        y_disp = delta(nf(i,2));         %
    end
    DISP(i,:) = [x_disp  y_disp];
end
%
%
nhp = 1;    % Calculate stresses at the centroid of the element
samp=hammer(nhp);
%
for i=1:nel
    [coord,g] = elem_T6(i);         % Retrieve coordinates and steering vector
    eld=zeros(eldof,1);             % Initialize element displacement to zero
    for m=1:eldof                   %
        if g(m)==0                  %
            eld(m)=0.;              %
        else                        %
            eld(m)=delta(g(m));     % Retrieve element displacement from
                                    % the global displacement vector
        end
    end
%
    for ig=1: nhp
        [der,fun] = fmT6_quad(samp,ig); % Derivative of shape functions in local coordinates
        jac=der*coord;                  % Compute Jacobian matrix
        jac1=inv(jac);                  % Compute inverse of the Jacobian
        deriv=jac1*der;                 % Derivative of shape functions in global coordinates
        [bee,radius]=formbee_axi(deriv,nne,fun, coord,eldof); %  Form matrix [B]
        eps=bee*eld;                    % Compute strains
        sigma=dee*eps ;                 % Compute stresses
    end
    SIGMA(i,:)=sigma ;              % Store stresses for all elements
end
%
[ZX, ZY,  Z_THETA, ZT] = Stresses_at_nodes_axi(SIGMA); U2 =
DISP(:,2);
%
%
% Plot stresses in the x_direction
%
cmin = min(ZT); cmax = max(ZT);
```

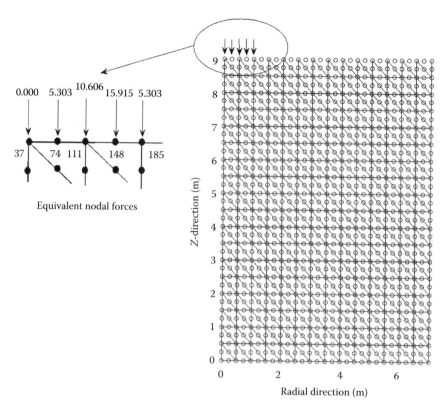

FIGURE 10.8 Finite element mesh using the 6-node triangle.

```
%
caxis([cmin cmax]); patch('Faces', connec, 'Vertices', geom,
'FaceVertexCData',ZT,...
                            'Facecolor','interp','Marker','.');
colorbar;
```

Figure 10.8 shows the finite element discretization of half the domain and the values of the equivalent nodal loads. The domain is meshed with 6-node triangle. In total the mesh consists of 1073 nodes and 504 elements. The nodes are numbered along the *y*-direction with the first node being at the origin. The loaded nodes are located using their coordinates, and the equivalent nodal loads calculated with the formulas given in Figure 10.4.

10.5.1.1 Numerical Integration of the Stiffness Matrix

The stiffness matrix is evaluated as

$$[K_e] = \sum_{i=1}^{nhp} W_i [B(\xi_i, \eta_i)]^T [D] [B(\xi_i, \eta_i)] r(\xi_i, \eta_i) det[J(\xi_i, \eta_i)] \tag{10.23}$$

1. For every element $i = 1$ to *nel*
2. Retrieve the coordinates of its nodes **coord(nne, 2)** and its steering vector **g(eldof)** using the function *elem_t6.m*
3. Initialize the stiffness matrix to zero
   a. Loop over the Hammer points $ig = 1$ to *nhp*
   b. Retrieve the weight **wi** as **samp(ig, 3)**

c. Use the function *fmT6_quad.m* to compute the shape functions, vector **fun**, and their local derivatives, **der**, at the local coordinates $\xi = $ **samp(ig, 1)** and $\eta = $ **samp(ig, 2)**

d. Evaluate the Jacobian **jac** = **der** $*$ **coord**

e. Evaluate the determinant of the Jacobian as **d** = **det(jac)**

f. Compute the inverse of the Jacobian as **jac1** = **inv(jac)**

g. Compute the derivatives of the shape functions with respect to the global coordinates x and y as **deriv** = **jac1** $*$ **der**

h. Use the function *formbee_axi* to form the strain matrix **bee** and calculate the radius r at the integration point as $r = \sum_{j}^{nne} N_j x_j$, where nne represents the number of nodes of the element

i. Compute the stiffness matrix as **ke** = **ke** + **wi** $*$ **bee'** $*$ **dee** $*$ **bee** $*$ **r** $*$ **d**$*$

4. Assemble the stiffness matrix **ke** into the global matrix **kk**

Note that the elasticity matrix $[D]$ is that given by Equation (10.7) for an axisymmetric conditions and has a dimension 4×4. It is formed using the function *formdax.m* listed in Appendix A. The strain matrix given by Equation (10.13) is evaluated using the function *formbee_axi.m* also listed in Appendix A.

10.5.1.2 Results

Figures 10.9 through 10.12 show respectively the contours of the vertical displacement v, the radial stress σ_{rr}, the vertical stress σ_{zz}, and the shear stress τ_{rz} obtained with the 6-node triangle element.

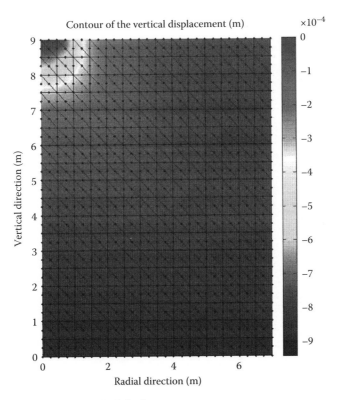

FIGURE 10.9   Contour plot of the vertical displacement.

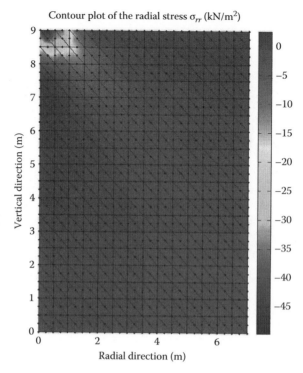

FIGURE 10.10   Contour plot of the radial stress.

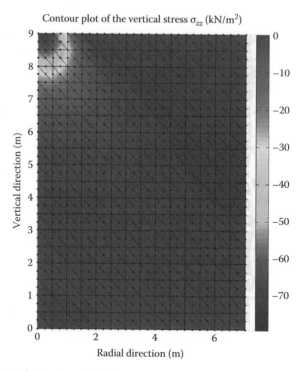

FIGURE 10.11   Contour plot of the vertical stress.

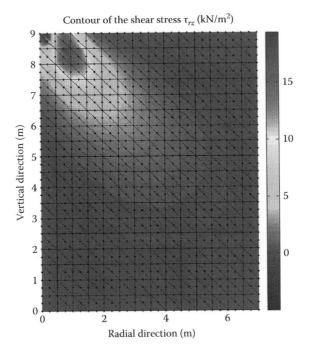

FIGURE 10.12   Contour plot of the shear stress.

10.5.2 COMPUTER CODE: AXI_SYM_Q8.m

The following program, *AXI_SYM_Q8.m*, is an adaptation of the plane stress/strain program *PLANE_Q8_MESH.m* to axisymmetric conditions. The program is listed next and includes the automatic mesh generation function *Q8_mesh.m*.

```
% THIS PROGRAM USES AN 8-NODDED QUADRILATERAL ELEMENT FOR THE LINEAR ELASTIC
% STATIC ANALYSIS OF AN AXISYMMETRIC PROBLEM. IT CONTAINS AN AUTOMATIC
% MESH GENERATION MODULE Q8_mesh.m
%
% Make these variables global so they can be shared by other functions
%
clc
clear all
global nnd nel nne  nodof eldof  n ngp
global geom connec dee nf Nodal_loads
global Length Width NXE NYE X_origin Y_origin dhx dhy
%
 format long g
%
% To change the size of the problem or change elastic properties
%     ALTER the q8_input_module.m
%
Length = 7.; % Length of the model
Width =9.;     % Width
NXE = 14;       % Number of rows in the x direction
NYE = 18;       % Number of rows in the y direction
dhx = Length/NXE; % Element size in the r direction
dhy = Width/NYE;  % Element size in the z direction
X_origin = 0. ;  % r origin of the global coordinate system
Y_origin = 0. ;   % z origin of the global coordinate system
%
nne = 8;
nodof = 2;
eldof = nne*nodof;
ngp = 3;
```

```
%
Q8_mesh       % Generate the mesh
%
E = 100000.;       % Elastic modulus in kN/m2
vu = 0.35;         % Poisson's ratio
%
% Form the elastic matrix for plane stress
%
dee = formdax(E,vu);
%
%
% Boundary conditions
%
nf = ones(nnd, nodof);    % Initialize the matrix nf to 1
%
% Restrain in all directions the nodes situated @
% (x = Length)
%
for i=1:nnd
    if geom(i,1) == 0 || geom(i,1) == Length
    nf(i,:) = [0  1]; % Restrain in direction r the nodes situated @
                     % (x = 0) and (x = Length)
    end
%
    if geom(i,2) == 0;
    nf(i,:) = [0 0]; % Restrain in all directions the nodes situated @
                     % (y = 0) Rock substratum
    end
end
%
% Counting of the free degrees of freedom
%
n=0;
for i=1:nnd
    for j=1:nodof
        if nf(i,j) ~= 0
            n=n+1;
            nf(i,j)=n;
        end
    end
end
%
% loading
%
Nodal_loads= zeros(nnd, 2);    % Initialize the matrix of nodal loads to 0
%
% Apply an equivalent nodal load to the nodes located at
% (r = 0, z = 9.), (r = 0.25, z = 9.), and (r = 0.5, z = 9.)
% (r = .75, z = 9.), (r = 1., z = 9.)
%
pressure = 63.662 ; % kN/m^2
%
for i=1:nnd
    if geom(i,1) == 0. && geom(i,2) == 9.
        Nodal_loads(i,:) = pressure*[0.   0.];
    elseif geom(i,1) == 0.25 && geom(i,2) == 9.
        Nodal_loads(i,:) = pressure*[0.  -0.0833];
    elseif geom(i,1) == 0.5 && geom(i,2) == 9.
        Nodal_loads(i,:) = pressure*[0.  (-0.0833-0.0833)];
    elseif geom(i,1) == 0.75 && geom(i,2) == 9.
        Nodal_loads(i,:) = pressure*[0.  -0.25];
    elseif geom(i,1) == 1. && geom(i,2) == 9.
        Nodal_loads(i,:) = pressure*[0.  -0.0833];
    end
end
%
%%%%%%%%%%%%%%%%%%%%%%%%%%%% End of input%%%%%%%%%%%%%%%%%%%%%%%%%%%%%%%%%
```

```
%
% Assemble the global force vector
%
fg=zeros(n,1);
for i=1: nnd
    if nf(i,1) ~= 0
        fg(nf(i,1))= Nodal_loads(i,1);
    end
    if nf(i,2) ~= 0
        fg(nf(i,2))= Nodal_loads(i,2);
    end
end
%
%  Form the matrix containing the abscissas and the weights of Gauss points
%
samp=gauss(ngp);
%
% Numerical integration and assembly of the global stiffness matrix
%
%  initialize the global stiffness matrix to zero
kk = zeros(n, n);
%
for i=1:nel
  [coord,g] = elem_q8(i) ;        % coordinates of the nodes of element i,
                                  % and its steering vector
  ke=zeros(eldof,eldof) ;         % Initialize the element stiffness
                                  % matrix to zero

  for ig=1: ngp
      wi = samp(ig,2);
  for jg=1: ngp
   wj=samp(jg,2);
   [der,fun] = fmquad(samp, ig,jg);  % Derivative of shape functions
                                     % in local coordinates
   jac=der*coord;                 % Compute Jacobian matrix
   d=det(jac);                    % Compute determinant of Jacobian matrix
   jac1=inv(jac);                 % Compute inverse of the Jacobian
   deriv=jac1*der;                % Derivative of shape functions in
                                  % global coordinates
   [bee,radius]=formbee_axi(deriv,nne,fun, coord,eldof); % Form matrix [B]
   ke=ke + d*wi*wj*bee'*dee*bee*radius; % Integrate stiffness matrix
   end
   end
   kk=form_kk(kk,ke, g);              % assemble global stiffness matrix
end
%
%
%%%%%%%%%%%%%%%%%%%%%%%%%  End of assembly %%%%%%%%%%%%%%%%%%%%%%%%%%%%%%%%%%
%
%
delta = kk\fg ;                             % solve for unknown displacements
disp('node        x_disp        y_disp ')  %
for i=1: nnd                                %
    if nf(i,1) == 0                         %
        x_disp =0.;                         %
    else
        x_disp = delta(nf(i,1));            %
    end
%
    if nf(i,2) == 0                         %
        y_disp = 0.;                        %
    else
        y_disp = delta(nf(i,2));            %
    end
disp([i  x_disp  y_disp]) ;                 % Display displacements of each node
DISP(i,:) = [x_disp  y_disp];
end
%
```

```
%
ngp=1;                                  % Calculate stresses and strains at
                                        % the center of each element
samp=gauss(ngp);
%
for i=1:nel
  [coord,g] = elem_q8(i);      % coordinates of the nodes of element i,
                               % and its steering vector
  eld=zeros(eldof,1);          % Initialize element displacement to zero
  for m=1:eldof                %
    if g(m)==0                 %
        eld(m)=0.;             %
    else                       %
        eld(m)=delta(g(m));    % Retrieve element displacement from the
                               % global displacement vector
    end
  end
%
  for ig=1: ngp
     wi = samp(ig,2);
  for jg=1: ngp
     wj=samp(jg,2);
     [der,fun] = fmquad(samp, ig,jg); % Derivative of shape functions in
                                      % local coordinates
     jac=der*coord;                   % Compute Jacobian matrix
     jac1=inv(jac);                   % Compute inverse of the Jacobian
     deriv=jac1*der;                  % Derivative of shape functions
                                      % in global coordinates
     [bee,radius]=formbee_axi(deriv,nne,fun, coord,eldof);%Form matrix [B]
     eps=bee*eld                      % Compute strains
     sigma=dee*eps                    % Compute stresses
  end
  end
  SIGMA(i,:)=sigma ;                % Store stresses for all elements
end
%
%
[ZX, ZY,  Z_THETA, ZT] = stresses_at_nodes_axi(SIGMA);
%
%
% Plot stresses in the x_direction
%
U2 = DISP(:,2);
cmin = min(ZT);
cmax = max(ZT);
caxis([cmin cmax]);
patch('Faces', connec, 'Vertices', geom, 'FaceVertexCData',ZT,...
                       'Facecolor','interp','Marker','.');
colorbar;
```

Figure 10.13 shows the finite element discretization of half the domain and the values of the equivalent nodal loads. The domain is meshed with 8-node quadrilaterals. In total the mesh consists of 821 nodes and 252 elements. The nodes are numbered along the y-direction with the first node being at the origin. The loaded nodes are located using their coordinates, and the equivalent nodal loads calculated with the formulas given in Figure 10.4.

10.5.2.1 Numerical Integration of the Stiffness Matrix

The stiffness matrix is evaluated using Gauss quadrature as

$$
\begin{aligned}
[K_e] &= \int_{-1}^{+1} \int_{-1}^{+1} [B(\xi,\eta)]^T [D][B(\xi,\eta)] r(\xi,\eta) det[J(\xi,\eta)] \, d\eta \, d\xi, \\
&= \sum_{i=1}^{ngp} \sum_{j=1}^{ngp} W_i W_j [B(\xi_i,\eta_j)]^T [D][B(\xi_i,\eta_j)] r(\xi_i,\eta_j) det[J(\xi_i,\eta_j)]
\end{aligned}
\tag{10.24}
$$

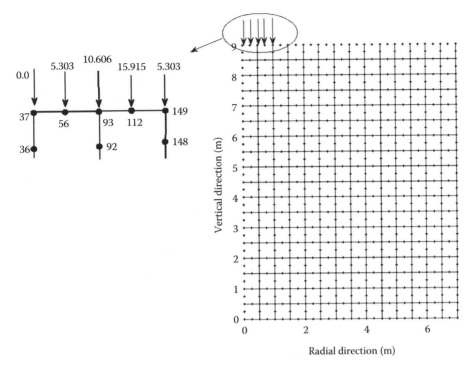

FIGURE 10.13 Finite element mesh using the 8-node quadrilateral.

For each element, it is evaluated as follows:

1. For every element $i = 1$ to *nel*
2. Retrieve the coordinates of its nodes **coord(nne, 2)** and its steering vector **g(eldof)** using the function *elem_Q8.m*
3. Initialize the stiffness matrix to zero
    a. Loop over the Gauss points $ig = 1$ to *ngp*
    b. Retrieve the weight **wi** as **samp(ig, 2)**
        i. Loop over the Gauss points $jg = 1$ to *ngp*
        ii. Retrieve the weight **wj** as **samp(jg, 2)**
        iii. Use the function *fmquad.m* to compute the shape functions, vector **fun**, and their derivatives, matrix **der**, in local coordinates, $\xi = $ **samp(ig, 1)** and $\eta = $ **samp(jg, 1)**.
        iv. Evaluate the Jacobian **jac** = **der** $*$ **coord**
        v. Evaluate the determinant of the Jacobian as **d** = **det(jac)**
        vi. Compute the inverse of the Jacobian as **jac1** = **inv(jac)**
        vii. Compute the derivatives of the shape functions with respect to the global coordinates x and y as **deriv** = **jac1** $*$ **der**
        viii. Use the function *formbee_axi* to form the strain matrix **bee** and calculate the radius r at the integration point as $r = \sum_{j}^{nne} N_j x_j$
        ix. Compute the stiffness matrix as **ke** = **ke** + **d** $*$ **wi** $*$ **wj** $*$ **bee'** $*$ **dee** $*$ **bee** $*$ **r**
4. Assemble the stiffness matrix **ke** into the global matrix **kk**

10.5.2.2 Results

Figures 10.14 through 10.17 show respectively the contours of the vertical displacement v, the radial stress σ_{rr}, the vertical stress σ_{zz}, and the shear stress τ_{rz} obtained with the 6-node triangle element.

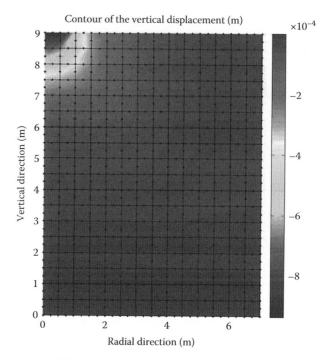

FIGURE 10.14   Contour plot of the vertical displacement.

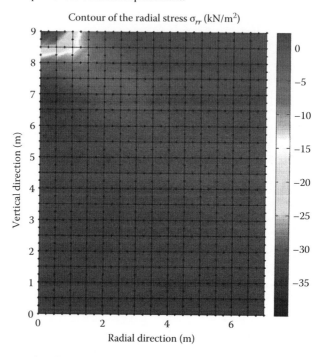

FIGURE 10.15   Contour plot of the radial stress.

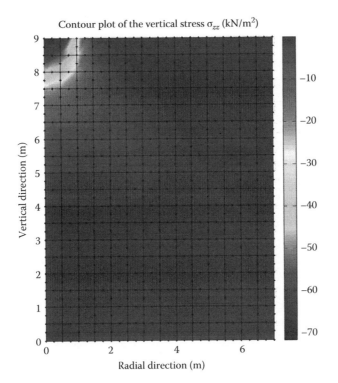

FIGURE 10.16   Contour plot of the vertical stress.

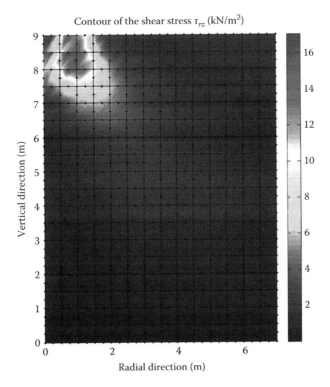

FIGURE 10.17   Contour plot of the shear stress.

10.6 ANALYSIS WITH ABAQUS USING THE 8-NODE QUADRILATERAL

In this section, we will analyze the circular footing shown in Figure 10.6 using the Abaqus interactive edition. Taking advantage of symmetry, only half the model is analyzed. We will use an element size of 0.5 m so that we could compare the results with those obtained previously.

Start **Abaqus CAE**. Click on **Create Model Database**. On the main menu, click on **File** and set **Set Work Directory** to choose your working directory. Click on **Save As** and name the file **FOOTING_Q8.cae**. On the left-hand-side menu, click on **Part** to begin creating the model. Name the part **FOOTING_Q8**, check **Axisymmetric**, check **Deformable** in the type. Choose **Shell** as the base feature. Enter an approximate size of 20 m and click on **Continue**. In the sketcher menu, choose the **Create-Lines Rectangle** icon to begin drawing the geometry of the footing. Make sure that the sketch is to the right or to the left of the centerline. Click on **Done** in the bottom-left corner of the viewport window (Figure 10.18).

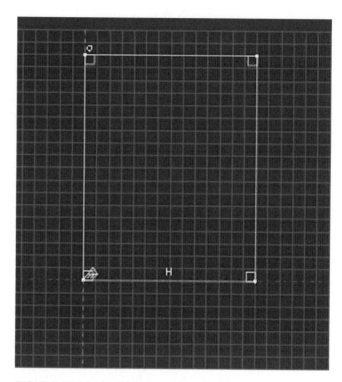

FIGURE 10.18　Creating the FOOTING_Q8 Part.

Define a material named **Dirt** with an elastic modulus of 100000 kN/m² and a Poisson's ratio of 0.35. Next, click on **Sections** to create a section named **Footing_section_Q8**. In the **Category** check **Solid**, and in the **Type**, check **Homogeneous**. Click on **Continue**. In the **Edit Section** dialog box, uncheck **Plane stress/strain thickness**. Click on **OK** (Figure 10.19).

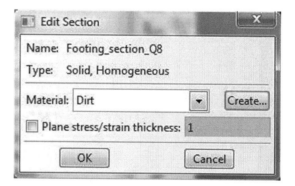

FIGURE 10.19　Creating an axisymmetric section.

Expand the menu under **Parts** and **FOOTING_Q8** and double click on **Section Assignments**. With the mouse select the whole part. In the **Edit Section Assignments** dialog box, select **Footing_section_Q8** and click on **OK** (Figure 10.20).

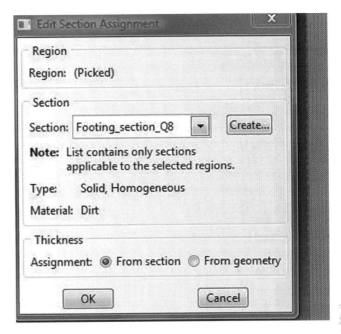

FIGURE 10.20   Editing section assignments.

It will be useful to partition the top edge so that we could apply the pressure load over a length of 2 m. Therefore, under the main menu, expand **Tools** and click on **Partition**. In the partition dialog box, select **Edge** for **Type**, and **Enter parameter** for **Method**. In the command line of the viewport enter $0.714285714285 = 5/7$ as shown in Figure 10.21. Click on **Create partition**.

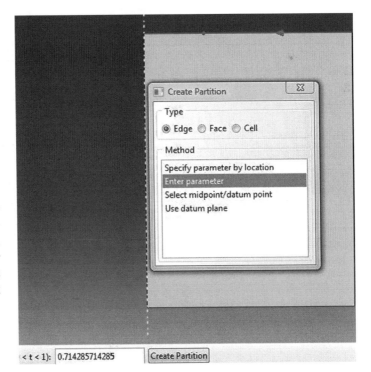

FIGURE 10.21   Edge partition.

In the model tree, double click on **Mesh** under the **FOOTING_Q8**. In the main menu, under **Mesh**, click on **Mesh Controls**. In the dialog box, check **Quad** for **Element shape** and **Structured** for **Technique**. Click on **OK**. Under **Mesh**, click on **Element Type**. In the dialog box, select **Standard** for element library, **Quadratic** for geometric order. In **Quad**, check **Reduced integration**. The description of the element **CAX8R: A 8-node biquadratic axisymmetric quadrilateral, reduced integration** can be seen in the dialog box. Click on **OK** (Figure 10.22).

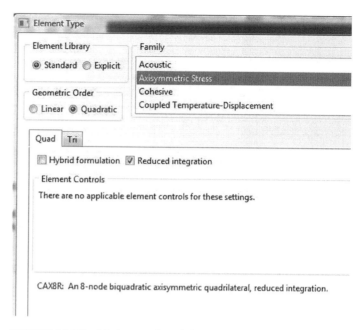

FIGURE 10.22   Mesh controls and element type.

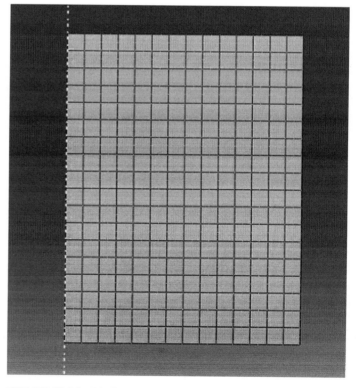

In the main menu, under **Seed**, click on **Part**. In the dialog box, enter 0.5 for **Approximate global size**. Click on **OK** and on **Done**. In the main menu, under **Mesh**, click on **Part**. In the prompt area, click on **Yes** (Figure 10.23).

FIGURE 10.23   Mesh.

In the model tree, expand the **Assembly** and double click on **Instances**. Select **FOOTING_Q8** for **Parts** and click **OK**. In the model tree, expand **Steps** and **Initial** and double click on **BC**. Name the boundary condition **Centerline**, select **Displacement/Rotation** for the type, and click on **Continue**. In the viewport, with the mouse select the centerline and click on **Continue**. In the **Edit Boundary Condition**, check **U1**. Click **OK**. Repeat the procedure again, this time select the right edge and click on **Continue**. In the **Edit Boundary Condition**, check **U1**. Click **OK**. Repeat the procedure again, this time select the bottom edge and click on **Continue**. In the **Edit Boundary Condition**, check **U1** and **U2**. Click **OK** (Figure 10.24).

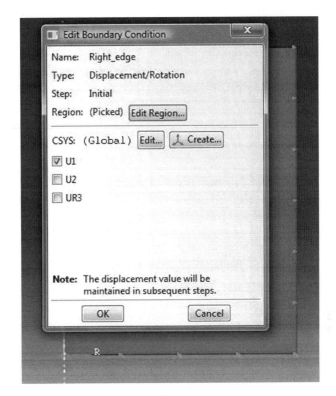

FIGURE 10.24 Imposing BC using geometry.

In the model tree, double click on **Steps**. Name the step **Apply_loads**. Set the procedure to **General** and select **Static, General**. Click on **Continue**. Give the step a description and click **OK**. In the model tree, under **steps**, and under **Apply_loads**, click on **Loads**. Name the load **Pressure** and select **Pressure** as the type. Click on **Continue**. In the viewport, with the mouse select the left part of the partitioned top edge. In the **Edit Load** dialog box, enter $63.662 \, \text{kN/m}^2$. Click **OK** (Figure 10.25).

FIGURE 10.25 Imposing loads using geometry.

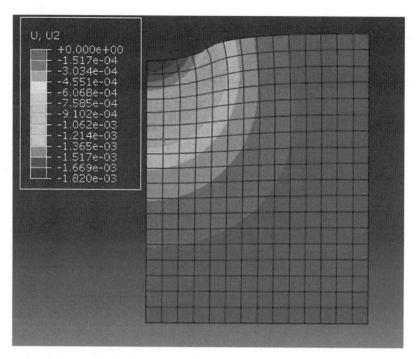

FIGURE 10.26   Contour of the vertical displacement.

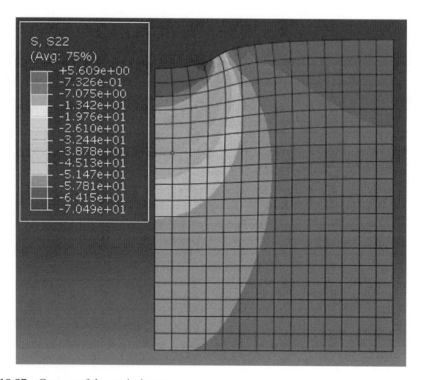

FIGURE 10.27   Contour of the vertical stress σ_{yy}.

Under **Analysis**, right click on **Jobs** and then click on **Create**.

In the **Create Job** dialog box, name the job **FOOTING_Q8** and click on **Continue**. In the **Edit Job** dialog box, enter a description for the job. Check **Full analysis**, select to run the job in **Background**, and check to start it **immediately**. Click **OK**. Expand the tree under **Jobs**, right click on **FOOTING_Q8**. Then, click on **Submit**. If you get the following message **FOOTING_Q8 completed successfully** in the bottom window, then your job is free of errors and was executed properly.

Under the top menu, in the **Module** scroll to **Visualization**, and click to load **Abaqus Viewer**. On the main menu, under **File**, click **Open**, navigate to your working directory and open the file **FOOTING_Q8.odb**. It should have the same name as the job you submitted. Click on the icon **Plot** on **Undeformed shape**. Under the main menu, select **U** and **U2** to plot the vertical displacement (Figure 10.26). It can be seen that the displacement contour is similar to that obtained with the MATLAB® code (Figures 10.9 and 10.14).

Under the main menu, select **S** and **S22** to plot σ_{yy} (Figure 10.27). Again, the contour is very similar to that shown in Figure 10.16.

11 Thin and Thick Plates

11.1 INTRODUCTION

Plates are very important structural elements. They are mainly used as slabs in buildings and bridge decks. They are structural elements that are bound by two lateral surfaces. The dimensions of the lateral surfaces are very large compared to the thickness of the plate. A plate may be thought of as the two-dimensional equivalent of a beam. Plates are also generally subject to loads normal to their plane.

11.2 THIN PLATES

11.2.1 DIFFERENTIAL EQUATION OF PLATES LOADED IN BENDING

The small deflection theory of plates attributed to Kirchhoff is based on the following assumptions:

1. The x–y plane coincides with the middle plane of the plate in the undeformed geometry.
2. The lateral dimension of the plate is at least 10 times its thickness.
3. The vertical displacement of any point of the plate can be taken equal to that of the point (below or above it) in the middle plane.
4. A vertical element of the plate before bending remains perpendicular to the middle surface of the plate after bending.
5. Strains are small: deflections are less than the order of (1/100) of the span length.
6. The strain of the middle surface is zero or negligible.

Considering the plate element shown in Figure 11.1, the in-plane displacements u and v, respectively in the directions x and y, can be expressed as

$$u = -z \frac{\partial w}{\partial x} \tag{11.1}$$

$$v = -z \frac{\partial w}{\partial y} \tag{11.2}$$

where w represents the vertical displacement of the plate mid-plane.

Because of the assumption number 4, that is, "a vertical element of the plate before bending remains perpendicular to the middle surface of the plate after bending," the transverse shear deformation is negligible. The in-plane strains can therefore be written in terms of the displacements as

$$\begin{Bmatrix} \epsilon_{xx} \\ \epsilon_{yy} \\ \gamma_{xy} \end{Bmatrix} = \begin{Bmatrix} \dfrac{\partial u}{\partial x} \\ \dfrac{\partial v}{\partial y} \\ \dfrac{\partial u}{\partial y} + \dfrac{\partial v}{\partial x} \end{Bmatrix} = \begin{Bmatrix} -z \dfrac{\partial^2 w}{\partial x^2} \\ -z \dfrac{\partial^2 w}{\partial y^2} \\ -2z \dfrac{\partial^2 w}{\partial x \partial y} \end{Bmatrix} = -z \begin{Bmatrix} \chi_x \\ \chi_y \\ \chi_{xy} \end{Bmatrix} \tag{11.3}$$

The vector $\{\chi\} = [\chi_x \ \chi_y \ \chi_{xy}]^T$ is called the vector of curvature or generalized strain.

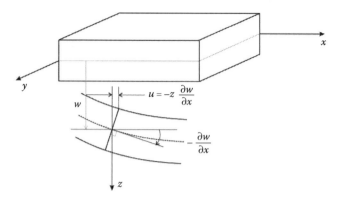

FIGURE 11.1   Deformed configuration of a thin plate in bending.

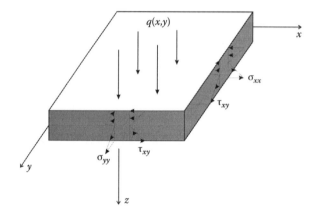

FIGURE 11.2   Internal stresses in a thin plate. Moments and shear forces due to internal stresses in a thin plate.

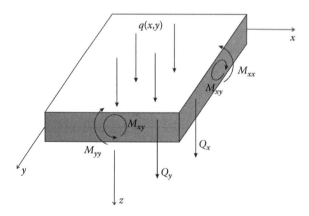

FIGURE 11.3   Moments and shear forces due to internal stresses in a thin plate.

Internal stresses in plates produce bending moments and shear forces as illustrated in Figures 11.2 and 11.3. The moments and shear forces are the resultants of the stresses and are defined as acting per unit length of plate. These internal actions are defined as

$$M_{xx} = \int\limits_{-h/2}^{h/2} \sigma_{xx} z \, dz \qquad (11.4)$$

$$M_{yy} = \int_{-h/2}^{h/2} \sigma_{yy} z \, dz \tag{11.5}$$

$$M_{xy} = \int_{-h/2}^{h/2} \tau_{xy} z \, dz \tag{11.6}$$

$$Q_{xx} = \int_{-h/2}^{h/2} \sigma_{xx} dz \tag{11.7}$$

$$Q_{yy} = \int_{-h/2}^{h/2} \sigma_{yy} dz \tag{11.8}$$

In general, the force and moment intensities vary with the coordinates x and y of the middle plane.

Assuming a state of plane stress conditions for plate bending,

$$\{\sigma\} = [D]\{\epsilon\} \tag{11.9}$$

with $[D]$ given as

$$[D] = \frac{E}{1-\nu^2} \begin{bmatrix} 1 & \nu & 0 \\ \nu & 1 & 0 \\ 0 & 0 & \dfrac{(1-\nu)}{2} \end{bmatrix} \tag{11.10}$$

and substituting for $\{\epsilon\}$ using Equation (11.3) yields the constitutive equation

$$\{\sigma\} = -z[D]\{\chi\} \tag{11.11}$$

Substituting for σ_{xx}, σ_{yy}, and τ_{xy} in Equation (11.4) and rearranging the results in a matrix notation yields

$$\{M\} = \frac{h^3}{12}[D]\{\chi\} \tag{11.12}$$

Consider the equilibrium of the free body of the differential plate element shown in Figure 11.4. Recalling that Q_x represents force per unit length along the edge dy and requiring force equilibrium in z direction results in

$$-Q_x dy - Q_y dx + \left(Q_x + \frac{\partial Q_x}{\partial x} dx\right) dy + \left(Q_y + \frac{\partial Q_y}{\partial y} dy\right) dx + q(x,y) dx dy = 0 \tag{11.13}$$

which upon dividing by $dxdy$ becomes

$$\frac{\partial Q_x}{\partial x} + \frac{\partial Q_y}{\partial y} + q(x,y) = 0 \tag{11.14}$$

Moment equilibrium about the x-axis leads to

$$\frac{\partial M_{xy}}{\partial x} + \frac{\partial M_{yy}}{\partial y} = Q_y \tag{11.15}$$

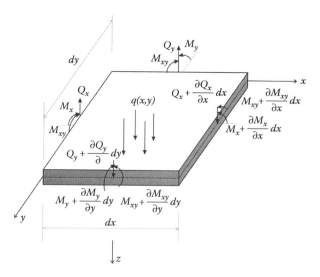

FIGURE 11.4 Free body diagram of a plate element.

Moment equilibrium about the y-axis leads to

$$\frac{\partial M_{xy}}{\partial y} + \frac{\partial M_{xx}}{\partial x} = Q_x \tag{11.16}$$

Substituting (11.15) and (11.16) in (11.14) results in the governing equation

$$\frac{\partial^2 M_{xx}}{\partial x^2} + \frac{\partial^2 M_{xy}}{\partial x \partial y} + \frac{\partial^2 M_{yy}}{\partial y^2} + q(x, y) = 0 \tag{11.17}$$

Since no relations regarding material behavior have entered Equation (11.17), it is valid for all types of materials.

11.2.2 GOVERNING EQUATION IN TERMS OF DISPLACEMENT VARIABLES

Substitution of Equation (11.12) into the equilibrium equation (11.17) leads to the general differential equation of simple rectangular plates:

$$\frac{\partial^4 w}{\partial x^4} + 2\frac{\partial^4 w}{\partial^2 x \partial^2 y} + \frac{\partial^4 w}{\partial y^4} = \frac{q(x, y)}{D_r} \tag{11.18}$$

which is often written as

$$\nabla^4 w = \frac{q}{D_r} \tag{11.19}$$

with

$$D_r = \frac{Eh^3}{12(1 - \nu^2)} \tag{11.20}$$

The solution of a simple rectangular plate in bending requires finding a function $w(x, y)$ that satisfies Equation (11.18) and also the boundary conditions of the specific problem.

11.3 THICK PLATE THEORY OR MINDLIN PLATE THEORY

As explained previously, the Kirchhoff plate theory does not include shear deformations. This is an acceptable assumption for very thin plates, but it can lead to errors, which are not negligible in thick plates; most of reinforced concrete slabs are classified in this latter category.

In thick plates, the assumption that a vertical element of the plate before bending remains perpendicular to the middle surface of the plate after bending is relaxed. Transverse normals may rotate without remaining normal to the mid-plane. A line originally normal to the middle plane will develop rotation components θ_x relative to the middle plane after deformation as shown in Figure 11.5. A similar definition holds for θ_y. Hence, the displacement field becomes

$$u = z\theta_x \tag{11.21}$$

$$v = z\theta_y \tag{11.22}$$

$$w = w(x, y) \tag{11.23}$$

The strains associated with these displacements are given as

$$\begin{Bmatrix} \epsilon_{xx} \\ \epsilon_{yy} \\ \gamma_{xy} \\ \gamma_{yz} \\ \gamma_{zx} \end{Bmatrix} = \begin{Bmatrix} z\dfrac{\partial \theta_x}{\partial x} \\ z\dfrac{\partial \theta_y}{\partial y} \\ z\left(\dfrac{\partial \theta_x}{\partial y} + \dfrac{\partial \theta_y}{\partial x}\right) \\ z\left(\theta_y - \dfrac{\partial w}{\partial y}\right) \\ z\left(\theta_x - \dfrac{\partial w}{\partial x}\right) \end{Bmatrix} \tag{11.24}$$

These equations are the main equations of the Mindlin plate theory. The theory accounts for transverse shear deformations and is applicable for moderately thick plates. Unlike in thin plate theory, it is important to notice that the transverse displacement $w(x, y)$ and slopes θ_x, θ_y are independent. Notice also that the thick plate theory reduces to thin plate theory if $\theta_x = -\frac{\partial w}{\partial x}$ and $\theta_y = -\frac{\partial w}{\partial y}$.

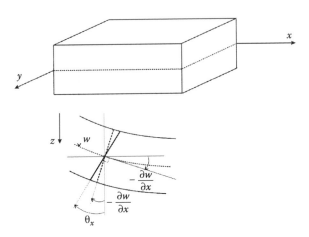

FIGURE 11.5 Deformed configuration of a thick plate in bending.

11.3.1   STRESS–STRAIN RELATIONSHIP

Assuming the material is homogeneous and isotropic, the plane stresses σ_{xx}, σ_{yy}, and τ_{xy} are related to the strains through the elasticity matrix $[D]$ given in Equation (11.10). The shear strains τ_{yz} and τ_{xz} are related to the shear strains γ_{yz} and γ_{xz} through

$$\begin{Bmatrix} \tau_{yz} \\ \tau_{xz} \end{Bmatrix} = \begin{bmatrix} G & 0 \\ 0 & G \end{bmatrix} \begin{Bmatrix} \gamma_{yz} \\ \gamma_{xz} \end{Bmatrix} \tag{11.25}$$

with

$$G = \frac{E}{2(1+\nu)} \tag{11.26}$$

The moment curvature relations for the Mindlin plate theory are obtained by combining (11.4), (11.9), (11.10), (11.24), and (11.25); that is

$$\begin{Bmatrix} M_x \\ M_y \\ M_{xy} \\ Q_y \\ Q_x \end{Bmatrix} = \begin{bmatrix} D_r & \nu \times D_r & 0 & 0 & 0 \\ \nu \times D_r & D_r & 0 & 0 & 0 \\ 0 & 0 & \dfrac{D_r(1-\nu)}{2} & 0 & 0 \\ 0 & 0 & 0 & Gh & 0 \\ 0 & 0 & 0 & 0 & Gh \end{bmatrix} \begin{Bmatrix} \dfrac{\partial \theta_x}{\partial x} \\ \dfrac{\partial \theta_y}{\partial y} \\ \left(\dfrac{\partial \theta_x}{\partial y} + \dfrac{\partial \theta_y}{\partial x}\right) \\ \left(\theta_y - \dfrac{\partial w}{\partial y}\right) \\ \left(\theta_x - \dfrac{\partial w}{\partial x}\right) \end{Bmatrix} \tag{11.27}$$

with

$$D_r = \frac{Eh^3}{12(1-\nu^2)} \tag{11.28}$$

Equation (11.27) can be written more compactly as

$$\{M\} = [D_M]\{\chi\} \tag{11.29}$$

The total strain energy of the plate is given as

$$U = \frac{1}{2} \int_A \{\chi\}^T [D_M]\{\chi\}\, dA \tag{11.30}$$

Equation (11.30) includes both the contributions from bending and shear energies. Hence, it can be decomposed as

$$U = U_B + U_S = \frac{1}{2} \int_A \{\chi_B\}^T [D_B]\{\chi_B\}\, dA + \frac{\kappa}{2} \int_A \{\chi_S\}^T [D_S]\{\chi_S\}\, dA \tag{11.31}$$

$$\{\chi_B\} = \begin{Bmatrix} \dfrac{\partial \theta_x}{\partial x} \\ \dfrac{\partial \theta_y}{\partial y} \\ \left(\dfrac{\partial \theta_x}{\partial y} + \dfrac{\partial \theta_y}{\partial x}\right) \end{Bmatrix} \tag{11.32}$$

$$\{\chi_S\} = \left\{ \begin{matrix} \left(\theta_y - \dfrac{\partial w}{\partial y} \right) \\ \left(\theta_x - \dfrac{\partial w}{\partial x} \right) \end{matrix} \right\} \tag{11.33}$$

$$[D_B] = D_r \begin{bmatrix} 1 & \nu & 0 \\ \nu & 1 & 0 \\ 0 & 0 & \dfrac{(1-\nu)}{2} \end{bmatrix} \tag{11.34}$$

$$[D_S] = G \begin{bmatrix} h & 0 \\ 0 & h \end{bmatrix} \tag{11.35}$$

and κ is the shear energy correction factor equal to 5/6.

11.4   LINEAR ELASTIC FINITE ELEMENT ANALYSIS OF PLATES

11.4.1   FINITE ELEMENT FORMULATION FOR THIN PLATES

The earliest finite elements for plates were based on the Kirchhoff theory, and their formulation required C^1 continuity. This required that the function $w(x, y)$ and its derivatives to be continuous across elements boundary to satisfy compatibility conditions, that is, the function $w(x, y)$ should satisfy the necessary identity of continuous functions:

$$\frac{\partial^2 w}{\partial x \partial y} = \frac{\partial^2 w}{\partial y \partial x} \tag{11.36}$$

A conventional plate element has three degrees of freedom per node: a vertical displacement w and two rotations. For small displacements, the rotations θ_x and θ_y are respectively the first derivatives of the vertical displacement w with respect to x and y:

$$\theta_x = \frac{\partial w}{\partial x} \tag{11.37}$$

$$\theta_y = \frac{\partial w}{\partial y} \tag{11.38}$$

The corresponding force components are the lateral force F_z and the moments M_x and M_y. The rotations θ_x and θ_y should be continuous all over the elements, otherwise the model will develop "kinks": no continuation in the slope.

11.4.1.1   Triangular Element

One of the earliest plate element is the three-node triangular plate bending element shown in Figure 11.6. It is important to note that θ_x and M_x are respectively the rotation and moment around the axis y. Any arbitrary point of the element has a deflection $w(x, y)$. Therefore, the displacement $w(x, y)$ is a continuous function of the variables x and y. At the nodes 1, 2, and 3, the function $w(x, y)$ should not only take on respectively the values w_1, w_2, and w_3, but should be continuous enough to have finite derivatives θ_{x1} and θ_{y1}, θ_{x2} and θ_{y2}, and θ_{x3} and θ_{y3}. To satisfy these requirements, a general approximation of the form shown in Equation (11.39) is used:

$$w(x, y) = \alpha_1 + \alpha_2 x + \alpha_3 y + \alpha_4 x^2 + \alpha_5 xy + \alpha_6 y^2 + \alpha_7 x^3 + \alpha_8 (x^2 y + xy^2) + \alpha_9 y^3 \tag{11.39}$$

Notice that there are nine parameters α_i as there are nine nodal variables. However, expression (11.39) does not constitute a complete polynomial, which contains 10 terms. The terms $x^2 y$ and xy^2

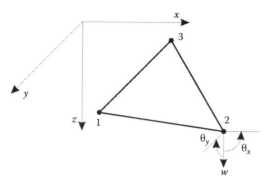

FIGURE 11.6   Three-node triangular plate bending element.

had to be grouped to make nine terms, and they do not vary independently. As a result, this element is known to behave badly, particularly when two sides of the triangle are parallel to the axes x and y [6].

The derivatives θ_x and θ_y are respectively obtained as

$$\frac{\partial w}{\partial x} = \alpha_2 + 2\alpha_4 x + \alpha_5 y + 3\alpha_7 x^2 + \alpha_8 (2xy + y^2) \tag{11.40}$$

$$\frac{\partial w}{\partial y} = \alpha_3 + \alpha_5 x + 2\alpha_6 y + \alpha_8 (x^2 + 2xy) + 3\alpha_9 y^2 \tag{11.41}$$

The general approximation (11.39) can be transformed into a nodal approximation using the method described in Chapter 7.

- At node 1, $x = x_1$, $y = y_1$, $w(x_1, y_1) = w_1$, $\theta_x = \theta_{x1}$ and $\theta_y = \theta_{x1}$
- At node 2, $x = x_2$, $y = y_2$, $w(x_2, y_2) = w_2$, $\theta_x = \theta_{x2}$ and $\theta_y = \theta_{x2}$
- At node 3, $x = x_3$, $y = y_3$, $w(x_3, y_3) = w_3$, $\theta_x = \theta_{x3}$ and $\theta_y = \theta_{x3}$

Substituting in Equations (11.39) and (11.40) yields

$$\begin{Bmatrix} w_1 \\ \theta_{x1} \\ \theta_{y1} \\ w_2 \\ \theta_{x2} \\ \theta_{y2} \\ w_3 \\ \theta_{x3} \\ \theta_{y3} \end{Bmatrix} = \begin{bmatrix} 1 & x_1 & y_1 & x_1^2 & x_1 y_1 & y_1^2 & x_1^3 & x_1^2 y_1 + x_1 y_1^2 & y_1^3 \\ 0 & 1 & 0 & 2x_1 & y_1 & 0 & 3x_1^2 & 2x_1 y_1 + y_1^2 & 0 \\ 0 & 0 & 1 & 0 & x_1 & 2y_1 & 0 & x_1^2 + 2x_1 y_1 & 3y_1^2 \\ 1 & x_2 & y_2 & x_2^2 & x_2 y_2 & y_2^2 & x_2^3 & x_2^2 y_2 + x_2 y_2^2 & y_2^3 \\ 0 & 1 & 0 & 2x_2 & y_2 & 0 & 3x_2^2 & 2x_2 y_2 + y_2^2 & 0 \\ 0 & 0 & 1 & 0 & x_2 & 2y_2 & 0 & x_2^2 + 2x_2 y_2 & 3y_2^2 \\ 1 & x_3 & y_3 & x_3^2 & x_3 y_3 & y_3^2 & x_3^3 & x_3^2 y_3 + x_3 y_3^2 & y_3^3 \\ 0 & 1 & 0 & 2x_3 & y_3 & 0 & 3x_3^2 & 2x_3 y_3 + y_3^2 & 0 \\ 0 & 0 & 1 & 0 & x_3 & 2y_3 & 0 & x_3^2 + 2x_3 y_3 & 3y_3^2 \end{bmatrix} \begin{Bmatrix} \alpha_1 \\ \alpha_2 \\ \alpha_3 \\ \alpha_4 \\ \alpha_5 \\ \alpha_6 \\ \alpha_7 \\ \alpha_8 \\ \alpha_9 \end{Bmatrix} \tag{11.42}$$

or in a more compact form as

$$\{a_e\} = [A]\{\alpha\} \tag{11.43}$$

Inverting Equation (11.43) and substituting in Equation (11.39) yields

$$w(x, y) = \begin{bmatrix} 1 & x & y & x^2 & xy & y^2 & x^3 & (x^2 y + xy^2) & y^3 \end{bmatrix} [A]^{-1}\{a_e\} \tag{11.44}$$

or simply as

$$w(x, y) = [N(x, y)]\{a_e\} \tag{11.45}$$

with

$$[N(x, y)] = \begin{bmatrix} 1 & x & y & x^2 & xy & y^2 & x^3 & (x^2y + xy^2) & y^3 \end{bmatrix} [A]^{-1} \tag{11.46}$$

Expression (11.45) is the equivalent nodal approximation. Substituting in Equation (11.3) for the in-plane strains yields

$$\{\epsilon\} = [B]\{a_e\} \tag{11.47}$$

with

$$= -z \begin{bmatrix} 0 & 0 & 0 & 2 & 0 & 0 & 6x & 2y & 0 \\ 0 & 0 & 0 & 0 & 0 & 2 & 0 & 2x & 6y \\ 0 & 0 & 0 & 0 & 2 & 0 & 0 & 4(x+y) & 0 \end{bmatrix} [A]^{-1} \tag{11.48}$$

The stiffness matrix is obtained in the usual manner as

$$[K_e] = \int_{A_e} \int_z [B]^T [D][B] dz dA \tag{11.49}$$

11.4.1.2 Rectangular Element

Consider the rectangular plate element shown in Figure 11.7. The element has four nodes and 12 dof in total. A trial function for the unknown $w(x, y)$ will contain 12 parameters:

$$w(x, y) = \alpha_1 + \alpha_2 x + \alpha_3 y + \alpha_4 x^2 + \alpha_5 xy + \alpha_6 y^2 + \alpha_7 x^3 + \alpha_8 x^2 y + \alpha_9 xy^2$$
$$+ \alpha_{10} y^3 + \alpha_{11} yx^3 + \alpha_{12} xy^3 \tag{11.50}$$

$$\frac{\partial w}{\partial x} = \alpha_2 + 2\alpha_4 x + \alpha_5 y + 3\alpha_7 x^2 + 2\alpha_8 xy + \alpha_9 y^2 + 3\alpha_{11} x^2 y + \alpha_{12} y^3 \tag{11.51}$$

$$\frac{\partial w}{\partial y} = \alpha_3 + \alpha_5 x + 2\alpha_6 y + \alpha_8 x^2 + 2\alpha_9 yx + 3\alpha_{10} y^2 + \alpha_{11} x^3 + 3\alpha_{12} y^2 x \tag{11.52}$$

It can be seen that both $w(x, y)$ and its derivatives are defined by cubic polynomials. As a cubic is uniquely defined by four constants, the two end values of the displacements and slopes will therefore define the displacements uniquely along any boundary. However, this is not the case for the derivatives, since only two end values of the slopes exist, the cubic is not specified uniquely. And in general a discontinuity of normal slope will occur. The function is therefore called "non-conforming."

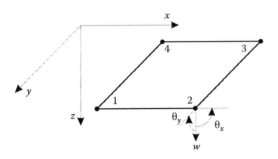

FIGURE 11.7 Four-node rectangular plate bending element.

11.4.2   FINITE ELEMENT FORMULATION FOR THICK PLATES

More recently, formulated elements use the Mindlin plate theory. From a FEM point of view, this is very important since the rotations θ_y and θ_x are independent from $w(x, y)$, therefore not requiring C^1 continuity. This greatly simplifies the formulation. As a result, C^0 isoparametric shape functions can be used for the thick plate element formulation. The transverse displacement and slopes can be interpolated independently using the same shape functions as

$$w = \sum_{i=1}^{n} N_i(\xi, \eta) w_i \tag{11.53}$$

$$\theta_x = \sum_{i=1}^{n} N_i(\xi, \eta) \theta_{xi} \tag{11.54}$$

$$\theta_y = \sum_{i=1}^{n} N_i(\xi, \eta) \theta_{yi} \tag{11.55}$$

where n represents the number of nodes.

There are three degrees of freedom at each node: w, θ_x, and θ_y. The curvatures in Equations (11.32) and (11.33) are defined in terms of the nodal unknowns as

$$\{\chi\}_B = [L_B][N]\{a\} \tag{11.56}$$

with

$$[L_B] = \begin{bmatrix} 0 & \dfrac{\partial}{\partial x} & 0 \\[2ex] 0 & 0 & \dfrac{\partial}{\partial y} \\[2ex] 0 & \dfrac{\partial}{\partial y} & \dfrac{\partial}{\partial x} \end{bmatrix} \tag{11.57}$$

and

$$\{\chi\}_S = [L_S][N]\{a\} \tag{11.58}$$

with

$$[L_S] = \begin{bmatrix} -\dfrac{\partial}{\partial y} & 0 & 1 \\[2ex] -\dfrac{\partial}{\partial x} & 1 & 0 \end{bmatrix} \tag{11.59}$$

The matrix of the shape functions $[N]$ is given as

$$[N] = \begin{bmatrix} N_1 & 0 & 0 & | & \dots & \dots & \dots & | & N_n & 0 & 0 \\ 0 & N_1 & 0 & | & \dots & \dots & \dots & | & 0 & N_n & 0 \\ 0 & 0 & N_1 & | & \dots & \dots & \dots & | & 0 & 0 & N_n \end{bmatrix} \tag{11.60}$$

and the vector of nodal unknowns as

$$\{a\} = \begin{bmatrix} w_1 & \theta_{x1} & \theta_{y1} & | & \dots & \dots & \dots & | & w_n & \theta_{xn} & \theta_{yn} \end{bmatrix}^T \tag{11.61}$$

Equations (11.56) and (11.58) can be rewritten in the usual manner as

$$\{\chi\}_B = [B_B]\{a\} \tag{11.62}$$

with

$$[B_B] = \begin{bmatrix} 0 & \dfrac{\partial N_1}{\partial x} & 0 & | & \cdots & \cdots & \cdots & | & 0 & \dfrac{\partial N_n}{\partial x} & 0 \\[2mm] 0 & 0 & \dfrac{\partial N_1}{\partial y} & | & \cdots & \cdots & \cdots & | & 0 & 0 & \dfrac{\partial N_n}{\partial y} \\[2mm] 0 & \dfrac{\partial N_1}{\partial y} & \dfrac{\partial N_1}{\partial x} & | & \cdots & \cdots & \cdots & | & 0 & \dfrac{\partial N_n}{\partial y} & \dfrac{\partial N_n}{\partial x} \end{bmatrix} \tag{11.63}$$

and

$$\{\chi\}_S = [B_S]\{a\} \tag{11.64}$$

with

$$[B_S] = \begin{bmatrix} -\dfrac{\partial N_1}{\partial y} & 0 & N_1 & | & \cdots & \cdots & \cdots & | & -\dfrac{\partial N_n}{\partial y} & 0 & N_n \\[2mm] -\dfrac{\partial N_1}{\partial x} & N_1 & 0 & | & \cdots & \cdots & \cdots & | & -\dfrac{\partial N_n}{\partial x} & N_n & 0 \end{bmatrix} \tag{11.65}$$

It follows therefore that the stiffness matrix is split into two matrices: one to model bending and the other to model shear:

$$[K_e] = [K_B] + [K_S] = \int_{A_e} [B_B]^T [D_B][B_B]\, dA + \kappa \int_{A_e} [B_S]^T [D_S][B_S]\, dA \tag{11.66}$$

Remark: It is important to note that the shear stiffness $[K_S]$ is a function of h since $[D_S]$ (Equation (11.35)) is a function of h, and the bending stiffness $[K_B]$ is a function of h^3 since $[D_B]$ (Equations (11.27) and (11.34)) is a function of h^3. A consequence of this is that the shear energy dominates as the thickness of the plate becomes very small compared to its side length. This is called shear locking. One way of resolving this problem is to under integrate the shear energy term. For example, if the 8 node quadrilateral is used, then the bending energy is to be integrated with 3×3 Gauss points, while the shear energy is to be integrated only with a 2×2 rule.

11.5 BOUNDARY CONDITIONS

Given a rectangular plate with dimensions $a \times b \times h$ as shown in Figure 11.8. The governing equation of the bending behavior of a thin plate is described by a fourth-order differential equation. Hence, two boundary conditions have to be specified on each edge.

11.5.1 SIMPLY SUPPORTED EDGE

If the edge $x = a$ is simply supported, the deflection $w_{(x=a)}$ along this edge must be zero. At the same time, the edge can rotate freely with respect to the support, that is, there is no bending moment M_{xx} along this edge:

$$(w)_{x=a} = 0 \quad \text{and} \quad (M_{xx})_{x=a} = -D_r\left(\frac{\partial^2 w}{\partial x^2} + \nu\frac{\partial^2 w}{\partial y^2}\right) = 0 \tag{11.67}$$

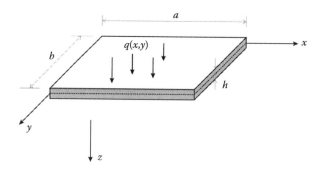

FIGURE 11.8   Plate boundary conditions.

The condition $w_{(x=a)} = 0$ along the edge $x = a$ means also that $\frac{\partial w}{\partial y} = \frac{\partial^2 w}{\partial y^2} = 0$ along that edge. The boundary conditions for a simply supported edge may also be written as

$$(w)_{x=a} = 0 \quad \text{and} \quad \frac{\partial^2 w}{\partial x^2} = 0 \tag{11.68}$$

The first boundary condition in (11.68) is a kinematic boundary condition and the second one is a dynamic or natural boundary condition. In FEA, only the kinematic boundary conditions needs to be imposed, the natural boundary condition is incorporated in the principle of virtual work.

11.5.2   BUILT-IN OR CLAMPED EDGE

If the edge $x = a$ is built-in or clamped, along this edge the deflection and the slope of the middle plane must be zero; that is,

$$(w)_{x=a} = 0 \quad \text{and} \quad \frac{\partial w}{\partial x} = 0 \tag{11.69}$$

These boundary conditions are both kinematic and need to be imposed.

11.5.3   FREE EDGE

If the edge $x = a$ is entirely free, it is natural to assume that along this edge there are no bending and twisting moments, and also no shear force; that is,

$$(M_{xx})_{x=a} = (M_{xy})_{x=a} = (Q_{xz})_{x=a} = 0 \tag{11.70}$$

Within the thin plate theory, these three conditions are combined into two conditions, namely,

$$(M_{xx})_{x=a} = 0 \quad \text{and} \quad \left(Q_{xz} + \frac{M_{xy}}{\partial y}\right)_{x=a} = 0 \tag{11.71}$$

The term $Q_{xz} + \frac{M_{xy}}{\partial y}$ is called the "effective shear force" or the "Kirchhoff shear force." The boundary conditions at a free edge are all natural and do not to be imposed.

11.6   COMPUTER PROGRAM FOR THICK PLATES USING THE 8-NODE QUADRILATERAL

11.6.1   MAIN PROGRAM: THICK_PLATE_Q8.m

Consider the simply supported square plate shown in Figure 11.9, which has an exact analytical solution [5]. Find the deflection at the center if the plate is subjected to a concentrated load of 1000 lb

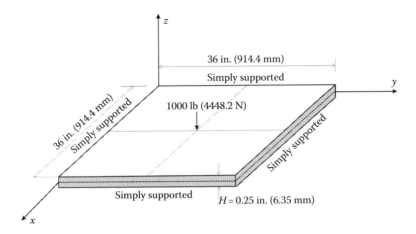

FIGURE 11.9 Simply supported plate on all edges.

(4448.2 N) at the center. The size of the plate is 36×36 in.^{2} (914.4×914.4 mm^2) and the thickness is 0.25 in. (6.35) mm. It is made of steel, $E = 30 \times 10^6$ psi (206843 MPa) and $\nu = 0.3$. The main program **Thick_plate_Q8.m** is listed next.

```
% THIS PROGRAM USES AN 8-NODDED QUADRILATERAL ELEMENT FOR THE LINEAR ELASTIC
% STATIC ANALYSIS OF A THICK PLATE IN BENDING
%
% Make these variables global so they can be shared by other functions
%
clc
clear all
%
global nnd nel nne nodof eldof n ngpb ngps global geom connec deeb
dees nf load dim
%
 format long g
%
% To cchange the size of the problem or change the elastic properties
%     ALTER the PlateQ8_input_module.m
%
dim = 2;
nne = 8;
nodof = 3;
eldof = nne*nodof;
%
% Plate_Q8_input_module
Length = 18.;      % Length of the in x-direction
Width =  18.;      % Width of the model in y-direction
NXE = 9;           % Number of rows in the x direction
NYE = 9;           % Number of rows in the y direction
dhx = Length/NXE; % Element size in the x direction
dhy = Width/NYE;  % Element size in the y direction
X_origin = 0. ;    % x origin of the global coordinate system
Y_origin = 0. ;    % y origin of the global coordinate system
%
thick = 0.25;  % Thickness of plate
ngpb = 3;             % number of Gauss points bending
ngps = 2;             % number of Gauss points for shear
%
Q8_mesh       % Generate the mesh
%
E = 30.e+6;      % Elastic modulus in kN/m2
vu = 0.3;        % Poisson's ratio
%
% Form the matrix of elastic properties
```

```
%
deeb=formdeeb(E,vu,thick); % Matrix of elastic properties for plate bending
dees=formdees(E,vu,thick); % Matrix of elastic properties for plate shear
%
% Boundary conditions
%
nf = ones(nnd, nodof);    % Initialize the matrix nf to 1
%
for i=1:nnd
if geom(i,1) == 0
    nf(i,1) = 0 ;    % Restrain in direction w
    nf(i,3) = 0 ;    % Restrain rotation theta_y (around x)
elseif geom(i,2) == 0
    nf(i,1) = 0. ;    % Restrain displacement  w
    nf(i,2) = 0. ;    % Restrain rotation theta_x (around y)
elseif geom(i,1) == Length
    nf(i,2) = 0. ;    % Restrain rotation theta_x (around y)
elseif geom(i,2) == Width
    nf(i,3) = 0. ;    % Restrain rotation theta_y (around x)
end
end
%
% Counting of the free degrees of freedom
%
n=0;
for i=1:nnd
    for j=1:nodof
        if nf(i,j) ~= 0
            n=n+1;
            nf(i,j)=n;
        end
    end
end
disp ('Nodal freedom')
nf
disp ('Total number of active degrees of freedom')
n
%
% loading
%
load = zeros(nnd, 3);
%
for i=1:nnd
    if geom(i,1) == Length && geom(i,2) == Width
    load(i,1) = - 1000/4; % Vertical load of 250 lb on the center node
    end
end
%
%
%
%%%%%%%%%%%%%%%%%%%%%%%%%%%% End of input%%%%%%%%%%%%%%%%%%%%%%%%%%%%%%%%%%%
%
% Assemble the global force vector
%
fg=zeros(n,1);
for i=1: nnd
    for j=1:nodof
    if nf(i,j) ~= 0
        fg(nf(i,j))= load(i,j);
    end
    end
end
%
% Form the matrix containing the abscissas and the weights of Gauss points
%
sampb=gauss(ngpb); samps=gauss(ngps);
%
```

```
% Numerical integration and assembly of the global stiffness matrix
%
%  initialize the global stiffness matrix to zero
kk = zeros(n, n);
%
for i=1:nel
    [coord,g] = platelem_q8(i) ;    % coordinates of the nodes of element i,
                                    % and its steering vector
    keb=zeros(eldof,eldof) ;        % Initialize the element bending
                                    % stiffness matrix to zero
    kes=zeros(eldof,eldof) ;        % Initialize the element Shear
                                    %  stiffness matrix to zero

    %
    % Integrate element bending stiffness and assemble it in global matrix
    %
    for ig=1: ngpb
        wi = sampb(ig,2);
    for jg=1: ngpb
        wj=sampb(jg,2);
        [der,fun] = fmquad(sampb, ig,jg); % Derivative of shape functions
                                          %  in local coordinates
        jac=der*coord;                    % Compute Jacobian matrix
        d=det(jac);                       % Compute the determinant of
                                          % Jacobian matrix
        jac1=inv(jac);                    % Compute inverse of the Jacobian
        deriv=jac1*der;                   % Derivative of shape functions
                                          % in global coordinates
        beeb=formbeeb(deriv,nne,eldof);   %  Form matrix [B]
        keb=keb + d*wi*wj*beeb'*deeb*beeb; % Integrate stiffness matrix
    end
    end
    kk=form_kk(kk,keb, g);                % assemble global stiffness matrix
    %
    % Integrate element Shear stiffness and assemble it in global matrix
    %
    for ig=1: ngps
        wi = samps(ig,2);
    for jg=1: ngps
        wj=samps(jg,2);
        [der,fun] = fmquad(samps, ig,jg); % Derivative of shape functions
                                          % in local coordinates
        jac=der*coord;                    % Compute Jacobian matrix
        d=det(jac);                       % Compute determinant of
                                          % Jacobian matrix
        jac1=inv(jac);                    % Compute inverse of the
                                          % Jacobian
        deriv=jac1*der;                   % Derivative of shape functions
                                          % in global coordinates
        bees=formbees(deriv,fun,nne,eldof);       %  Form matrix [B]
        kes=kes + (5/6)*d*wi*wj*bees'*dees*bees; % Integrate stiffness matrix
    end
    end
    kk=form_kk(kk,kes, g);                % assemble global stiffness matrix
end
%
%
%%%%%%%%%%%%%%%%%%%%%%%  End of assembly %%%%%%%%%%%%%%%%%%%%%%%%%%%%%%%%%%%
%
%
delta = kk\fg            % solve for unknown displacements

format short e
disp('node         w_disp        x_slope          y_slope ')      %
for i=1: nnd                                    %
    if nf(i,1) == 0                             %
        w_disp =0.;                             %
    else
```

```
            w_disp = delta(nf(i,1));              %
      end
%
      if nf(i,2) == 0                             %
            x_slope = 0.;                          %
      else
            x_slope = delta(nf(i,2));             %
      end
 %
      if nf(i,3) == 0                             %
            y_slope = 0.;                          %
      else
            y_slope = delta(nf(i,3));             %
      end

disp([i  w_disp   x_slope  y_slope])  % Display displacements of each node
DISP(i,:) = [ w_disp   x_slope  y_slope];
end
%
%
%
ngp=1;                                 % Calculate moments and shear forces
                                       % the center of each element
samp=gauss(ngp);
%
for i=1:nel
  [coord,g] =  platelem_q8(i);  % coordinates of the nodes of element i,
                                % and its steering vector
  eld=zeros(eldof,1);                  % Initialize element displacement to zero
  for m=1:eldof                        %
      if g(m)==0                       %
            eld(m)=0.;                 %
      else                             %
            eld(m)=delta(g(m));  % Retrieve element displacement from the
                                 % global displacement vector
      end
  end
  %
  for ig=1: ngp
        wi = samp(ig,2);
     for jg=1: ngp
        wj=samp(jg,2);
        [der,fun] = fmquad(samp, ig,jg);  % Derivative of shape functions
                                          %  in local coordinates
        jac=der*coord;                    % Compute Jacobian matrix
        d=det(jac);                       % Compute the determinant of
                                          % Jacobian matrix
        jac1=inv(jac);                    % Compute inverse of the Jacobian
        deriv=jac1*der;                   % Derivative of shape functions
                                          % in global coordinates
                                          %
        beeb=formbeeb(deriv,nne,eldof);   %  Form matrix [B_b]
        chi_b = beeb*eld  ;               % compute bending curvatures
        Moment = deeb*chi_b ;             % Compute moments
        bees=formbees(deriv,fun,nne,eldof); %  Form matrix [B_s]
        chi_s = bees*eld  ;               % compute shear curvatures
        Shear = dees*chi_s ;              % Compute shera forces
    end
  end
  Element_Forces(i,:)=[Moment' Shear'];
end
%
W = DISP(:,1);
[MX, MY,  MXY, QX, QY] = Forces_at_nodes_plate(Element_Forces);
%
cmin = min(W);
cmax = max(W);
```

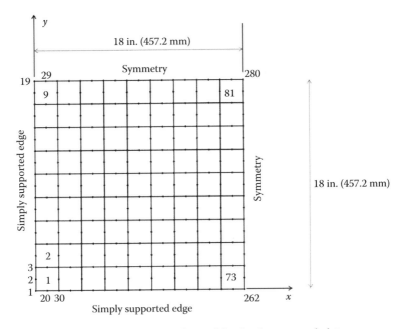

FIGURE 11.10   Finite element mesh of one quadrant of the simply supported plate.

```
caxis([cmin cmax]);
patch('Faces',
connec, 'Vertices', geom, 'FaceVertexCData',W,...
                        'Facecolor','interp','Marker','.');
colorbar;
```

Because of symmetry only one quadrant of the plate is discretized. The finite element mesh is shown in Figure 11.10 and generated using the function **mesh_Q8**.m. Both the nodes and the elements are numbered in the y-direction. In total there are 282 nodes and 81 elements.

11.6.2 DATA PREPARATION

11.6.2.1 Stiffness Matrices

Note two different integrations schemes are used: one consisting of a 3×3 rule, $ngpb = 3$, to integrate the flexural matrix *deeb*, and the other consisting of a 2×2 rule, $ngps = 2$, to integrate the shear stiffness matrix *dees*. The matrices are respectively formed with the functions *formdeeb.m* and *formdees.m* listed in Appendix A.

11.6.2.2 Boundary Conditions

The boundary conditions are given as follows:

$$Edge \quad x = 0 \quad w = 0 \quad \theta_y = 0 \qquad Edge \quad x = 18 \text{ in.} \quad \theta_x = 0$$
$$Edge \quad y = 0 \quad w = 0 \quad \theta_x = 0 \qquad Edge \quad y = 18 \text{ in.} \quad \theta_y = 0$$

They are introduced as follows:

- For all the nodes located at $x = 0$, restrain the degree of freedom No. 1 corresponding to the vertical translation, and the degree of freedom No. 3 corresponding to the rotation θ_y around the axis x.

- For all the nodes located at $y = 0$, restrain the degree of freedom No. 1 corresponding to the vertical translation, and the degree of freedom No. 2 corresponding to the rotation θ_x around the axis y.
- For all the nodes located at $x = Length$, the length of the quarter plate, that is, restrain the degree of freedom No. 2 corresponding to the rotation θ_x around the axis y.
- For all the nodes located at $y = Width$, the width of the quarter plate, that is, restrain the degree of freedom No. 3 corresponding to the rotation θ_y around the axis x.

11.6.2.3 Loading

A quarter of the 1000. lb load is applied at node 282 in the opposite z-direction. This node is located by its coordinates.

File: Plate_Q8_input_module.m

```
%
%%%%%%%%%%%%%%%%%%%%%%%%%%%%% Beginning of data input %%%%%%%%%%%%%%%%%%%%%%%%%
%
global nnd nel nne nodof eldof  n ngpb ngps
global geom connec deeb dees nf  load dim

%
dim=2;                        % Dimension
nnd = 21 ;                    % Number of nodes:
nel = 4;                      % Number of elements:
nne = 8 ;                     % Number of nodes per element:
nodof =3;                     % Number of degrees of freedom per node
ngpb = 3;                      % number of Gauss points bending
ngps = 2;                     % number of Gauss points shear
eldof = nne*nodof;            % Number of degrees of freedom per element
%
% Thickness of the domain
thick = 0.25;
%
% Nodes coordinates x and y
geom = [0.0    18; ...         %   x and y coordinates of node 1
        0.0    13.5; ...       %   x and y coordinates of node 2
        0.0    9; ...          %   x and y coordinates of node 3
        0.0    4.5; ...        %   x and y coordinates of node 4
        0.0    0; ...          %   x and y coordinates of node 5
        4.5    18; ...         %   x and y coordinates of node 6
        4.5    9.; ...         %   x and y coordinates of node 7
        4.5    0; ...          %   x and y coordinates of node 8
        9      18; ...         %   x and y coordinates of node 9
        9      13.5; ...       %   x and y coordinates of node 10
        9       9; ...         %   x and y coordinates of node 11
        9      4.5; ...        %   x and y coordinates of node 12
        9       0.; ...        %   x and y coordinates of node 13
        13.5   18; ...         %   x and y coordinates of node 14
        13.5   9; ...          %   x and y coordinates of node 15
        13.5    0.; ...        %   x and y coordinates of node 16
        18     18; ...         %   x and y coordinates of node 17
        18     13.5; ...       %   x and y coordinates of node 18
        18     9; ...          %   x and y coordinates of node 19
        18     4.5; ...        %   x and y coordinates of node 20
        18      0.];           %   x and y coordinates of node 21
%
disp ('Nodes X-Y coordinates')
geom
%
% Element connectivity
connec= [  1    2    3    7    11   10    9     6;...   % Element 1
           3    4    5    8    13   12   11     7;...   % Element 2
           9   10   11   15   19   18   17    14;...   % Element 3
          11   12   13   16   21   20   19    15];     % Element 4
```

```
disp ('Elements connectivity')
connec
%
% Material properties
%
E=30.e+6; vu=0.3;              % Young's modulus and Poisson's ration
%
% Form the matrix of elastic properties
%
deeb=formdeeb(E,vu,thick); % Matrix of elastic properties for plate bending
dees=formdees(E,vu,thick); % Matrix of elastic properties for plate shear
%
% Boundary conditions
%
nf = ones(nnd, nodof);                % Initialize the matrix nf to 1
nf(1,1) = 0; nf(1,3)=0;
nf(2,1) = 0; nf(2,3)=0;
nf(3,1) = 0; nf(3,3)=0;
nf(4,1) = 0; nf(4,3)=0;
nf(5,1) = 0; nf(5,2)=0; nf(5,3)=0;
nf(6,3)=0;
nf(8,1)=0; nf(8,2)=0;
nf(9,3)=0;
nf(13,1)=0; nf(13,2)=0;
nf(14,3)=0
nf(16,1)=0; nf(16,2)=0;
nf(17,2)=0;nf(17,3)=0;
nf(18,2)=0;
nf(19,2)=0;
nf(20,2)=0;
nf(21,1)=0;nf(21,2)=0;
%
% Counting of the free degrees of freedom
%
n=0;
for i=1:nnd
    for j=1:nodof
        if nf(i,j) ~= 0
            n=n+1;
            nf(i,j)=n;
        end
    end
end
disp ('Nodal freedom')
nf
disp ('Total number of active degrees of freedom')
n
%
% loading
%
load = zeros(nnd, 3);
load(17,1) = 1000/4;          % Vertical load of 250 lb on node 17
%
% End input
```

11.6.2.4 Numerical Integration of the Stiffness Matrix

The stiffness matrix is given by Equation (11.66). For each element, it is computed as follows:

1. For every element $i = 1$ to *nel*
2. Retrieve the coordinates of its nodes **coord(nne, 2)** and its steering vector **g(eldof)** using the function *platelem_q8.m*
3. Initialize the stiffness matrices to zero

a. Loop over the Gauss points $ig = 1$ to *ngpb*

b. Retrieve the weight **wi** as **sampb(ig, 2)**

     i. Loop over the Gauss points $jg = 1$ to *ngpb*

     ii. Retrieve the weight **wj** as **sampb(jg, 2)**

     iii. Use the function *fmquad.m* to compute the shape functions, vector **fun**, and their derivatives, matrix **der**, in local coordinates, $\xi = $ **sampb(ig, 1)** and $\eta = $ **sampb(jg, 1)**

     iv. Evaluate the Jacobian **jac = der ∗ coord**

     v. Evaluate the determinant of the Jacobian as **d = det(jac)**

     vi. Compute the inverse of the Jacobian as **jac1 = inv(jac)**

     vii. Compute the derivatives of the shape functions with respect to the global coordinates x and y as **deriv = jac1 ∗ der**

     viii. Use the function *formbeeb.m* to form the strain matrix **beeb**

     ix. Compute the stiffness matrix as **keb = keb + d ∗ wi ∗ wj ∗ beeb′ ∗ deeb ∗ beeb**

4. Assemble the stiffness matrix **keb** into the global matrix **kk**

a. Loop over the Gauss points $ig = 1$ to *ngps*

b. Retrieve the weight **wi** as **samps(ig, 2)**

     i. Loop over the Gauss points $jg = 1$ to *ngps*

     ii. Retrieve the weight **wj** as **samps(jg, 2)**

     iii. Use the function *fmquad.m* to compute the shape functions, vector **fun**, and their derivatives, matrix **der**, in local coordinates, $\xi = $ **samps(ig, 1)** and $\eta = $ **samps(jg, 1)**.

     iv. Evaluate the Jacobian **jac = der ∗ coord**

     v. Evaluate the determinant of the Jacobian as **d = det(jac)**

     vi. Compute the inverse of the Jacobian as **jac1 = inv(jac)**

     vii. Compute the derivatives of the shape functions with respect to the global coordinates x and y as **deriv = jac1 ∗ der**

     viii. Use the function *formbees.m* to form the strain matrix **bees**

     ix. Compute the stiffness matrix as **kes = kes + d ∗ wi ∗ wj ∗ bees′ ∗ dees ∗ bees**

5. Assemble the stiffness matrix **kes** into the global matrix **kk**

The functions *formbeeb.m* and *formbees.m*, which form the flexural and shear strain matrices, are listed in Appendix A.

11.6.3 Results

11.6.3.1 Determination of the Resulting Moments and Shear Forces

Once the global equations are solved or the global displacement, for each element we retrieve its nodal displacements and calculate the resulting moments and shear forces at its center. For such we use only one Gauss point as detailed next:

1. For every element $i = 1$ to *nel*

2. Retrieve the coordinates of its nodes **coord(nne, 2)** and its steering vector **g(eldof)** using the function *platelem_q8.m*

3. Retrieve its vector of nodal displacements **eld(eldof)**

a. Loop over the Gauss points $ig = 1$ to *ngp*

b. Retrieve the weight **wi** as **samp(ig, 2)**

     i. Loop over the Gauss points $jg = 1$ to *ngp*

     ii. Retrieve the weight **wj** as **samp(jg, 2)**

iii. Use the function *fmquad.m* to compute the shape functions, vector **fun**, and their derivatives, matrix **der**, in local coordinates, $\xi = $ **samp(ig, 1)** and $\eta = $ **samp(jg, 1)**

iv. Evaluate the Jacobian **jac** = **der** ∗ **coord**

v. Evaluate the determinant of the Jacobian as **d** = **det(jac)**

vi. Compute the inverse of the Jacobian as **jac1** = **inv(jac)**

vii. Compute the derivatives of the shape functions with respect to the global coordinates *x* and *y* as **deriv** = **jac1** ∗ **der**

viii. Use the function *formbeeb.m* to form the strain matrix **beeb**

ix. Compute the flexural curvature $\chi_b = beeb * eld$ and the corresponding moments as *Moment = deeb* ∗ χ_b

x. Use the function *formbees.m* to form the strain matrix **bees**

xi. Compute the shear curvature $\chi_s = bees * eld$ and the corresponding shear forces as *Shear = dees* ∗ χ_s

4. Store the moments and shear forces in the array *Element_Forces(nel, 5)*

Using the data stored in the array *Element_Forces(nel, 5)*, the function **Forces_at_nodes_plate.m** calculates the moments and shear forces at the nodes, and returns them as arrays for plotting using the MATLAB® function **patch**.

11.6.3.2 Contour Plots

Figure 11.11 shows the contour plot of the vertical displacement. The program predicts a vertical displacement of −0.35239 in. at node 282, which is the center of the plate, that is very close to the exact solution of −0.35022 in. Figures 11.12 and 11.13 show the contour plots of the moments M_{xx} and M_{xy}. It is very interesting to note in Figure 11.13 that the corner of the plate tends to rise. Indeed, it is well known that the corners of a flat plate under transverse load have the tendency to rise when upward displacements are not restricted as shown in Figure 11.14.

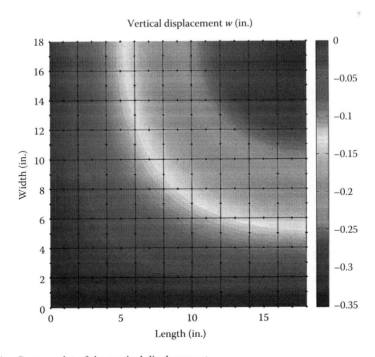

FIGURE 11.11   Contour plot of the vertical displacement.

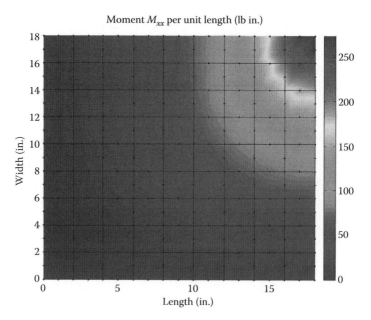

FIGURE 11.12   Contour plot of the moment M_{xx}.

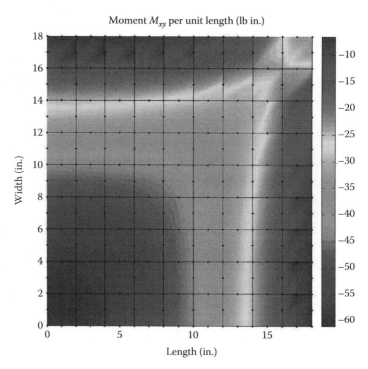

FIGURE 11.13   Contour plot of the moment M_{xy}.

11.7 ANALYSIS WITH ABAQUS

11.7.1 PRELIMINARY

Abaqus does not have plate elements as such. Instead it uses shell elements. In Abaqus, a plate is merely considered as a flat shell. A shell element can be considered as a sophisticated version of

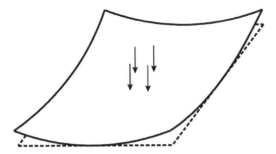

FIGURE 11.14   Lifting of corners of a plate.

a plate element that can carry in-plane forces. Abaqus offers two types of three-dimensional shell elements: conventional shell elements and continuum shell elements. Detailed descriptions of these elements can be found in the Abaqus manual and in Ref. [6].

In Abaqus, shell elements are named as in the following sections.

11.7.1.1   Three-Dimensional Shell Elements
S8R5W

- **S**, conventional stress/displacement shell; **SC**, continuum stress/displacement shell; **STRI**, triangular stress/displacement thin shell; **DS**, heat transfer shell
- **8**, number of nodes
- **R**, reduced integration (optional)
- **5**, number of degrees of freedom per node (optional)
- **W**, warping considered in small-strain formulation

11.7.1.2   Axisymmetric Shell Elements
SAX2T

- **S**, stress/displacement shell; **DS**, heat transfer shell
- **AX**, axisymmetric; **AXA**, axisymmetric with nonlinear, asymmetric deformation
- **2**, order of interpolation
- **T**, coupled temperature displacement

11.7.1.3   Thick versus Thin Conventional Shell

Before choosing a shell element in Abaqus, it is worthwhile to check whether it is suitable for thin shells only, thick shells only, or both.

The following elements are suitable for both: S3, S3R, S3RS, S4, S4R, S4RSW, SAX1, SAX2, SAX2T, SC6R, and SC8R. They include the transverse shear deformation, which becomes very small as the shell thickness decreases.

The following elements S8R and S8RT are only for use in thick shell problems.

Elements STRI3, S4R5, STRI65, S8R, S9R5, SAXA1n, and SAXA2n should not be used for thick shells where transverse shear deformation is important.

11.7.2   SIMPLY SUPPORTED PLATE

In this section, we will analyze the simply supported square plate shown in Figure 11.9. As before, we will only analyze a quarter for reasons of symmetry in both geometry and loading. We will use the S4R element, which is suitable for both thin and thick shells.

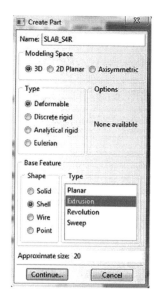

Start **Abaqus CAE**. Click on **Create Model Database**. On the main menu, click on **File** and set **Set Work Directory** to choose your working directory. Click on **Save As** and name the file **SLAB_S4R.cae**. On the left-hand-side menu, click on **Part** to begin creating the model. Name the part **SLAB_S4R**, check **3D**, check **Deformable** in the type. Choose **Shell** as the shape, and **Extrusion** for type. Enter an approximate size of 20 in. and click on **Continue** (Figure 11.15).

FIGURE 11.15   Creating the Slab_S4R Part.

In the sketcher menu, choose the **Create-Lines connected** icon to draw a straight line 18 in. long. In the prompt area in the bottom-left corner of the viewport window, click on **Sketch the section for the shell extrusion**. In the **Edit base extrusion** dialog box, enter 18 in. for depth and click **OK** (Figure 11.16).

FIGURE 11.16   Sketching the Slab_S4R Part.

Define a material named steel with an elastic modulus of 30000000 psi and a Poisson's ratio of 0.3. Next, click on **Sections** to create a section named **Slab_section_S4R**. In the **Category** check **Shell**, and in the **Type**, check **Homogeneous**. Click on **Continue**. In the **Edit Section** dialog box, enter 0.25 in. as the thickness. Check **Simpson** for thickness integration rule. Click on **OK** (Figure 11.17).

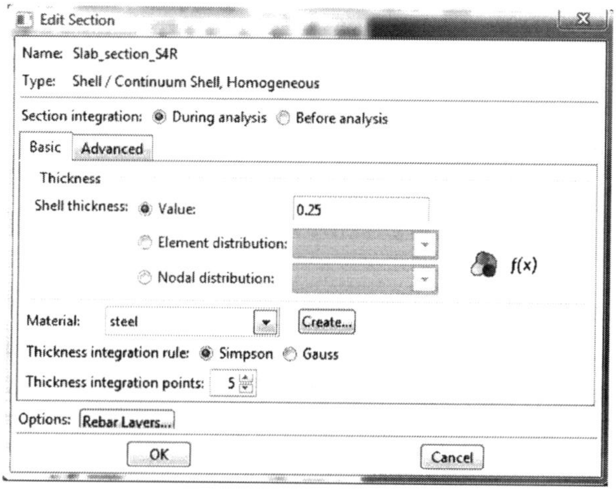

FIGURE 11.17   Creating a homogeneous shell section.

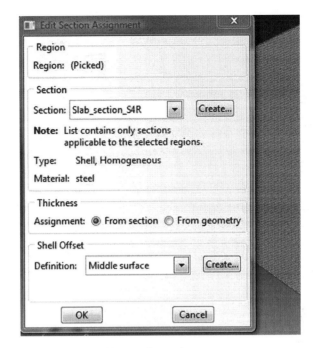

FIGURE 11.18   Editing section assignments.

Expand the menu under **Parts** and **SLAB_S4R** and double click on **Section Assignments**. With the mouse select the whole part. In the **Edit Section Assignments** dialog box, select **Slab_section_S4R**, and click on **OK** (Figure 11.18).

In the model tree, double click on **Mesh** under the **SLAB_S4R**. In the main menu, under **Mesh**, click on **Mesh Controls**. In the dialog box, check **Quad** for **Element shape** and **Structured** for **Technique**. Click on **OK**. Under **Mesh**, click on **Element Type**. In the dialog box, select **Standard** for element library, **Linear** for geometric order. In **Quad**, check **Reduced integration**. The description of the element **S4R: A 4-node doubly curved thin or thick shell, reduced integration, hour glass control, finite membrane strains** can be seen in the dialog box. Click on **OK** (Figure 11.19).

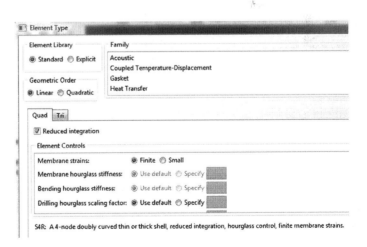

FIGURE 11.19   Mesh Controls and element type.

In the main menu, under **Seed**, click on **Part**. In the dialog box, enter 2 in. for **Approximate global size**. Click on **OK** and on **Done**. In the main menu, under **Mesh**, click on **Part**. In the prompt area, click on **Yes** (Figure 11.20).

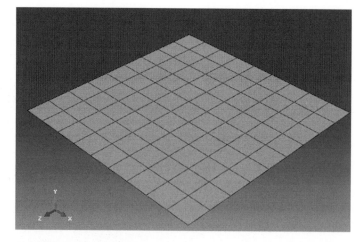

FIGURE 11.20 Mesh.

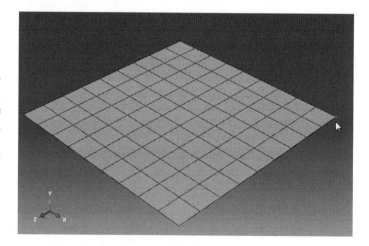

Under **Part**, in the left-hand-side menu, click on **Sets**. In the dialog box, name the set **Loaded_node**, and check **Node** for **Type**. Click on **Continue**. In the viewport, locate the central node as shown in Figure 11.21. Click on **Done**.

FIGURE 11.21 Creating a node set.

In the model tree, expand the **Assembly** and double click on **Instances**. Select **SLAB_S4R** for **Parts** and click **OK**. In the model tree, expand **Steps** and **Initial** and double click on **BC**. Name the boundary condition **Edge_X0**, select **Displacement/Rotation** for the type, and click on **Continue**. With the mouse select edge having $X = -9$ in. as shown in Figure 11.22 and click on **Done** in the prompt area. In the **Edit Boundary Condition**, check **U2, UR1, UR2**: no displacement is allowed along Y, and no rotations are allowed around X and Y. Click **OK**.

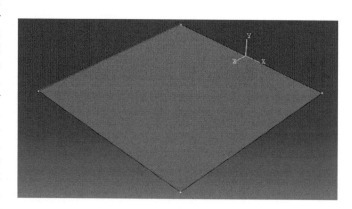

FIGURE 11.22 Imposing BC **Edge_X0** using geometry.

Repeat the procedure and this time name the boundary condition **Edge_Z18**, select **Displacement/Rotation** for the type, and click on **Continue**. With the mouse select edge having $Z = 18$ in. as shown in Figure 11.23 and click on **Done** in the prompt area. In the **Edit Boundary Condition**, check **U2, UR2, UR3**: no displacement is allowed along Y, and no rotations are allowed around Y and Z. Click **OK**.

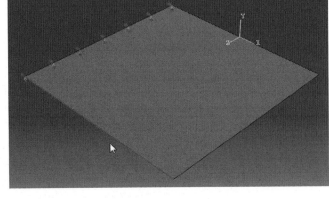

FIGURE 11.23   Imposing BC **Edge_Z18** using geometry.

Repeat the procedure and this time name the boundary condition **Edge_Z0**, select **Displacement/Rotation** for the type, and click on **Continue**. With the mouse select edge having $Z = 0$ in. as shown in Figure 11.24 and click on **Done** in the prompt area. In the **Edit Boundary Condition**, check **U3, UR1, UR2**: because of symmetry no displacement is allowed along Z, and no rotations are allowed around X and Y. Click **OK**.

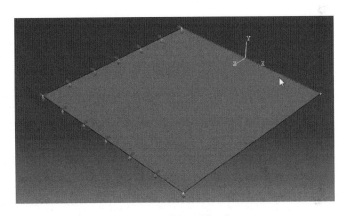

FIGURE 11.24   Imposing BC **Edge_Z0** using geometry.

Repeat the procedure and this time name the boundary condition **Edge_X9**, select **Displacement/Rotation** for the type, and click on **Continue**. With the mouse select edge having $X = 9$ in. as shown in Figure 11.25 and click on **Done** in the prompt area. In the **Edit Boundary Condition**, check **U1, UR2, UR3**: because of symmetry no displacement is allowed along X, and no rotations are allowed around Y and Z. Click **OK**.

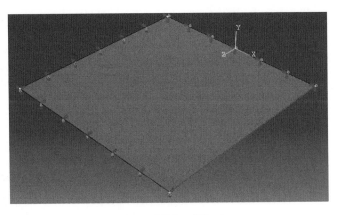

FIGURE 11.25   Imposing BC **Edge_X9** using geometry.

In the model tree, double click on **Steps**. Name the step **Apply_loads**. Set the procedure to **General** and select **Static, General**. Click on **Continue**. Give the step a description and click **OK**. In the model tree, under **steps**, and under **Apply_loads**, click on **Loads**. Name the load **Point_Load** and select **Concentrated Force** as the type. Click on **Continue**. In the bottom-right corner of the viewport, click on **Sets** and select **SLAB_S4R-1.Loaded_node**. In the **Edit load** dialog box, enter −250, a quarter of the load, for **CF2**. Click **OK** (Figure 11.26).

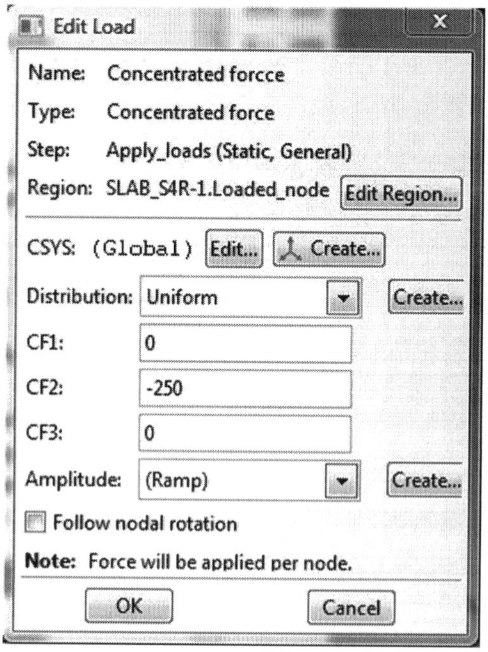

FIGURE 11.26 Imposing a concentrated force using a node set.

In the model tree, expand the **Field Output Requests** and then double click on **F-Output-1**. **F-Output-1** is the default and is automatically generated when creating the step. Uncheck the variables **Contact** and select any other variable you wish to add to the field output. Click on **OK**. Under **Analysis**, right click on **Jobs** and then click on **Create**.

In the **Create Job** dialog box, name the job **SLAB_S4R** and click on **Continue**. In the **Edit Job** dialog box, enter a description for the job. Check **Full analysis**, select to run the job in **Background**, and check to start it **immediately**. Click **OK**. Expand the tree under **Jobs**, right click on **SLAB_S4R**. Then, click on **Submit**. If you get the following message **SLAB_S4R completed successfully** in the bottom window, then your job is free of errors and was executed properly. Under the top menu, in the **Module** scroll to **Visualization**, and click to load **Abaqus Viewer**. On the main menu, under **File**, click **Open**, navigate to your working directory, and open the file **SLAB_S4R.odb**. It should have the same name as the job you submitted. Click on the **Common options** icon to display the **Common Plot options** dialog box. Under **labels**, check **Show Element labels** and **Show Node labels** to display elements and nodes' numbering. Click on the icon **Plot Contours on both shapes** to display the deformed shape of the beam. Under the main menu, select **U** and **U2** to plot the vertical displacement. It can be seen that the displacement of center of the plate is equal to −**0.351** in., which is very close to the analytical solution (Figure 11.27).

11.7.3 THREE-DIMENSIONAL SHELLS

In this section, we will show some more features of modeling with Abaqus. We will analyze a castellated beam as an assembly of three-dimensional shell elements. Castellated beams such as the one shown in Figure 11.28 are widely used in the steel construction industry. They are fabricated from standard universal beam sections. The beam is initially split along its length in a zigzag cut. The two halves of the beam are then separated, displaced by one profile to join the peaks, and welded together to increase the depth of the beam.

From a universal beam section such as the one shown in Figure 11.29, we will make a castellated beam whose cross section is shown in Figure 11.30. Notice that we will only model

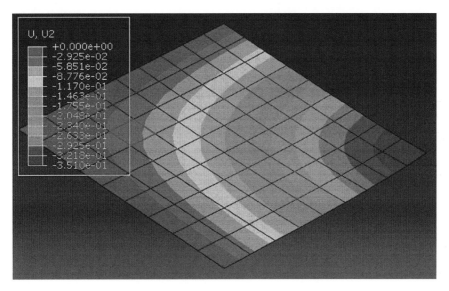

FIGURE 11.27   Plotting displacements on deformed shape.

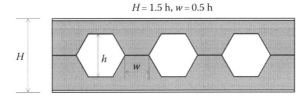

FIGURE 11.28   Castellated beam.

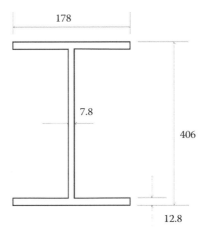

FIGURE 11.29   Base profile.

the middle plane as the behavior of a conventional shell element is described by that of its middle plane.

Figure 11.31 shows the castellated beam over a length of 12 m. There are 19 hexagons through the length spaced at 203 mm. The beam will be fixed at both ends and subject to uniformly distributed load of 178 kN/m, as shown in Figure 11.32.

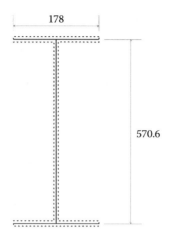

FIGURE 11.30   Castellated beam profile.

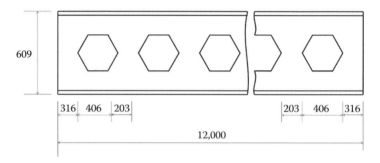

FIGURE 11.31   Geometrical details of the castellated beam.

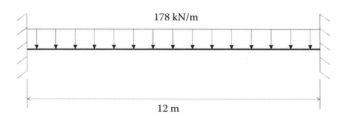

FIGURE 11.32   Loading and boundary conditions.

Start **Abaqus CAE**. Click on **Create Model Database**. On the main menu, click on **File** and set **Set Work Directory** to choose your working directory. Click on **Save As** and name the file **Castellated_beam.cae**. On the left-hand-side menu, click on **Part** to begin creating the model. Name the part **Castellated_beam**, check **3D**, check **Deformable** in the type. Choose **Shell** as the shape and **Extrusion** for type. Enter an approximate size of 1000 mm and click on **Continue**. In the sketcher menu, choose the **Create Lines: connected** icon to begin drawing the profile of the beam. Draw an I profile as shown in Figure 11.33 without paying too much attention to the dimensions.

FIGURE 11.33   Sketching the I profile.

Click on the **Add Dimension** icon. With the mouse click on the first vertice of the flange and on the second vertice representing the middle of the flange as shown. In the command line of the viewport, enter 89 mm as shown. Click on **Return** (Figure 11.34).

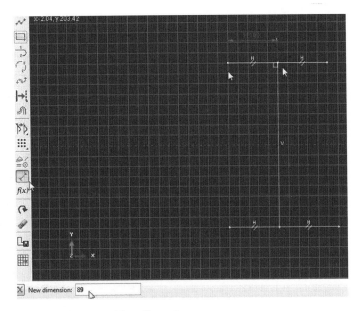

FIGURE 11.34   Adding dimensions.

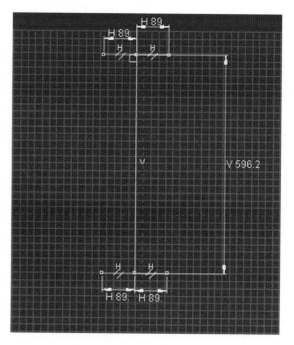

Repeat the operation for the parts of the flanges and enter 596.2 mm for the web. Click on **Return**. The result should look like the one shown in Figure 11.35.

FIGURE 11.35   Finishing dimensioning the profile.

When finished, the **Add dimension** too, and click on **Done** in the prompt area to sketch the section for the shell extrusion. In the **Edit base extrusion**, enter 12000 mm as shown in Figure 11.36, and click **OK**.

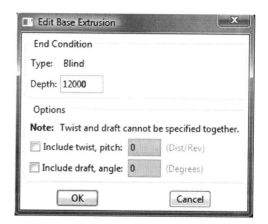

FIGURE 11.36   Editing shell extrusion.

Under the main menu, click on **Shape**, **Cut** and **Extrude**. Select the web as the plane for the extruded cut. Next select the right-hand end of the beam as the edge or the axis that will appear vertical on the right (Figure 11.37).

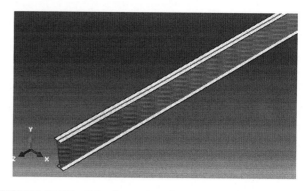

FIGURE 11.37   Selecting a plane for an extruded cut.

The sketcher is loaded again. This time we will use it to sketch the hexagon. Use the **Magnify View** tool to increase the size of the sketch (Figure 11.38).

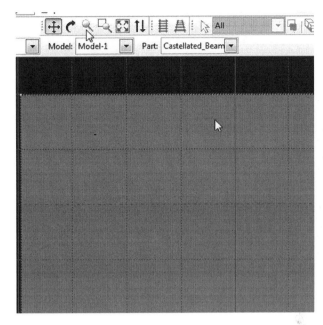

FIGURE 11.38 Magnify view tool.

Draw a circle, and using the **Add dimension** tool, enter its radius as 203 mm. Then draw two other circles as shown in Figure 11.39, each having a radius of 203 mm. Then using the **Create Lines: connected** tool, join the intersecting points as shown to create a perfect hexagon.

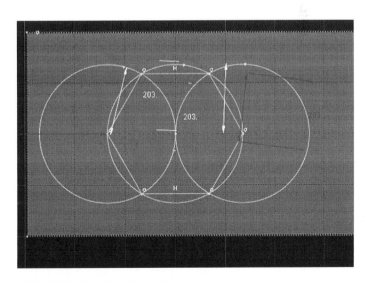

FIGURE 11.39 Sketching a hexagon.

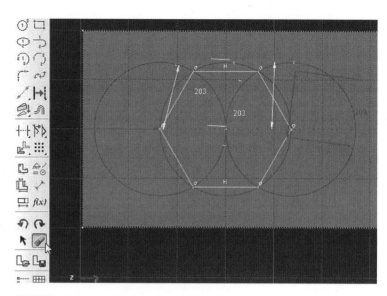

Select the **Delete** tool. By keeping the **Shift** key down, select all the circles. When finished click in the prompt area on **Done**. All that is left is a hexagon (Figure 11.40).

FIGURE 11.40   Delete tool.

Next, we need to position the hexagon at exactly 316 mm from the edge. Using the **Add dimension** tool, enter 316 as the distance from the left vertex to the edge (Figure 11.41).

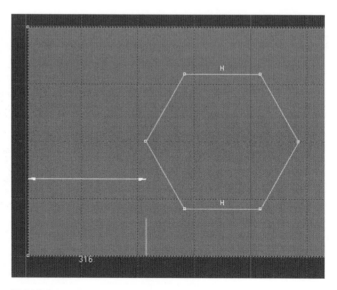

FIGURE 11.41   Dimension tool.

Next, we need to copy the hexagon along the length of the beam. Click on the **Linear Pattern** tool and select the hexagon. Click on **Done** (Figure 11.42).

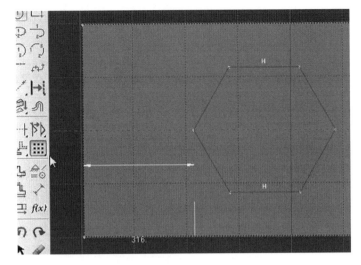

FIGURE 11.42 Linear pattern tool.

In the **Edit Linear Pattern** dialog box, enter 19 for direction 1, and 1 for direction 2. Enter the distance from vertice to vertice as 609 mm. Click on **OK** (Figure 11.43).

FIGURE 11.43 Editing a linear pattern.

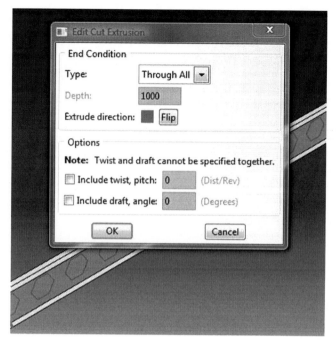

Then in the prompt area of the viewport, click on **Sketch the section for the extruded cut**. In the **Edit Cut Extrusion** dialog box, select **Through all** for the type and click **OK**. The result should be an image of a castellated beam (Figure 11.44).

FIGURE 11.44   Edit cut extrusion.

Define a material named **Steel** with an elastic modulus of 200000 MPa and a Poisson's ratio of 0.27. Next, click on **Sections** to create two sections: one for the web and the other for the flanges. Name the first one **Web_section**. In the **Category** check **Shell**, and in the **Type**, check **Homogeneous**. Click on **Continue**. In the **Edit Section** dialog box, enter the web thickness as 7.8 mm and the material as steel. Click on **OK**. Create another section named **Flange_section**. Enter the shell thickness as 12.8 mm as shown in Figure 11.45. Click on **OK**.

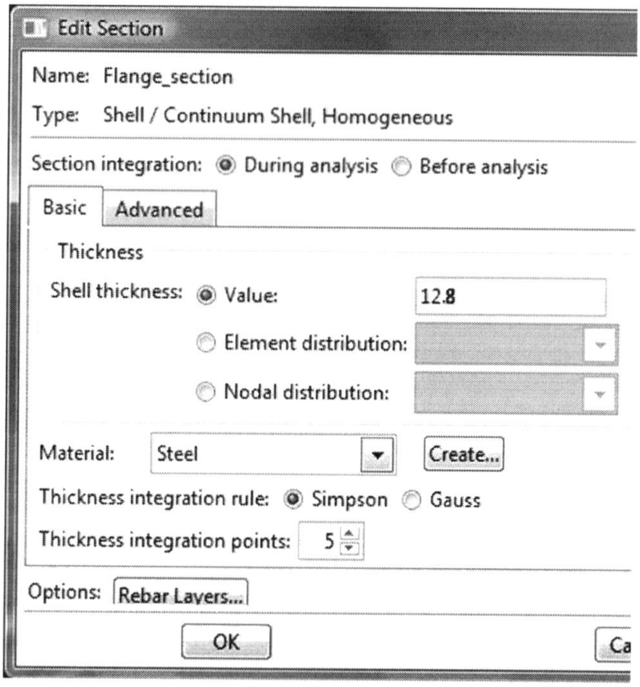

FIGURE 11.45   Creating a shell section.

Expand the menu under **Parts** and **Castellated_beam** and double click on **Section Assignments**. With the mouse select the web. In the **Edit Section Assignments** dialog box, select **Web_section** and click on **OK**. Double click on **Section Assignments** again, select the flanges. In the **Edit Section Assignments** dialog box, select **Flange_section** and click on **OK**. In the prompt area, click on **Done** (Figure 11.46).

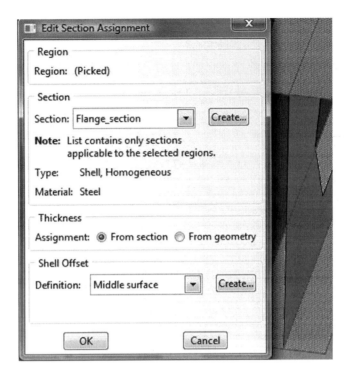

FIGURE 11.46   Editing section assignments.

In the model tree, double click on **Mesh** under the **Castellated_beam**. In the main menu, under **Mesh**, click on **Mesh Controls**, select all the regions, and click on **Done**. In the dialog box, check **Quad** for **Element shape** and **Structured** for **Technique**. A pop-up will appear stating that the web is too complex to be meshed with a structured technique. As a result, select the flanges only for a structured mesh and the web on its own for a free mesh (Figure 11.47).

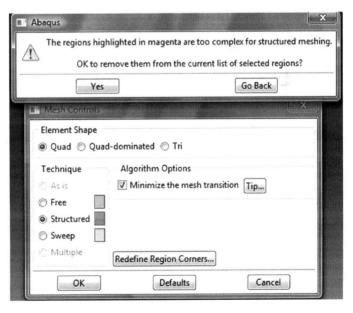

FIGURE 11.47   Mesh controls and element type.

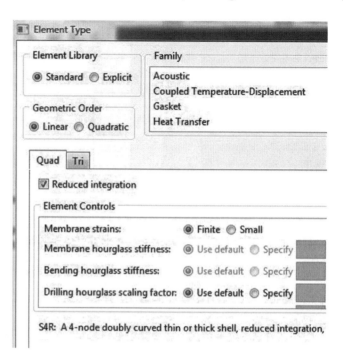

Under **Mesh**, click on **Element Type**. In the dialog box, select **Standard** for element library, **Linear** for geometric order. In **Quad**, check **Reduced integration**. The description of the element **S4R** can be seen. Click on **OK** (Figure 11.48).

FIGURE 11.48   Element type.

In the main menu, under **Seed**, click on **Part**. In the dialog box, enter 40 for **Approximate global size**. Click on **OK** and on **Done**. In the main menu, under **Mesh**, click on **Part**. In the prompt area, click on **Yes** (Figure 11.49).

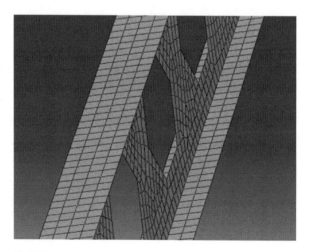

FIGURE 11.49   Mesh.

In the model tree, expand the **Assembly** and double click on **Instances**. Select **Castellated_beam** for **Parts**, and click **OK**. In the model tree, expand **Steps** and **Initial** and double click on **BC**. Name the boundary condition **FIXED**, select **Symmetry/Antisymmetry/Encastre** for the type, and click on **Continue**. In the viewport, select the two ends of the beam and click on **Continue**. In the **Edit Boundary Condition**, check **Encastre**. Click **OK** (Figure 11.50).

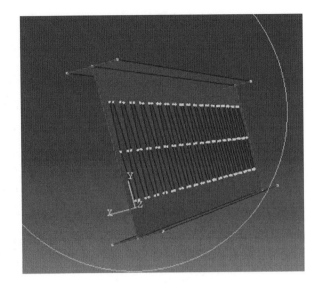

FIGURE 11.50 Imposing BC using geometry.

In the model tree, double click on **Steps**. Name the step **Apply_loads**. Set the procedure to **General** and select **Static, General**. Click on **Continue**. Click on **OK**. In the model tree, under **steps**, and under **Apply_loads**, click on **Loads**. Name the load **Pressure** and select **Pressure** as the type. Click on **Continue**. In the viewport, select the two top surfaces. If any of the surface appears brown, select it and flip the color to purple in the prompt area. In the **Edit Load** dialog box, enter $1 \, N/mm^2$ for **magnitude**. Click **OK** (Figure 11.51).

FIGURE 11.51 Applying a pressure load on a shell surface.

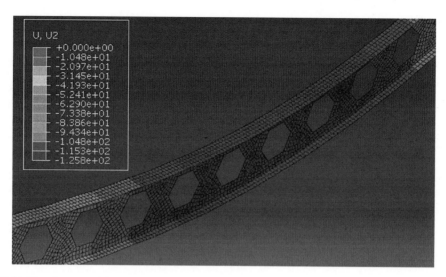

FIGURE 11.52   Contour of the vertical displacement.

Under **Analysis**, right click on **Jobs** and then click on **Create**.

In the **Create Job** dialog box, name the job **Castellated_Beam** and click on **Continue**. In the **Edit Job** dialog box, enter a description for the job. Check **Full analysis**, select to run the job in **Background**, and check to start it **immediately**. Click **OK**. Expand the tree under **Jobs**, right click on **Castellated_Beam**. Then, click on **Submit**. If you get the following message **Castellated_Beam completed successfully** in the bottom window, then your job is free of errors and was executed properly.

Under the top menu, in the **Module** scroll to **Visualization**, and click to load **Abaqus Viewer**. On the main menu, under **File**, click **Open**, navigate to your working directory, and open the file **Castellated_Beam**. It should have the same name as the job you submitted. Click on the icon **Plot** on **Undeformed shape**. Under the main menu, select **U** and **U2** to plot the vertical displacement (Figure 11.52).

Under the main menu, select **S** and **Mises** to plot the von Mises stress (Figure 11.53).

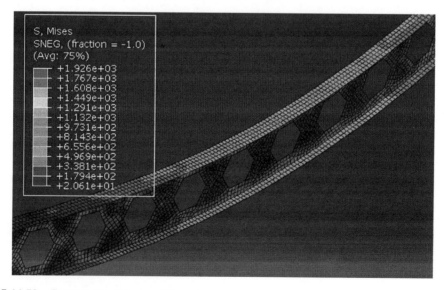

FIGURE 11.53   Contour plot of the von Mises stress.

Appendix A: List of MATLAB® Modules and Functions

A.1 Assem_Elem_loads.m

```
function[F] = Assem_Elem_loads(F , fg, g)
%
% This function assemble the global force vector
%
global eldof
%
% This function assembles the global force vector
%
for idof=1:eldof
    if (g(idof))~= 0
        F(g(idof))= F(g(idof))+  fg(idof);
    end
end
%
% end function Assem_Elem_loads
```

A.2 Assem_Joint_Loads.m

```
function[F] = Assem_Joint_Loads(F)
%
% This function assembles the joints loads
% to the global force vector
%
global nnd nodof
global nf Joint_loads
%
for i=1:nnd
    for j=1:nodof
        if nf(i,j)~= 0
            F(nf(i,j)) = Joint_loads(i,j);
        end
    end
end
end
%%%%%%%% End function form_beam_F  %%%%%%%%%%%%%%%%%
```

A.3 beam_column_C.m

```
function[C] = beam_column_C(i)
%
% This function forms the transformation between
% local and global coordinates
%
global nnd nel nne nodof eldof
global geom connec prop nf load
%
% retrieve the nodes of element i
%
node1=connec(i,1);
node2=connec(i,2);
%
%
% Retrieve the x and y coordinates of nodes 1 and 2
```

```
%
x1=geom(node1,1); y1=geom(node1,2);
x2=geom(node2,1); y2=geom(node2,2);
%
% Evaluate the angle that the member makes with the global axis X
%
if(x2-x1)==0
    if(y2>y1)
    theta=2*atan(1);
    else
    theta=-2*atan(1);
    end
else
    theta=atan((y2-y1)/(x2-x1));
end
%
% Construct the transformation matrix
%
C = [cos(theta)    -sin(theta)    0      0             0            0 ; ...
      sin(theta)    cos(theta)    0      0             0            0 ; ...
          0             0         1      0             0            0 ; ...
          0             0         0    cos(theta)   -sin(theta)     0 ; ...
          0             0         0    sin(theta)    cos(theta)     0 ; ...
          0             0         0      0             0            1 ];
%
% end function beam_column_C
```

A.4   beam_column_g.m

```
function[g] = beam_column_g(i)
%
% This function forms the steering vector for element i
%
global nnd nel nne nodof eldof
global geom connec prop nf load
%
% retrieve the nodes of element i
%
node1=connec(i,1);
node2=connec(i,2);
%
% Retrieve the element degrees of freedom to be stored
% in the steering vector
%
g=[nf(node1,1); ...
   nf(node1,2); ...
   nf(node1,3); ...
   nf(node2,1); ...
   nf(node2,2); ...
   nf(node2,3)];
%
% end function beam_column_g
```

A.5   beam_column_k.m

```
function[kl] = beam_column_k(i)
%
% This function forms the beam-column element stiffness in local coordinates
%
global nnd nel nne nodof eldof
global geom connec prop nf load Hinge
%
% retrieve the nodes of element i
%
node1=connec(i,1);
node2=connec(i,2);
```

```
%
%
% Retrieve the x and y coordinates of nodes 1 and 2
%
x1=geom(node1,1); y1=geom(node1,2);
x2=geom(node2,1); y2=geom(node2,2);
%
% Evaluate length of element i
%
L = sqrt((x2-x1)^2 + (y2-y1)^2);
%
%  Retrieve section properties of element i
%
E = prop(i,1); A = prop(i,2); I = prop(i,3);
%
EA=E*A; EI=E*I;
%
%Calculate element stiffness matrix in its local coordinates
%
if Hinge(i,1) == 0
kl=[EA/L        0            0        -EA/L          0             0       ; ...
     0       3*EI/L^3        0          0       -3*EI/L^3     3*EI/L^2   ; ...
     0          0            0          0            0             0       ; ...
   -EA/L        0            0        EA/L           0             0       ; ...
     0      -3*EI/L^3        0          0        3*EI/L^3     -3*EI/L^2   ; ...
     0       3*EI/L^2        0          0       -3*EI/L^2      3*EI/L    ];

elseif Hinge(i,2) == 0
kl=[EA/L        0            0        -EA/L          0          0 ; ...
     0       3*EI/L^3     3*EI/L^2      0       -3*EI/L^3      0 ; ...
     0       3*EI/L^2     3*EI/L        0       -3*EI/L^2      0 ; ...
   -EA/L        0            0        EA/L           0         0 ; ...
     0      -3*EI/L^3    -3*EI/L^2      0        3*EI/L^3      0 ; ...
     0          0            0          0            0         0];

else
kl=[EA/L        0            0        -EA/L          0             0       ; ...
     0      12*EI/L^3     6*EI/L^2      0       -12*EI/L^3     6*EI/L^2 ; ...
     0       6*EI/L^2     4*EI/L        0        -6*EI/L^2     2*EI/L   ; ...
   -EA/L        0            0        EA/L           0             0       ; ...
     0     -12*EI/L^3    -6*EI/L^2      0        12*EI/L^3    -6*EI/L^2 ; ...
     0       6*EI/L^2     2*EI/L        0        -6*EI/L^2     4*EI/L    ];
end
 %
% End function beam_column_k
```

A.6   beam_g.m

```
function[g] = beam_g(i)
%
% This function forms the steering vector for element i
%
global connec nf
%
% retrieve the nodes of element i
%
node_1=connec(i,1);
node_2=connec(i,2);
%
% Form the steering vector from element's degrees
% of freedom
%
g=[nf(node_1,1); nf(node_1,2); nf(node_2,1);nf(node_2,2)];
%
%%%%%%%%%%%%%%%%%%% end function beam_g %%%%%%%%%%%%%%%%%
```

A.7   beam_k.m

```
function[kl] = beam_k(i)
%
% This function forms the element stiffness in local coordinates
%
global nnd nel nne nodof eldof
global geom connec prop nf load Hinge
%
% retrieve the nodes of element i
%
node1=connec(i,1);
node2=connec(i,2);
%
%
% Retrieve the x and y coordinates of nodes 1 and 2
%
x1=geom(node1); x2=geom(node2);
%
% Evaluate length of element i
%
L = abs(x2-x1);
%
%  Retrieve section properties of element i
%
EI = prop(i,1)*prop(i,2);
%
%Calculate element stiffness matrix in its local coordinates
%
if Hinge(i, 1) == 0
kl=[ 3*EI/L^3        0      -3*EI/L^3      3*EI/L^2  ; ...
        0            0          0             0      ; ...
    -3*EI/L^3        0       3*EI/L^3     -3*EI/L^2  ; ...
     3*EI/L^2        0      -3*EI/L^2      3*EI/L    ];

elseif Hinge(i, 2) == 0
kl=[ 3*EI/L^3      3*EI/L^2    -3*EI/L^3     0 ; ...
     3*EI/L^2      3*EI/L      -3*EI/L^2     0 ; ...
    -3*EI/L^3     -3*EI/L^2     3*EI/L^3     0 ; ...
        0            0            0          0 ] ;
else
kl=[ 12*EI/L^3     6*EI/L^2    -12*EI/L^3     6*EI/L^2  ; ...
      6*EI/L^2     4*EI/L       -6*EI/L^2     2*EI/L    ; ...
    -12*EI/L^3    -6*EI/L^2     12*EI/L^3    -6*EI/L^2  ; ...
      6*EI/L^2     2*EI/L       -6*EI/L^2     4*EI/L    ];
end
%
% End function beam_k
```

A.8   coord_q8.m

```
function[coord] = coord_q8(k,nne, geom, connec )
%
% This function returns the coordinates of the nodes of element k
%
coord=zeros(nne,2);
for i=1: nne
    coord(i,:)=geom(connec(k,i),:);
end
%
% End function coord_q8
```

A.9   elem_q4.m

```
function[coord,g] = elem_q4(i)
%
% This function returns the coordinates of the nodes of
% element i and its steering vector g
%
global nnd nel nne nodof eldof n ngp
global geom connec dee nf load
%
l=0;
coord=zeros(nne,nodof);
for k=1: nne
    for j=1:nodof
    coord(k,j)=geom(connec(i,k),j);
    l=l+1;
    g(l)=nf(connec(i,k),j);
    end
end
%
% End function elem_q4
```

A.10   Elem_q8.m

```
function[coord,g] = elem_q8(i)
%
% This function returns the coordinates of the nodes of element i
% and its steering vector
%
global nnd nel nne nodof eldof n ngp
global geom connec dee nf load
%
l=0;
coord=zeros(nne,nodof);
for k=1: nne
    for j=1:nodof
    coord(k,j)=geom(connec(i,k),j);
    l=l+1;
    g(l)=nf(connec(i,k),j);
    end
end
%
% End function elem_q8
```

A.11   elem_T3.m

```
function[bee,g,A] = elem_T3(i)
%
% This function returns the coordinates of the nodes of element i
% and its steering vector
%
global nnd nel nne nodof eldof n
global geom connec dee nf load
%
x1 = geom(connec(i,1),1);    y1 = geom(connec(i,1),2);
x2 = geom(connec(i,2),1);    y2 = geom(connec(i,2),2);
x3 = geom(connec(i,3),1);    y3 = geom(connec(i,3),2);
%
A = (0.5)*det([1    x1    y1; ...
               1    x2    y2; ...
               1    x3    y3]);
%
 m11 = (x2*y3 - x3*y2)/(2*A);
 m21 = (x3*y1 - x1*y3)/(2*A);
```

```
m31 = (x1*y2 - y1*x2)/(2*A);
m12 = (y2 - y3)/(2*A);
m22 = (y3 - y1)/(2*A);
m32 = (y1 - y2)/(2*A);
m13 = (x3 - x2)/(2*A);
m23 = (x1 - x3)/(2*A);
m33 = (x2 -x1)/(2*A);
%
bee = [ m12    0    m22    0    m32       0; ...
          0   m13     0   m23    0     m33; ...
        m13   m12   m23   m22   m33    m32] ;
%
l=0;
for k=1:nne
    for j=1:nodof
    l=l+1;
    g(l)=nf(connec(i,k),j);
    end
end
%
% End function elem_T3
```

A.12   elem_T6.m

```
function[coord,g] = elem_T6(i)
%
% This function returns the coordinates of the nodes of element i
% and its steering vector
%
global nnd nel nne nodof eldof n
global geom connec dee nf load
%
l=0;
coord=zeros(nne,nodof);
for k=1: nne
    for j=1:nodof
    coord(k,j)=geom(connec(i,k),j);
    l=l+1;
    g(l)=nf(connec(i,k),j);
    end
end
%
% End function elem_T6
```

A.13   fmlin.m

```
function[der,fun] = fmlin(samp, ig,jg)
%
% This function returns the vector of the shape function and their
% derivatives with respect to xi and eta
%
xi=samp(ig,1);
eta=samp(jg,1);
%
fun = 0.25*[(1.- xi - eta + xi*eta);...
            (1.+ xi - eta - xi*eta);...
            (1.+ xi + eta + xi*eta);...
            (1.- xi + eta - xi*eta)];
%
der = 0.25*[-(1-eta)    (1-eta)    (1+eta)   -(1+eta);...
            -(1-xi)    -(1+xi)    (1+xi)     (1-xi)];
% end function fmlin
```

A.14   fmquad.m

```
function[der,fun] = fmquad(samp, ig,jg)
%
% This function returns the vector of the shape function and their
% derivatives with respect to xi and eta at the gauss points for
% an 8-nodded quadrilateral
%
xi=samp(ig,1);
eta=samp(jg,1);
etam=(1.-eta);
etap=(1.+eta);
xim=(1.-xi);
xip=(1.+xi);
%
fun(1) = -0.25*xim*etam*(1.+ xi + eta);
fun(2) = 0.5*(1.- xi^2)*etam;
fun(3) = -0.25*xip*etam*(1. - xi + eta);
fun(4) = 0.5*xip*(1. - eta^2);
fun(5) = -0.25*xip*etap*(1. - xi - eta);
fun(6) = 0.5*(1. - xi^2)*etap;
fun(7) = -0.25*xim*etap*(1. + xi - eta);
fun(8) = 0.5*xim*(1. - eta^2);
%
der(1,1)=0.25*etam*(2.*xi + eta);  der(1,2)=-1.*etam*xi;
der(1,3)=0.25*etam*(2.*xi-eta);  der(1,4)=0.5*(1-eta^2);
der(1,5)=0.25*etap*(2.*xi+eta);  der(1,6)=-1.*etap*xi;
der(1,7)=0.25*etap*(2.*xi-eta);  der(1,8)=-0.5*(1.-eta^2);
%
der(2,1)=0.25*xim*(2.*eta+xi);  der(2,2)=-0.5*(1. - xi^2);
der(2,3)=-0.25*xip*(xi-2.*eta);  der(2,4)=-1.*xip*eta;
der(2,5)=0.25*xip*(xi+2.*eta);  der(2,6)=0.5*(1.-xi^2);
der(2,7)=-0.25*xim*(xi-2.*eta);  der(2,8)=-1.*xim*eta;
%
% end function fmquad
```

A.15   fmT6_quad.m

```
function[der,fun] = fmT6_quad(samp, ig)
%
% This function returns the vector of the shape function and their
% derivatives with respect to xi and eta at the gauss points for
% an 8-nodded quadrilateral
%
xi=samp(ig,1);
eta=samp(ig,2);
lambda = 1. - xi - eta;
%
fun(1) = -lambda*(1.-2*lambda);
fun(2) = 4.*xi*lambda;
fun(3) = -xi*(1.-2*xi);
fun(4) = 4.*xi*eta;
fun(5) = -eta*(1.-2*eta);
fun(6) = 4.*eta*lambda;
%
der(1,1)=1.-4*lambda;  der(1,2)=4.*(lambda-xi);
der(1,3)=-1.+4*xi;  der(1,4)=4.*eta;
der(1,5)=0.;  der(1,6)=-4.*eta;
%
der(2,1)=1.-4*lambda;  der(2,2)=-4.*xi;
der(2,3)=0.;  der(2,4)=4.*xi;
der(2,5)=-1.+4.*eta;  der(2,6)=4.*(lambda-eta);
%
% end function fmT6_quad
```

A.16   Forces_at_nodes_plate.m

```
function[MX, MY,  MXY, QX, QY]=Forces_at_nodes_plate(Element_Forces)
%
% This function averages the stresses at the nodes
%
global nnd nel nne connec
%
for k = 1:nnd
    mx = 0. ; my = 0.; mxy = 0.; qx = 0.; qy = 0.;
    ne = 0;
    for iel = 1:nel;
        for jel=1:nne;
            if connec(iel,jel) == k;
                ne=ne+1;
                mx = mx + Element_Forces(iel,1);
                my = my + Element_Forces(iel,2);
                mxy = mxy + Element_Forces(iel,3);
                qx = qx + Element_Forces(iel,4);
                qy = qy + Element_Forces(iel,5);
            end
        end
    end
        MX(k,1) = mx/ne;
        MY(k,1) = my/ne;
        MXY(k,1) = mxy/ne;
        QX(k,1) = qx/ne;
        QY(k,1) = qy/ne;
end
```

A.17   File:form_beam_F.m

```
function[F] = form_beam_F(F)
%
% This function forms the global force vector
%
global nnd nodof nel eldof
global nf Element_loads Joint_loads
%
for i=1:nnd
    for j=1:nodof
        if nf(i,j)~= 0
            F(nf(i,j)) = Joint_loads(i,j);
        end
    end
end
%
%
for i=1:nel
    g=beam_g(i) ;        % Retrieve the element steering vector
    for j=1:eldof
        if g(j)~= 0
        F(g(j))= F(g(j)) + Element_loads(i,j);
        end
    end
end
%%%%%%%%% End function form_beam_F  %%%%%%%%%%%%%%%%
```

A.18   File:form_ff.m

```
function[ff]=form_ff(ff,fg, g)
%
% This function assemble the global force vector
%
global nodof nne eldof
%
```

```
% This function assembles the global force vector
%
for idof=1:eldof
    if (g(idof))~= 0
        ff(g(idof))= ff(g(idof))+  fg(idof);
    end
end
%
% end function form_ff
```

A.19   File:form_KK.m

```
function[KK]=form_KK(KK, kg, g)
%
% This function assembles the global stiffness matrix
%
global  eldof
%
% This function assembles the global stiffness matrix
%
for i=1:eldof
    if g(i) ~= 0
        for j=1: eldof
            if g(j) ~= 0
            KK(g(i),g(j))= KK(g(i),g(j)) + kg(i,j);
            end
        end
    end
end
%
%%%%%%%%%%%%% end function form_KK %%%%%%%%%%%%%%%%%%
```

A.20   form_truss_F.m

```
function[F] = form_truss_F(F)
%
% This function forms the global force vector
%
global nnd nodof
global nf load
%
for i=1:nnd
    for j=1:nodof
        if nf(i,j)~= 0
            F(nf(i,j)) = load(i,j);
        end
    end
end
%%%%%%%%% End function form_truss_F  %%%%%%%%%%%%%%%%%
```

A.21   formbee.m

```
function[bee] = formbee(deriv,nne,eldof)
%
%  This function assembles the matrix [bee] from the
%  derivatives of the shape functions in global coordinates
%
bee=zeros(3,eldof);
for m=1:nne
    k=2*m;
    l=k-1;
    x=deriv(1,m);
    bee(1,l)=x;
    bee(3,k)=x;
    y=deriv(2,m);
```

```
    bee(2,k)=y;
    bee(3,l)=y;
end
%
% End function formbee
```

A.22   formbee_axi.m

```
function[bee, radius] = formbee_axi(deriv,nne,fun, coord,eldof)
%
% This function assembles the matrix [bee] for an axisymmetric
% problem from the derivatives of the shape functions in global
% coordinates
%
bee=zeros(4,eldof);
%
radius = dot(fun,coord(:,1));
%
for m=1:nne
    k=2*m;
    l=k-1;
    x=deriv(1,m);
    bee(1,l)=x;
    bee(4,k)=x;
    y=deriv(2,m);
    bee(2,k)=y;
    bee(4,l)=y;
    bee(3,l) = fun(m)/radius;
end
%
% End function formbee_axi
```

A.23   formbeeb.m

```
function[beeb] = formbeeb(deriv,nne,eldof)
%
%  This function assembles the matrix [beeb] from the
%  derivatives of the shape functions in global coordinates
%  for a thick plate element (bending action)
%
beeb=zeros(3,eldof);
for m=1:nne
    k=3*m;
    j=k-1;
    x=deriv(1,m);
    beeb(1,j)=x;
    beeb(3,k)=x;
    y=deriv(2,m);
    beeb(2,k)=y;
    beeb(3,j)=y;
end
%
% End function formbeeb
```

A.24   formbees.m

```
function[bees] = formbees(deriv,fun, nne,eldof)
%
%  This function assembles the matrix [bees] from the
%  derivatives of the shape functions in global coordinates
%  for the shear action in a plate element
%
bees=zeros(2,eldof);
for m=1:nne
    k=3*m;
```

```
        j=k-1;
        i=k-2;
        x=deriv(1,m); y=deriv(2,m);
        bees(2,i)=-x;
        bees(1,i)=-y;
        bees(1,k) = fun(m);
        bees(2,j) = fun(m);
end
%
% End function formbees
```

A.25 formdax.m

```
function[dee] = formdax(E,vu)
%
% This function forms the elasticity matrix for a plane stress problem
%
v1 = 1. - vu;
c = E/((1. + vu)*(1. - 2.*vu));
%
dee = c*[v1          vu          vu          0;...
         vu          v1          vu          0;...
         vu          vu          v1          0;...
         0.          0.          0.     .5*(1.-vu)];
%
% end function fromdeps
```

A.26 formdeeb.m

```
function[deeb] = formdeeb(E,vu,thick)
%
% This function forms the elasticity matrix for a bending
% action in a plate element
%
DR= E*(thick^3)/(12*(1.-vu*vu));
%
deeb=DR*[1          vu          0.          ;...
         vu          1          0.          ;...
         0.          0.     (1.-vu)/2    ;
%
% end function fromdeeb
```

A.27 formdees.m

```
function[dees] = formdees(E,vu,thick)
%
% This function forms the elasticity matrix for the shear
% action in a thick plate element
%
G= E/(2*(1.+vu));
%
dees=G*[thick      0  ;...
          0      thick];
%
% end function fromdees
```

A.28 formdeps.m

```
function[dee] = formdeps(E,vu)
%
% This function forms the elasticity matrix for a plane strain problem
%
v1=1.-vu
c=E/((1.+vu)*(1.-2.*vu))
```

```
%
dee=[v1*c          vu*c            0.          ;...
     vu*c          v1*c            0.          ;...
      0.            0.          .5*c*(1.-2.*vu)];
%
% end function fromdeps
```

A.29 formdsig.m

```
function[dee] = formdsig(E,vu)
%
% This function forms the elasticity matrix for a plane stress problem
%
c=E/(1.-vu*vu);
%
dee=c*[1           vu             0.           ;...
       vu           1             0.           ;...
       0.           0.          .5*(1.-vu)];
%
% end function formdsig
```

A.30 gauss.m

```
function[samp]=gauss(ngp)
%
% This function returns the abscissas and weights of the Gauss
% points for ngp equal up to 4
%
%
samp=zeros(ngp,2);
%
if ngp==1
    samp=[0.  2];
elseif ngp==2
    samp=[-1./sqrt(3)    1.;...
           1./sqrt(3)    1.];
elseif ngp==3
    samp= [-.2*sqrt(15.)    5./9; ...
            0.              8./9.;...
           .2*sqrt(15.)     5./9];
elseif ngp==4
    samp= [-0.861136311594053       0.347854845137454; ...
           -0.339981043584856       0.652145154862546; ...
            0.339981043584856       0.652145154862546; ...
            0.861136311594053       0.347854845137454];
end
%
% End function Gauss
```

A.31 hammer.m

```
function[samp]=hammer(npt)
%
% This function returns the abscissae and weights of the
% integration points for npt equal up to 7
%
%
samp=zeros(npt,3);
%
if npt==1
    samp=[1/3.  1/3.  1/2.];
elseif (npt==2 | npt==3)
    npt=3;
    samp=[1/6.  1/6.   1/6.; ...
          2/3   1./6   1/6.; ...
```

```
          1/6.   2./3.   1/6];
elseif (npt==4 | npt==5)
    npt=4;
    samp= [1/3      1/3    -27/96; ...
           1/5      1/5.    25/96;...
           3/5      1/5.    25/96;...
           1/5      3/5     25/96];
elseif npt==6
    a = 0.445948490915965;  b = 0.091576213509771;
    samp= [  a        a      0.111690794839005; ...
           1-2*a      a      0.111690794839005; ...
             a      1-2*a    0.111690794839005; ...
             b        b      0.054975871827661; ...
           1-2*b      b      0.054975871827661; ...
             b      1-2*b    0.054975871827661];
elseif npt==7
    a = (6+sqrt(15))/21 ; b = 4/7 -a;
    A = (155+sqrt(15))/2400; B = (31/240 -A);
samp= [   1/3        1/3       9/80; ...
           a          a          A   ; ...
         1-2*a        a          A   ; ...
           a        1-2*a        A   ; ...
           b          b          B   ; ...
         1-2*b        b          B   ; ...
           b        1-2*b        B];
end
%
% End function hammer
```

A.32   platelem_q8.m

```
function[coord,g] = platelem_q8(i)
%
% This function returns the coordinates of the nodes of element i
% and its steering vector
%
global nne nodof geom connec  nf   dim
%
coord=zeros(nne,dim);
for k=1: nne
    for j=1:dim
    coord(k,j)=geom(connec(i,k),j);
    end
end
%
l=0;
for k=1: nne
    for j=1:nodof
    l=l+1;
    g(l)=nf(connec(i,k),j);
    end
end
%
% End function platelem_q8
```

A.33   prepare_contour_data.m

```
function[ZX, ZY, ZT, Z1, Z2]=prepare_contour_data(SIGMA)
%
% This function averages the stresses at the nodes
% and rearrange the values in the matrices Z for contouring
%
global nnd nel nne  geom  connec XIG YIG NXE NYE
%
for k = 1:nnd
    sigx = 0. ;sigy = 0.; tau = 0.;
```

```
    ne = 0;
    for iel = 1:nel;
        for jel=1:nne;
            if connec(iel,jel) == k;
                ne=ne+1;
                sigx = sigx+SIGMA(iel,1);
                sigy = sigy + SIGMA(iel,2);
                tau = tau + SIGMA(iel,3);
            end
        end
    end
    xc = geom(k,1); yc = geom(k,2);
    for i = 1:2*NXE+1;
      for j=1:2*NYE +1;
        if xc == XIG(i) && yc == YIG(j);
        ZX(j,i) = sigx/ne;
        ZY(j,i) = sigy/ne;
        ZT(j,i)=tau/ne;
        Z1(j,i)= ((sigx+sigy)/2 + sqrt(((sigx+sigy)/2)^2 +tau^2))/ne;
        Z2(j,i)= ((sigx+sigy)/2 - sqrt(((sigx+sigy)/2)^2 +tau^2))/ne;
        end
      end
    end
end
```

A.34   print_beam_model.m

```
%
fprintf(fid, ' ******* PRINTING MODEL DATA *************\n\n\n');
%
% Print Nodal coordinates
%
fprintf(fid, '------------------------------------------------------ \n');
fprintf(fid, 'Number of nodes:                            %g\n', nnd );
fprintf(fid, 'Number of elements:                         %g\n', nel );
fprintf(fid, 'Number of nodes per element:                %g\n', nne );
fprintf(fid, 'Number of degrees of freedom per node:      %g\n', nodof);
fprintf(fid, 'Number of degrees of freedom per element:   %g\n\n\n', eldof);
%
%
%
fprintf(fid, '------------------------------------------------------ \n');
fprintf(fid, 'Node        X\n');
for i=1:nnd
fprintf(fid,' %g,      %07.2f\n',i, geom(i));
end
fprintf(fid,'\n');
%
% Print element connectivity
%
fprintf(fid, '------------------------------------------------------ \n');
fprintf(fid, 'Element       Node_1        Node_2 \n');
for i=1:nel
fprintf(fid,'    %g,           %g,              %g\n',i, connec(i,1), connec(i,2));
end
fprintf(fid,'\n');
%
% Print element property
%
fprintf(fid, '------------------------------------------------------ \n');
fprintf(fid, 'Element          E               I \n');
for i=1:nel
fprintf(fid,'    %g,        %g,        %g\n',i, prop(i,1), prop(i,2));
end
fprintf(fid,'\n');
%
% Print Nodal freedom
```

```
%
fprintf(fid, '------------------------------------------------------ \n');
fprintf(fid, '-------------Nodal freedom-------------------------- \n');
fprintf(fid, 'Node      disp_w      Rotation \n');
for i=1:nnd
fprintf(fid,'  %g,          %g,            %g\n',i, nf(i,1), nf(i,2));
end
fprintf(fid,'\n');
%
% Print Nodal loads
%
fprintf(fid, '------------------------------------------------------ \n');
fprintf(fid, '----------------Applied Nodal Loads------------------ \n');
fprintf(fid, 'Node      load_Y          Moment\n');
for i=1:nnd
    for j=1:nodof
        node_force(i,j) = 0;
        if nf(i,j)~= 0;
        node_force(i,j) = F(nf(i,j))
        end
    end
fprintf(fid,'  %g,       %07.2f,          %07.2f\n',i, node_force(i,1), node_force(i,2));
end
%
fprintf(fid, '------------------------------------------------------ \n');
fprintf(fid,'\n');
fprintf(fid,'Total number of active degrees of freedom, n = %g\n',n);
fprintf(fid,'\n');
%
```

A.35   print_beam_results.m

```
%
fprintf(fid, '------------------------------------------------------ \n');
fprintf(fid, ' \n\n\n ******* PRINTING ANALYSIS RESULTS *************\n\n\n');
%
% Print global force vector
%
fprintf(fid, '------------------------------------------------------ \n');
fprintf(fid,'Global force vector  F \n');
fprintf(fid,'   %g\n',F);
fprintf(fid,'\n');
%
%
% Print Displacement solution vector
%
fprintf(fid, '------------------------------------------------------ \n');
fprintf(fid,'Displacement solution vector:  delta \n');
fprintf(fid,' %8.5f\n',delta);
fprintf(fid,'\n');
%
% Print nodal displacements
%
fprintf(fid, '------------------------------------------------------ \n');
fprintf(fid, 'Nodal displacements \n');
fprintf(fid, 'Node      disp_y          rotation\n');
for i=1:nnd
fprintf(fid,' %g,       %8.5f,         %8.5f\n',i, node_disp(i,1), node_disp(i,2));
end
fprintf(fid,'\n');
%
% Print Members actions
%
fprintf(fid, '------------------------------------------------------ \n');
fprintf(fid, 'Members actions \n');
fprintf(fid, 'element      fy1          M1          Fy2          M2\n');
for i=1:nel
```

```
        fprintf(fid,' %g,        %9.2f,      %9.2f,      %9.2f,      %9.2f\n',i, ...
                force(i,1),force(i,2),force(i,3),force(i,4));
end
```

A.36   print_CST_results.m

```
%
fprintf(fid, '-------------------------------------------------------- \n');
fprintf(fid, ' \n ******* PRINTING ANALYSIS RESULTS ************\n\n');
%
% Print nodal displacements
%
fprintf(fid, '-------------------------------------------------------- \n');
fprintf(fid, 'Nodal displacements \n');
fprintf(fid, 'Node        disp_x            disp_y \n');

for i=1:nnd
fprintf(fid,' %g,       %8.5e,       %8.5e\n', ...
              i, node_disp(i,1), node_disp(i,2));
end
fprintf(fid,'\n');
%
% Print element stresses
%
fprintf(fid, '-------------------------------------------------------- \n');
fprintf(fid, '                         Element stresses \n');
fprintf(fid, 'element      sigma_(xx)          sigma_(yy)          tau_(xy)\n');
%
for i=1:nel
    fprintf(fid,' %g,        %7.4e,        %7.4e,        %7.4e\n',i, ...
              SIGMA(i,1),SIGMA(i,2),SIGMA(i,3));
end
%
%
% Print element strains
%
fprintf(fid, '-------------------------------------------------------- \n');
fprintf(fid, '                         Element strains \n');
fprintf(fid, 'element      epsilon_(xx)          epsilon_(yy)          gamma_(xy)\n');
%
for i=1:nel
    fprintf(fid,' %g,        %7.4e,        %7.4e,        %7.4e\n',i, ...
              EPS(i,1),EPS(i,2),EPS(i,3));
end
```

A.37   print_frame_model.m

```
%
fprintf(fid, ' ******* PRINTING MODEL DATA *************\n\n\n');

% Print Nodal coordinates
%
fprintf(fid, '-------------------------------------------------------- \n');
fprintf(fid, 'Number of nodes:                            %g\n', nnd );
fprintf(fid, 'Number of elements:                         %g\n', nel );
fprintf(fid, 'Number of nodes per element:                %g\n', nne );
fprintf(fid, 'Number of degrees of freedom per node:      %g\n', nodof);
fprintf(fid, 'Number of degrees of freedom per element:   %g\n\n\n', eldof);
%
%
%
fprintf(fid, '-------------------------------------------------------- \n');
fprintf(fid, 'Node        X         Y\n');
for i=1:nnd
fprintf(fid,' %g,        %07.2f        %07.2f\n',i, geom(i,1), geom(i,2));
end
```

```
fprintf(fid,'\n');
%
% Print element connectivity
%
fprintf(fid, '----------------------------------------------------- \n');
fprintf(fid, 'Element        Node_1        Node_2 \n');
for i=1:nel
fprintf(fid,'    %g,             %g,                 %g\n',i, connec(i,1), connec(i,2));
end
fprintf(fid,'\n');
%
% Print element property
%
fprintf(fid, '----------------------------------------------------- \n');
fprintf(fid, 'Element        E            A            I \n');
for i=1:nel
fprintf(fid,'    %g,          %g,          %g              %g\n', ...
                  i, prop(i,1), prop(i,2), prop(i,3));
end
fprintf(fid,'\n');
%
% Print Nodal freedom
%
fprintf(fid, '----------------------------------------------------- \n');
fprintf(fid, '-------------Nodal freedom------------------------- \n');
fprintf(fid, 'Node       disp_u         disp_u         Rotation \n');
for i=1:nnd
fprintf(fid,'   %g,           %g,           %g,            %g\n', ...
                    i, nf(i,1), nf(i,2),nf(i,3) );
end
fprintf(fid,'\n');
%
% Print joint loads
%
fprintf(fid, '----------------------------------------------------- \n');
fprintf(fid, '----------------Applied joint Loads------------------ \n');
fprintf(fid, 'Node       load_X        load_Y         Moment\n');
for i=1:nnd
    for j=1:nodof
        node_force(i,j) = 0;
        if nf(i,j)~= 0;
        node_force(i,j) = F(nf(i,j));
        end
    end
fprintf(fid,'   %g,        %07.2f,         %07.2f,            %07.2f\n', ...
           i, node_force(i,1), node_force(i,2), node_force(i,3));
end
%
fprintf(fid, '----------------------------------------------------- \n');
fprintf(fid,'\n');
fprintf(fid,'Total number of active degrees of freedom, n = %g\n',n);
fprintf(fid,'\n');
%
```

A.38   print_frame_results.m

```
%
fprintf(fid, '----------------------------------------------------- \n');
fprintf(fid, ' \n\n\n ******* PRINTING ANALYSIS RESULTS ***********\n\n\n');
%
% Print global force vector
%
fprintf(fid, '----------------------------------------------------- \n');
fprintf(fid,'Global force vector  F \n');
fprintf(fid,'   %g\n',F);
fprintf(fid,'\n');
%
```

```
%
% Print Displacement solution vector
%
fprintf(fid, '-------------------------------------------------------- \n');
fprintf(fid,'Displacement solution vector:  delta \n');
fprintf(fid,' %8.5f\n',delta);
fprintf(fid,'\n');
%
% Print nodal displacements
%
fprintf(fid, '-------------------------------------------------------- \n');
fprintf(fid, 'Nodal displacements \n');
fprintf(fid, 'Node       disp_x      disp_y        rotation\n');
for i=1:nnd
fprintf(fid,' %g,     %8.5e,     %8.5e,       %8.5e\n',i, ...
             node_disp(i,1), node_disp(i,2),node_disp(i,3));
end
fprintf(fid,'\n');
%
% Print Members actions
%
fprintf(fid, '-------------------------------------------------------- \n');
fprintf(fid, 'Members actions in local coordinates \n');
fprintf(fid, 'element    fx1      fy1       M1      fx2      Fy2       M2\n');
for i=1:nel
    fprintf(fid,' %g,    %7.4f,    %7.4f,    %7.4f,    %7.4f,    %7.4f,    %9.4f\n',i, ...
               force_l(i,1),force_l(i,2),force_l(i,3),force_l(i,4),force_l(i,5),force_l(i,6));
end
%
fprintf(fid, '-------------------------------------------------------- \n');
fprintf(fid, 'Members actions in global coordinates \n');
fprintf(fid, 'element    fx1      fy1       M1      fx2      Fy2       M2\n');
for i=1:nel
    fprintf(fid,' %g,    %7.4f,    %7.4f,    %7.4f,    %7.4f,    %7.4f,    %9.4f\n',i, ...
    force_g(i,1),force_g(i,2),force_g(i,3),force_g(i,4),force_g(i,5),force_g(i,6));
end
```

A.39   print_truss_model.m

```
%
fprintf(fid, ' ******* PRINTING MODEL DATA **************\n\n\n');

% Print Nodal coordinates
%
fprintf(fid, '-------------------------------------------------------- \n');
fprintf(fid, 'Number of nodes:                                %g\n', nnd );
fprintf(fid, 'Number of elements:                             %g\n', nel );
fprintf(fid, 'Number of nodes per element:                    %g\n', nne );
fprintf(fid, 'Number of degrees of freedom per node:          %g\n', nodof);
fprintf(fid, 'Number of degrees of freedom per element:       %g\n\n\n', eldof);
%
%
%
fprintf(fid, '-------------------------------------------------------- \n');
fprintf(fid, 'Node       X            Y \n');
for i=1:nnd
fprintf(fid,' %g,      %07.2f,      %07.2f\n',i, geom(i,1), geom(i,2));
end
fprintf(fid,'\n');
%
% Print element connectivity
%
fprintf(fid, '-------------------------------------------------------- \n');
fprintf(fid, 'Element       Node_1      Node_2 \n');
for i=1:nel
fprintf(fid,'    %g,          %g,            %g\n',i, connec(i,1), connec(i,2));
end
```

```
fprintf(fid,'\n');
%
% Print element property
%
fprintf(fid, '-------------------------------------------------------- \n');
fprintf(fid, 'Element        E              A \n');
for i=1:nel
fprintf(fid,'   %g,        %g,          %g\n',i, prop(i,1), prop(i,2));
end
fprintf(fid,'\n');
%
% Print Nodal freedom
%
fprintf(fid, '-------------------------------------------------------- \n');
fprintf(fid, 'Node      disp_U     disp_V\n');
for i=1:nnd
fprintf(fid,'  %g,         %g,              %g\n',i, nf(i,1), nf(i,2));
end
fprintf(fid,'\n');
%
% Print Nodal loads
%
fprintf(fid, '-------------------------------------------------------- \n');
fprintf(fid, 'Node      load_X         load_Y\n');
for i=1:nnd
fprintf(fid,'  %g,      %07.2f,        %07.2f\n',i, load(i,1), load(i,2));
end
%
fprintf(fid, '-------------------------------------------------------- \n');
fprintf(fid,'\n');
fprintf(fid,'Total number of active degrees of freedom, n = %g\n',n);
fprintf(fid,'\n');
%
```

A.40   print_truss_results.m

```
%
fprintf(fid, '-------------------------------------------------------- \n');
fprintf(fid, ' \n\n\n ******* PRINTING ANALYSIS RESULTS **************\n\n\n');
%
%
%
% Print global force vector
%
fprintf(fid, '-------------------------------------------------------- \n');
fprintf(fid,'Global force vector  F \n');
fprintf(fid,'   %g\n',F);
fprintf(fid,'\n');
%
%
% Print Displacement solution vector
%
fprintf(fid, '-------------------------------------------------------- \n');
fprintf(fid,'Displacement solution vector:  delta \n');
fprintf(fid,' %8.5f\n',delta);
fprintf(fid,'\n');
%
% Print nodal displacements
%
fprintf(fid, '-------------------------------------------------------- \n');
fprintf(fid, 'Nodal displacements \n');
fprintf(fid, 'Node       disp_X         disp_Y\n');
for i=1:nnd
fprintf(fid,' %g,      %8.5f,        %8.5f\n',i, node_disp(i,1), node_disp(i,2));
end
fprintf(fid,'\n');
%
```

```
% Print Members actions
%
fprintf(fid, '----------------------------------------------------- \n');
fprintf(fid, 'Members actions \n');
fprintf(fid, 'element        force          action\n');
for i=1:nel
    if force(i) > 0
        fprintf(fid,' %g,        %9.2f,        %s\n',i, force(i), 'Tension');
    else
        fprintf(fid,' %g,        %9.2f,        %s\n',i, force(i), 'Compression');
    end
end
```

A.41 Q4_mesh.m

```
% This module generates a mesh of linear quadrilateral elements
%
global nnd nel nne  nodof eldof  n
global geom connec dee nf Nodal_loads
global Length Width NXE NYE X_origin Y_origin dhx dhy
%
%
nnd = 0;
k = 0;
for i = 1:NXE
    for j=1:NYE
        k = k + 1;
        n1 = j + (i-1)*(NYE + 1);
        geom(n1,:) = [(i-1)*dhx - X_origin    (j-1)*dhy - Y_origin ];
        n2 = j + i*(NYE+1);
        geom(n2,:) = [i*dhx - X_origin        (j-1)*dhy - Y_origin ];
        n3 = n1 + 1;
        geom(n3,:) = [(i-1)*dhx - X_origin       j*dhy - Y_origin ];
        n4 = n2 + 1;
        geom(n4,:) = [i*dhx- X_origin        j*dhy - Y_origin        ];
        nel = k;
        connec(nel,:) = [n1  n2   n4   n3];
        nnd = n4;
        end
end
%
```

A.42 Q8_mesh.m

```
% This function module a mesh of 8-nodded quadrilateral elements
%
global nnd nel nne  nodof eldof  n
global geom connec dee nf Nodal_loads
global Length Width NXE NYE X_origin Y_origin dhx dhy
%
%
nnd = 0;
k = 0;
for i = 1:NXE
    for j=1:NYE
    k = k + 1;
%
    n1 = (i-1)*(3*NYE+2)+2*j - 1;
    n2 = i*(3*NYE+2)+j - NYE - 1;
    n3 = i*(3*NYE+2)+2*j-1;
    n4 = n3 + 1;
    n5 = n3 + 2;
    n6 = n2 + 1;
    n7 = n1 + 2;
    n8 = n1 + 1;
%
```

```
    geom(n1,:) = [(i-1)*dhx - X_origin     (j-1)*dhy - Y_origin ];
    geom(n3,:) = [i*dhx - X_origin         (j-1)*dhy - Y_origin ];
    geom(n2,:) = [(geom(n1,1)+geom(n3,1))/2  (geom(n1,2)+geom(n3,2))/2];
    geom(n5,:) = [i*dhx- X_origin          j*dhy - Y_origin ];
    geom(n4,:) = [(geom(n3,1)+ geom(n5,1))/2  (geom(n3,2)+ geom(n5,2))/2];
    geom(n7,:) = [(i-1)*dhx - X_origin       j*dhy - Y_origin ];
    geom(n6,:) = [(geom(n5,1)+ geom(n7,1))/2  (geom(n5,2)+ geom(n7,2))/2];
    geom(n8,:) = [(geom(n1,1)+ geom(n7,1))/2  (geom(n1,2)+ geom(n7,2))/2];
%
    nel = k;
    nnd = n5;
    connec(k,:) = [n1 n2 n3 n4 n5 n6 n7 n8];
  end
```

A.43   Stresses_at_nodes_axi.m

```
function[ZX, ZY, Z_THETA, ZT]=Stresses_at_nodes_axi(SIGMA)
%
% This function averages the stresses at the nodes
%
global nnd nel nne  geom  connec
%
for k = 1:nnd
    sigx = 0. ;sigy = 0.; sig_theta = 0.; tau = 0.;
    ne = 0;
    for iel = 1:nel;
        for jel=1:nne;
            if connec(iel,jel) == k;
                ne=ne+1;
                sigx = sigx+SIGMA(iel,1);
                sigy = sigy + SIGMA(iel,2);
                sig_theta = sig_theta + SIGMA(iel,3);
                tau = tau + SIGMA(iel,4);
            end
        end
    end
        ZX(k,1) = sigx/ne;
        ZY(k,1) = sigy/ne;
        ZT(k,1)=tau/ne;
        Z_THETA(k,1) = sig_theta/ne;
end
```

A.44   Stresses_at_nodes_Q4.m

```
function[ZX, ZY, ZT, Z1, Z2]=Stresses_at_nodes_Q4(SIGMA)
%
% This function averages the stresses at the nodes
%
%
global nnd nel nne  geom  connec XIG YIG NXE NYE
%
for k = 1:nnd
    sigx = 0. ;sigy = 0.; tau = 0.;
    ne = 0;
    for iel = 1:nel;
        for jel=1:nne;
            if connec(iel,jel) == k;
                ne=ne+1;
                sigx = sigx+SIGMA(iel,1);
                sigy = sigy + SIGMA(iel,2);
                tau = tau + SIGMA(iel,3);
            end
        end
    end
        ZX(k,1) = sigx/ne;
        ZY(k,1) = sigy/ne;
```

```
            ZT(k,1)=tau/ne;
            Z1(k,1)= ((sigx+sigy)/2 + sqrt(((sigx+sigy)/2)^2 +tau^2))/ne;
            Z2(k,1)= ((sigx+sigy)/2 - sqrt(((sigx+sigy)/2)^2 +tau^2))/ne;
end
```

A.45   Stresses_at_nodes_Q8.m

```
function[ZX, ZY, ZT, Z1, Z2]=Stresses_at_nodes_Q8(SIGMA)
%
% This function averages the stresses at the nodes
%
global nnd nel nne  geom  connec
%
for k = 1:nnd
    sigx = 0. ;sigy = 0.; tau = 0.;
    ne = 0;
    for iel = 1:nel;
        for jel=1:nne;
            if connec(iel,jel) == k;
                ne=ne+1;
                sigx = sigx+SIGMA(iel,1);
                sigy = sigy + SIGMA(iel,2);
                tau = tau + SIGMA(iel,3);
            end
        end
    end
        ZX(k,1) = sigx/ne;
        ZY(k,1) = sigy/ne;
        ZT(k,1)=tau/ne;
        Z1(k,1)= ((sigx+sigy)/2 + sqrt(((sigx+sigy)/2)^2 +tau^2))/ne;
        Z2(k,1)= ((sigx+sigy)/2 - sqrt(((sigx+sigy)/2)^2 +tau^2))/ne;
end
```

A.46   T3_mesh.m

```
% This module generates a mesh of triangular elements
%
global nnd nel nne  nodof eldof  n
global geom connec dee nf Nodal_loads
global Length Width NXE NYE X_origin Y_origin dhx dhy
%
nnd = 0;
k = 0;
for i = 1:NXE
    for j=1:NYE
        k = k + 1;
        n1 = j + (i-1)*(NYE + 1);
        geom(n1,:) = [(i-1)*dhx - X_origin     (j-1)*dhy - Y_origin ];
        n2 = j + i*(NYE+1);
        geom(n2,:) = [i*dhx - X_origin         (j-1)*dhy - Y_origin ];
        n3 = n1 + 1;
        geom(n3,:) = [(i-1)*dhx - X_origin       j*dhy - Y_origin ];
        n4 = n2 + 1;
        geom(n4,:) = [i*dhx- X_origin        j*dhy - Y_origin       ];
        nel = 2*k;
        m = nel -1;
        connec(m,:) = [n1  n2  n3];
        connec(nel,:) = [n2  n4  n3];
        nnd = n4;
    end
end
%
for i=1:nel
    x =[geom(connec(i,1),1) ; geom(connec(i,2),1); geom(connec(i,3),1)];
    y =[geom(connec(i,1),2) ; geom(connec(i,2),2); geom(connec(i,3),2)];
end
```

A.47   T6_mesh.m

```
% This module generates a mesh of the linear strain triangular element
%
global nnd nel geom connec XIG YIG
global Length Width NXE NYE X_origin Y_origin dhx dhy
%
%
nnd = 0;
k = 0;
for i = 1:NXE
    for j=1:NYE
            k = k + 1;
            n1 = (2*j-1) + (2*i-2)*(2*NYE+1) ;
            n2 = (2*j-1) + (2*i-1)*(2*NYE+1);
            n3 = (2*j-1) + (2*i)*(2*NYE+1);
            n4 = n1 + 1;
            n5 = n2 + 1;
            n6 = n3 + 1 ;
            n7 = n1 + 2;
            n8 = n2 + 2;
            n9 = n3 + 2;
            %
            geom(n1,:) = [(i-1)*dhx - X_origin          (j-1)*dhy - Y_origin];
            geom(n2,:) = [((2*i-1)/2)*dhx - X_origin    (j-1)*dhy - Y_origin ];
            geom(n3,:) = [i*dhx - X_origin              (j-1)*dhy - Y_origin ];
            geom(n4,:) = [(i-1)*dhx - X_origin          ((2*j-1)/2)*dhy - Y_origin ];
            geom(n5,:) = [((2*i-1)/2)*dhx - X_origin    ((2*j-1)/2)*dhy - Y_origin ];
            geom(n6,:) = [i*dhx - X_origin              ((2*j-1)/2)*dhy - Y_origin ];
            geom(n7,:) = [(i-1)*dhx - X_origin          j*dhy - Y_origin];
            geom(n8,:) = [((2*i-1)/2)*dhx - X_origin    j*dhy - Y_origin];
            geom(n9,:) = [i*dhx - X_origin              j*dhy - Y_origin];
            %
            nel = 2*k;
            m = nel -1;
            connec(m,:) = [n1   n2   n3     n5     n7     n4];
            connec(nel,:) = [n3    n6     n9    n8     n7    n5];
            max_n = max([n1   n2   n3    n4   n5   n6   n7    n8     n9]);
            if(nnd <= max_n); nnd = max_n; end;
            %
            % XIN and YIN are two vectors that holds the coordinates X and Y
            % of the grid necessary for the function contourf (XIN,YIN, stress)
            %
            XIG(2*i-1) = geom(n1,1); XIG(2*i) = geom(n2,1); XIG(2*i+1) = geom(n3,1);
            YIG(2*j-1) = geom(n1,2); YIG(2*j) = geom(n4,2); YIG(2*j+1) = geom(n7,2);
    end
end
%
```

A.48   truss_C.m

```
function[C] = truss_C(i)
%
% This function forms the transformation between
% local and global coordinates
%
global geom connec
%
% retrieve the nodes of element i
%
node_1=connec(i,1);
node_2=connec(i,2);
%
% Retrieve the x and y coordinates of nodes 1 and 2
%
x1=geom(node_1,1); y1=geom(node_1,2);
```

```
x2=geom(node_2,1); y2=geom(node_2,2);
%
% Evaluate the angle that the member makes with the
% global axis X
%
if(x2-x1)==0
    if(y2>y1)
        theta=2*atan(1);
    else
        theta=-2*atan(1);
    end
else
    theta=atan((y2-y1)/(x2-x1));
end
%
% Construct the transformation matrix
%
C = [cos(theta) -sin(theta)     0         0 ; ...
     sin(theta)  cos(theta)     0         0 ; ...
         0           0      cos(theta) -sin(theta) ; ...
         0           0      sin(theta) cos(theta) ];
%
%%%%%%%%%%%%%%%% end function truss_C %%%%%%%%%%%%%
```

A.49   truss_g.m

```
function[g] = truss_g(i)
%
% This function forms the steering vector for element i
%
global connec nf
%
% retrieve the nodes of element i
%
node_1=connec(i,1);
node_2=connec(i,2);
%
% Form the steering vector from element's degrees
% of freedom
%
g=[nf(node_1,1); nf(node_1,2); nf(node_2,1);nf(node_2,2)];
%
%%%%%%%%%%%%%%%%%%%% end function truss_g %%%%%%%%%%%%%%%%
```

A.50   truss_kl.m

```
function[kl] = truss_kl(i)
%
% This function forms the element stiffness matrix
% in local coordinates
%
global geom connec prop
%
% retrieve the nodes of element i
%
node_1=connec(i,1);
node_2=connec(i,2);
%
%
% Retrieve the x and y coordinates of nodes 1 and 2
%
x1=geom(node_1,1); y1=geom(node_1,2);
x2=geom(node_2,1); y2=geom(node_2,2);
%
% Evaluate length of element i
%
```

```
L = sqrt((x2-x1)^2 + (y2-y1)^2);
%
% Retrieve section properties of element i
%
E= prop(i,1); A=prop(i,2);
%
% Calculate element stiffness matrix in its
% local coordinates
%
kl=[E*A/L 0 -E*A/L 0 ; ...
0 0 0 0 ; ...
-E*A/L 0 E*A/L 0 ; ...
0 0 0 0 ];
%
%%%%%%%%%%%%%%%%%%% End function truss_kl%%%%%%%%%%%%%%
```

Appendix B: Statically Equivalent Nodal Forces

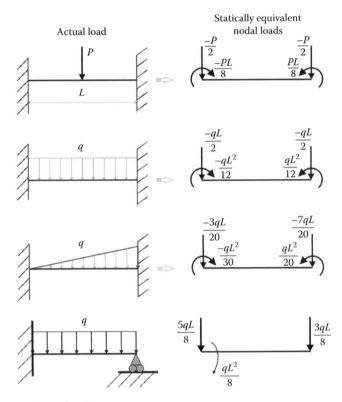

FIGURE B.1 Common beam loadings.

Appendix C: Index Notation and Transformation Laws for Tensors

C.1 INDEX NOTATION FOR VECTORS AND TENSORS

C.1.1 VECTOR AND TENSOR COMPONENTS

Operations on Cartesian components of vectors and tensors can be expressed very efficiently and clearly using index notation. The index notation refers to vectors or tensors by their general term, with the indices ranging over the dimensions of the vector or the tensor.

Let $\vec{u}$ be a vector and $\mathbf{a}$ a second-order tensor defined in a Cartesian basis. Using matrix notation, they can be represented by their Cartesian components as

$$\vec{u} = \begin{Bmatrix} u_1 \\ u_2 \\ u_3 \end{Bmatrix} \qquad \mathbf{a} = \begin{pmatrix} a_{11} & a_{12} & a_{13} \\ a_{21} & a_{22} & a_{23} \\ a_{31} & a_{32} & a_{33} \end{pmatrix} \tag{C.1}$$

Using index notation, the vector $\vec{u}$ and the tensor $\mathbf{a}$ can be expressed in a compact manner as

$$\vec{u} = u_i \qquad \mathbf{a} = a_{ij} \tag{C.2}$$

C.1.2 EINSTEIN SUMMATION CONVENTIONS

Under the rules of index notation, if an index is repeated in a product of vectors or tensors, summation is implied over the range of the repeated index. For example, for a range from 1 to 3, the following expressions can be developed as

$$a_i b_i = a_1 b_1 + a_2 b_2 + a_3 b_3 \tag{C.3}$$

$$c_i = a_{ik} x_k = \begin{Bmatrix} a_{11} x_1 + a_{12} x_2 + a_{13} x_3 \\ a_{21} x_1 + a_{22} x_2 + a_{23} x_3 \\ a_{31} x_1 + a_{32} x_2 + a_{33} x_3 \end{Bmatrix} \tag{C.4}$$

$$\lambda = a_{ij} b_{ij} = a_1 b_1 + a_1 b_2 + a_1 b_3 + a_2 b_1 + a_2 b_2 + a_2 b_3 + a_3 b_1 + a_3 b_2 + a_3 b_3 \tag{C.5}$$

$$c_{ij} = a_{ik} b_{kj} = a_{i1} b_{1j} + a_{i2} b_{2j} + a_{i3} b_{3j} \equiv [C] = [A] \times [B] \tag{C.6}$$

$$a_{ij} = b_{ji} \equiv [A] = [B]^T \tag{C.7}$$

Expression (C.6) is equivalent to the product of two matrices.

C.1.3   THE KRONECKER DELTA AND THE PERMUTATION SYMBOL

In the index notation, two special quantities, the Kronecker delta and the permutation factor, must be defined for use in the various operations involving vectors and tensors.

The Kronecker δ_{ij} is defined as

$$\delta_{ij} = \begin{Bmatrix} 1 & i = j \\ 0 & i \neq j \end{Bmatrix} \tag{C.8}$$

Thus

$$\delta_{11} = \delta_{22} = \delta_{33} = 1 \tag{C.9}$$

and

$$\delta_{12} = \delta_{21} = \delta_{13} = \delta_{31} = \delta_{23} = \delta_{32} = 0 \tag{C.10}$$

In matrix notation, the equivalent of the Kronecker delta is the identity matrix.

The Kronecker delta can be used as a substitution operator, since

$$\delta_{ij}b_j = \delta_{i1}b_1 + \delta_{i2}b_2 + \delta_{i3}b_3 = b_i \tag{C.11}$$

$$\delta_{ij}a_{ik} = \delta_{1j}a_{1k} + \delta_{2j}a_{2k} + \delta_{3j}a_{3k} = a_{jk} \tag{C.12}$$

The permutation factor e_{ijk} is defined as

$$e_{ijk} = \begin{Bmatrix} 1 & i,j,k = 1,2,3 & 2,3,1 & 3,1,2 \\ -1 & i,j,k = 3,2,1 & 2,1,3 & 1,3,2 \\ & 0 & \text{otherwise} & \end{Bmatrix} \tag{C.13}$$

We can observe that

$$e_{ijk} = e_{jki} = e_{kij} = -e_{ikj} = -e_{kji} = e_{jik} \tag{C.14}$$

$$e_{kki} = 0 \tag{C.15}$$

$$e_{ijk}e_{imn} = \delta_{jm}\delta_{kn} - \delta_{jn}\delta_{mk} \tag{C.16}$$

Using these definitions, the cross-product of two vectors can be written as

$$\vec{u} \times \vec{v} = e_{ijk}u_jv_k \tag{C.17}$$

C.1.4   RULES OF INDEX NOTATION

There three important rules in index notation, which are as follows:

- An index may occur either once or twice in a given term. When an index occurs unrepeated in a term, that index is understood to take all the values of its range. Unrepeated indices are known as free indices. Free indices appearing on each term must agree.
- When an index appears twice in a term, that index is understood to take all the values of its range and the resulting terms summed. Repeated indices are often referred to as dummy indices.

- Free and dummy indices may be changed without altering the meaning of the equation. The number and the location of the free indices reveal the exact tensorial rank of the quantity expressed.

The following expressions are valid

$$A_{ik}x_k, \quad A_{ij}B_{ik}C_{nk}, \quad a_i = A_{ki}B_{kj}x_j + C_{ik}u_k$$

but

$$x_i = A_{ij}, \quad x_j = a_{ik}u_k, \quad x_j = A_{ki}B_{kj}u_j$$

are meaningless.

C.2 COORDINATE TRANSFORMATIONS

C.2.1 TRANSFORMATION LAWS FOR VECTORS

Given two arbitrary coordinate systems $\vec{e}_1, \vec{e}_2, \vec{e}_3$ and $\vec{e}_1', \vec{e}_2', \vec{e}_3'$ in a three-dimensional Euclidean space. Any change of coordinate system is characterized by a Jacobian matrix $[J]$, which helps express the vectors of the new base in terms of the ones in the old base:

$$[J] = \left\| \frac{\partial e_i'}{\partial e_j} \right\| = \begin{bmatrix} \dfrac{\partial e_1'}{\partial e_1} & \dfrac{\partial e_1'}{\partial e_2} & \dfrac{\partial e_1'}{\partial e_3} \\[2mm] \dfrac{\partial e_2'}{\partial e_1} & \dfrac{\partial e_2'}{\partial e_2} & \dfrac{\partial e_2'}{\partial e_3} \\[2mm] \dfrac{\partial e_3'}{\partial e_1} & \dfrac{\partial e_3'}{\partial e_2} & \dfrac{\partial e_3'}{\partial e_3} \end{bmatrix} \tag{C.18}$$

If the Jacobian does not vanish, expression (C.18) possesses a unique inverse.

The coordinates systems represented by $\vec{e}_1, \vec{e}_2, \vec{e}_3$ and $\vec{e}_1', \vec{e}_2', \vec{e}_3'$ are completely general and may be any curvilinear or Cartesian systems. In the case of Cartesian systems as shown in Figure C.1, the Jacobian takes the form of a constant tensor l_{ij} or, because of the identity between second-order

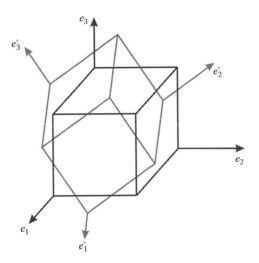

FIGURE C.1 Transformation of coordinates.

tensors and square matrices, a constant matrix $[Q]$, which is called the transition matrix from the old basis to the new basis. In index notation, the transformation takes the form

$$e'_i = l_{ij} e_j \tag{C.19}$$

with

$$l_{ij} = \cos\left(\vec{e}'_i, \vec{e}_j\right) = \begin{pmatrix} l_{11} & l_{12} & l_{13} \\ l_{21} & l_{22} & l_{23} \\ l_{31} & l_{32} & l_{33} \end{pmatrix} \tag{C.20}$$

In matrix notation, it takes the form:

$$\{e'\} = [Q]\{e\} \tag{C.21}$$

with

$$[Q] = \left[\cos\left(\vec{e}'_i, \vec{e}_j\right) \right] = \begin{bmatrix} Q_{11} & Q_{12} & Q_{13} \\ Q_{21} & Q_{22} & Q_{23} \\ Q_{31} & Q_{32} & Q_{33} \end{bmatrix} \tag{C.22}$$

Example: Anticlockwise Rotation around the Axis 3

In the case of an anticlockwise rotation as shown in Figure C.2, the relation between the bases is written as

$$\vec{e}'_1 = \cos(\psi)\vec{e}_1 + \sin(\psi)\vec{e}_2 + 0 \times \vec{e}_3 \tag{C.23}$$

$$\vec{e}'_2 = -\sin(\psi)\vec{e}_1 + \cos(\psi)\vec{e}_2 + 0 \times \vec{e}_3 \tag{C.24}$$

$$\vec{e}'_3 = 0 \times \vec{e}_1 + 0 \times \vec{e}_2 + 1 \times \vec{e}_3 \tag{C.25}$$

The matrix $[Q]$ takes the form

$$[Q] = \begin{bmatrix} \cos(\psi) & \sin(\psi) & 0 \\ -\sin(\psi) & \cos(\psi) & 0 \\ 0 & 0 & 1 \end{bmatrix} \tag{C.26}$$

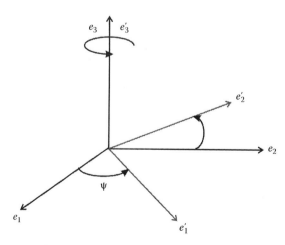

FIGURE C.2   Rotation around the third axis.

The matrix [Q] is an orthonormal matrix and has the following properties:

$$[Q]^T = [Q]^{-1} \tag{C.27}$$

In index notation, the relationship (C.27) is given as

$$l_{ik}l_{jk} = \delta_{ij} \tag{C.28}$$

Note that in index notation, and by analogy to matrix notation, you cannot write (C.28) as $l_{ji}l_{ij} = \delta_{ij}$. This is completely erroneous in index notation since the repeated indices in the first term imply summation; therefore, the first term is a scalar and the second a tensor.

Given an arbitrary vector $\vec{v}$ represented in the base $(\vec{e}_1, \vec{e}_2, \vec{e}_3)$ as

$$\vec{v} = v_1\vec{e}_1 + v_2\vec{e}_2 + v_3\vec{e}_3 = v_j\vec{e}_j \tag{C.29}$$

The same vector can also be represented in the base $(\vec{e'}_1, \vec{e'}_2, \vec{e'}_3)$ as

$$\vec{v} = v'_1\vec{e'}_1 + v'_2\vec{e'}_2 + v'_3\vec{e'}_3 = v'_i\vec{e'}_i \tag{C.30}$$

Using Equation (C.19), Equation (C.30) is rewritten as:

$$\vec{v} = v'_i l_{ij}\vec{e}_j \tag{C.31}$$

Comparing Equations (C.29) and (C.31) reveal that the vector components in the primed and unprimed basis are related by

$$v_j = v'_i l_{ij} = l_{ij}v'_i \tag{C.32}$$

in matrix notation

$$\{v\} = [Q]^T\{v'\} \tag{C.33}$$

The inverse transformation is defined as

$$v'_i = v_j l_{ij} \tag{C.34}$$

or in matrix notation as

$$\{v'\} = [Q]\{v\} \tag{C.35}$$

C.2.2 TRANSFORMATION LAWS FOR TENSORS

Given two arbitrary vectors $\vec{u}$ and $\vec{v}$ represented in the base $(\vec{e}_1, \vec{e}_2, \vec{e}_3)$ respectively as

$$\vec{u} = u_1\vec{e}_1 + u_2\vec{e}_2 + u_3\vec{e}_3 \tag{C.36}$$
$$\vec{v} = v_1\vec{e}_1 + v_2\vec{e}_2 + v_3\vec{e}_3 \tag{C.37}$$

Now suppose the existence of a linear application between the two vectors defined by $\vec{u} = f(\vec{u})$ and expressed in index notation as

$$u_i = a_{ij}v_j \tag{C.38}$$

or in matrix notation as

$$\{u\} = [a]\{v\} \tag{C.39}$$

In another base, say $(\vec{e}_1', \vec{e}_2', \vec{e}_3')$, the vectors $\vec{u}$ and $\vec{v}$ are expressed as

$$\vec{u} = u_1'\vec{e}_1' + u_2'\vec{e}_2' + u_3'\vec{e}_3' \tag{C.40}$$

$$\vec{v} = v_1'\vec{e}_1' + v_2'\vec{e}_2' + v_3'\vec{e}_3' \tag{C.41}$$

and the relationship $\vec{u} = f(\vec{u})$ is expressed in index notation as

$$u_i' = a_{ij}'v_j' \tag{C.42}$$

and in matrix notation as

$$\{u'\} = [a']\{v'\} \tag{C.43}$$

The problem is to find a relationship between the tensors **a** and **a'**.

Using (C.32), Equation (C.38) is rewritten as

$$u_i = a_{ij}l_{mj}v_m' = l_{mj}a_{ij}v_m' \tag{C.44}$$

Substituting in (C.44) for u_i using (C.32) leads to

$$l_{ki}u_k' = a_{ij}l_{mj}v_m' = l_{mj}a_{ij}v_m' \tag{C.45}$$

Multiplying both sides of the equations by l_{ni}, and noting that $l_{ni}l_{ki} = \delta_{nk}$, Equation C.46 becomes

$$\delta_{nk}u_k' = l_{ki}l_{mj}a_{ij}v_m' \tag{C.46}$$

That is,

$$u_k' = l_{ki}l_{mj}a_{ij}v_m' \tag{C.47}$$

Comparing (C.42) and (C.47), it follows

$$a_{km}' = l_{ki}l_{mj}a_{ij} \tag{C.48}$$

Using matrix notation, and after substituting Equation (C.33), Equation (C.39) becomes

$$\{u\} = [a][Q]^T\{v'\} \tag{C.49}$$

Replacing the vector $\{u\}$ by $[Q]T\{u'\}$, Equation (C.49) becomes

$$[Q]T\{u'\} = [a][Q]^T\{v'\} \tag{C.50}$$

and premultiplying both sides of the equation by the matrix $[Q]$, and noting $[Q][Q]^T = [I]$, yields the result

$$\{u'\} = [Q][a][Q]^T\{v'\} \tag{C.51}$$

Comparing Equations (C.43) and (C.51) yields the result

$$[a'] = [Q][a][Q]^T \tag{C.52}$$

The inverse relation is expressed as

$$[a] = [Q]^T[a'][Q] \tag{C.53}$$

References and Bibliography

In the course of writing this present work, many books on matrix structural analysis and the theories of elasticity and finite element methods have been consulted. Some of these books have been explicitly cited while others not. An exhaustive list of all the books consulted is given below.

REFERENCES

1. Dhatt G. and Touzot G. *Une Présentation de la Méthode des Éléments Finis,* Deuxième édition. Maloine S.A. Editeurs, Paris, France, 1984.
2. Hammer P.C., Marlowe O.J., and Stroud A.H. Mathematical tables and other aids to computation, *American Mathematical Society*, 10(55), 130–136, 1956.
3. Smith I.M. and Griffiths D.V. *Programming the Finite Element Method*, 2nd edn. Wiley, Chichester, U.K., 1988.
4. Timoshenko S. and Goodier J. *Theory of Elasticity*, 3rd edn. McGraw-Hill, New York, 1970.
5. Timoshenko S. and Woinowsky-Krieger S. *Theory of Plates and Shells*. McGraw-Hill, New York, 1959.
6. Zienkiewicz O.C. *The Finite Element Method*, 3rd edn. McGraw-Hill, York, London, 1977.

BIBLIOGRAPHY

Chandrupatla T.R. and Belegundu A.D. *Introduction to Finite Elements in Engineering*, 3rd edn. Prentice-Hall, Upper Saddle River, NJ, 2002.
Cook R.D. *Finite Element Modeling for Stress Analysis*. Wiley, New York, 1995.
Kwon Y.W. and Bang H. *The Finite Element Method Using Matlab*, 2nd edn. CRC Press, London, U.K., 2000.
Logan D.L. *A First Course in the Finite Element Method Using Algor*, 2nd edn. Brooks/Cole Thompson Learning, Pacific Groove, CA, 2001.
Mase G.E. *Schaum's Outline Series: Theory and Problems of Continuum Mechanics*. McGraw-Hill, New York, 1970.
McGuire M., Gallagher G.H., and Ziemian R.D. *Matrix Structural Analysis,* 2nd edn. Wiley, New York, 2000.
Meek J.L. *Computer Methods in Structural Analysis*. E & FN SPON, London, U.K., 1991.
Reddy J.N. *An Introduction to the Finite Element Method*, 3rd edn. McGraw-Hill, New York, 2006.
Saada A.S. *Elasticity: Theory and Applications*, 2nd edn. Krieger Publishing, Melbourne, FL, 1993.

Index

A

Abaqus
  beam orientation, 94, 105, 108, 127, 155
  CAE, 2, 36–37, 91, 101, 125, 131, 254, 258, 274, 296,
      321, 372, 402, 409
  encastré, 45, 47, 74, 109, 111
  field output, 54–57, 59–60, 101–102, 130–132, 134,
      258–259, 262, 278, 299–300, 304, 406
  history output, 59–60, 105
  input file, 24, 27, 32, 36, 57, 77, 81, 84, 87, 103, 115,
      120, 131–132, 244, 249, 258, 260–261,
      267, 278, 287, 302–303, 329, 333, 337,
      339, 359
  instance, 44, 98, 145, 216
  keyword, 57–61, 103–105, 132, 134, 260–262, 278,
      302–304
  keyword edition, 57, 103–104, 132, 134, 260, 278,
      302, 304
  mesh generation, 249, 252, 266, 271, 291, 295, 315,
      319, 329, 333, 337, 359, 365
  odb file, 53, 57, 60, 102, 105, 131–132, 134,
      258–259, 262, 278–279, 299–300, 304, 326,
      377, 406
  part, 1–2, 35, 37–38, 43–44, 57, 91, 95, 97, 102, 125,
      127–128, 132, 179, 182, 254–257, 259,
      274–276, 296–298, 300, 321–323, 372–375,
      402–404, 409, 416
  pinned, 46, 59–60, 129, 133
  seed, 43, 96, 128, 256, 276, 298, 323, 374, 404, 416
  step, 49, 78–79, 87, 89, 132, 134–135, 162–164, 289,
      291–292, 330, 333–334
  viewer, 53, 61, 102, 105, 131, 134, 258, 262, 278–279,
      299, 304, 326, 377, 406, 418
  visualization, 36, 53, 57, 61, 102, 105, 131, 134, 258,
      262, 278, 299, 304, 326, 377, 406, 418
Assembly
  global force vector assembly, 14
  stiffness matrix
    assembly, 9, 109
    in local coordinates, 21, 112
    in global coordinates, 22, 113
  transformation matrix, 9, 12, 21–25, 109, 112–115,
      422, 444
Axisymmetric
  axisymmetric equivalent nodal loads, 356
  axisymmetric problems, 3, 353–354, 357, 359,
      361, 363, 365, 367, 369, 371, 373, 375,
      377

B

Bibliography
  Dhatt, G., 419
  Griffiths, D.V., 19, 419
  Timoshenko, S., 419
  Touzot, G., 419
  Zienkiewicz, O.C., 419
Boundary conditions
  built-in/clamped edge, 390
  encastré, 45, 47, 74, 109, 111
  essential, 1, 135, 178–179, 183
  free edge, 390
  natural, 178–179, 390
  pinned, 15, 46, 59–60, 129, 133
  simply supported edge, 389–390, 395

C

Castellated beam, 406–408, 414
  shell element, 400–401, 407
  three-dimensional shells, 406
Constitutive equations
  Hooke's law, 154–155, 160, 354
  isotropic, 158, 172, 174, 354
  Lamé, 159–160

E

Elements
  beam column element, 108–109
  beam element, 63–65, 73–77, 79, 81, 83, 85, 87, 89, 91,
      93–97, 99, 101, 103–105, 107, 109, 186, 191,
      195, 211
  bilinear quadrilateral, 207, 217, 279, 302–303
  brick element, 209–210
  constant strain triangle, 235, 237
  linear strain triangle, 263, 272, 282
  Q4 quadrilateral, 295
  Q8 quadrilateral, 321
  quadrilateral element, 203, 207, 280, 287, 291, 304,
      311, 314–315, 337, 357, 365, 391
  rectangular element, 387
  rod element, 7, 75
  tetrahedra, 209
Energy principles
  Castigliano, 63, 65–66
  potential energy, 1, 185–186, 191
  Rayleigh Ritz, 183, 186
  virtual work, 1, 182, 191, 211, 233, 390

F

Finite element approximation
  compatibility, 152, 195–197, 232, 385
  completeness, 196–197
  isoparametric, 202, 207, 217, 221, 223, 228, 280, 388
  Jacobian, 217, 223–225, 229, 265, 268–269, 271–272,
      287–290, 293–294, 312–313, 317–318,
      335–336, 361, 363, 367–369, 393–394, 398–399

shape functions, 193, 195, 197, 199, 203, 207, 217–218, 224–225, 227–229, 233, 235, 238, 263, 265–266, 268–269, 271–272, 279–280, 282, 287–290, 293–294, 304, 312–313, 317–318, 335–336, 361, 363, 367–369, 388, 393–394, 398–399, 429–430

trial function, 176–177, 179, 193, 196–197, 199, 202, 226–228, 387

Finite element formulation

body forces, 186, 233, 238, 282, 356

constant strain triangle, 235, 237

linear strain triangle, 263, 272, 282

mesh generation, 249, 252, 266, 271, 291, 295, 315, 319, 329, 333, 337, 359, 365

numerical integration, 266, 270, 287, 292, 312, 317, 362, 367–368, 393, 397

parent element, 199–200, 202–203, 228, 280

reference element, 199–200, 202–204, 216–218, 228

strain matrix, 233, 237, 244–246, 251, 265, 268, 271–272, 289–290, 330–331, 335, 360, 363, 369, 398–399

traction forces, 233, 238, 356

I

Index notation

Kronecker delta, 450

tensor, 447–452

the permutation symbol, 450

transformation laws, 449–450, 452–454

M

MATLAB programs

AXI_SYM_Q8.M, 365

AXI_SYM_T6.M, 359

beam.m, 73, 77, 81, 109, 246

CST_PLANE_STRESS.m, 243, 247

CST_PLANE_STRESS_MESH.m, 249, 266

CST_STRIP_FOOTING.m, 329

Eight_Q8.m, 223–225

frame.m, 109, 114, 246

IXX.m, 223

LST_PLANE_STRESS_MESH.m, 266, 359

LST_STRIP_FOOTING.m, 333

modules and functions, 421–445

Plane_Q8_MESH.m, 315, 365

precipitation_T3.m, 227

precipitation_T6.m, 229

Q4_mesh.m, 291, 294, 440

Q4_PLANE_STRESS.m, 287, 307

Q4_PLANE_STRESS_MESH.m, 291

Q8_mesh.m, 315, 318, 337, 365, 440

Q8_PLANE_STRESS.m, 307, 311, 336, 338, 351

T3_mesh.m, 249, 252, 328, 442

T6_mesh.m, 266, 269, 332, 359, 443

Thick_plate_Q8.m, 390–391

truss.m, 19–21, 24, 27, 57, 73, 75–76, 109, 113, 246

Two_Q8.m, 223–224

N

Numerical integration

area coordinates, 202–203, 217–218

Gauss quadrature, 211, 213–214, 216, 218–220, 223, 225, 282, 289, 368

Hammer, 3, 218, 229, 266, 268–269, 271–272, 334–335, 360–362, 419, 432–433

Jacobian, 217, 223–225, 229, 265, 268–269, 271–272, 287–290, 293–294, 312–313, 317–318, 335–336, 361, 363, 367–369, 393–394, 398–399

numerical integration, 3, 5, 211, 213, 217–219, 221, 223, 225, 227, 229, 266, 270, 287, 292, 312, 317, 362, 367–368, 393, 397

P

Plane problems

plane strain, 231, 327, 329, 340, 431

plane stress, 3, 231, 242–243, 250, 255–256, 260–262, 267, 274–275, 286, 291, 297–298, 302–303, 311, 316, 321–322, 333, 337, 353, 359, 365–366, 372, 381, 431–432

S

Stress and strain

engineering representation

of stress, 144

of strain, 153

equilibrium equations, 15, 20, 182

Green Lagrange strain, 148

Mohr's circle, 143

principal strain, 152

principal stress, 140, 278–279, 327, 331–332, 336, 350–351

small deformation theory, 149

strain tensor, 150, 152, 154, 171, 174

stress tensor, 135–137, 140–141, 152, 154, 164–165

von Mises stress, 141, 418

T

Thin and thick plates

built-in/clamped edge, 390

free edge, 390

Kirchhoff, 379, 385, 390

Mindlin, 388

shear stiffness, 389, 393, 395

simply supported edge, 389–390, 395

thick plate, 388, 391, 430–431

thin plate, 380, 389–390

Transformation of coordinates

global coordinates, 10, 12, 21–25, 73, 76, 105, 109, 112–115, 117–118, 121, 124, 130, 268–269, 271–272, 288–290, 293–294, 312–313, 317–318, 335–336, 361, 363, 367–369, 393–394, 398–399, 421, 429–430, 438, 443

local coordinates, 9–10, 12, 21, 24–25, 78, 112–114,
    118, 124, 130, 228, 265, 269, 271–272,
    287–290, 293, 312–313, 317–318, 336, 361,
    363, 367–369, 393–394, 398–399, 422–424,
    438, 444–445
parent element, 199–200, 202–203, 228, 280
reference element, 199–200, 202–204,
    216–218, 228
transformation matrix, 9, 12, 21–25, 109, 112–115,
    422, 444

W

Weighted residual methods
    equilibrium equations, 182
    functional, 183, 186
    Galerkin, 176–177, 179, 183
    Green theorem, 179
    Rayleigh Ritz, 183, 186
    virtual work, 182
    weak form, 178–179, 182